# TECHNOLOGY HOW INFLUENCES ORGANIZATIONAL AND SOCIAL BEHAVIORAL CHANGE

## SOCIAL BEHAVIORAL CHANGE

JOHN LOK

Copyright © John Lok
All Rights Reserved.

This book has been published with all efforts taken to make the material error-free after the consent of the author. However, the author and the publisher do not assume and hereby disclaim any liability to any party for any loss, damage, or disruption caused by errors or omissions, whether such errors or omissions result from negligence, accident, or any other cause.

While every effort has been made to avoid any mistake or omission, this publication is being sold on the condition and understanding that neither the author nor the publishers or printers would be liable in any manner to any person by reason of any mistake or omission in this publication or for any action taken or omitted to be taken or advice rendered or accepted on the basis of this work. For any defect in printing or binding the publishers will be liable only to replace the defective copy by another copy of this work then available.

# Contents

# Preface

Technology how influence human behavioral change? This book brings readers to enter our future technological world to let you to feel how we will encounter in future predictive technological innovation or invention to influence our daily life. This is future US and UK technological fiction to explain what images we will feel when our societies were become technological life. I write this book aims to explain how future technology change will influence human living of standard. I suppose that human living of standard will become better if future human technology will be improved. Otherwise, human living of standard will become worse if future human technology won't be improved , even will be fallen down. I shall give US and UK technologies development past and present evidences to explain how their technologies development will improve human living of standard. I hope readers can make personal judgement to evaluate whether technology development will improve human living of standard.

This book divides four parts. The first part explains what future technology factors will influence US economy development. The second part explains what future technology factors will influence UK economy development. The final part that I shall give some economy theories to explain why these kinds of technological factors can influence future these both countries' economy growth.

This book first part concern to be given on my opinions and some economists' opinions to UK and US both countries government and businessmen what will be the most urgent important needs to develop on the technological product investments aspect. I shall indicate the online teaching technology and environment protect technology, such as natural energy, farming environment protect technology etc. as well as automatic technology in manufacturing industry , such as human intelligence products to give my reasons and some economists' reasons to explain to support why these three technological developments will be valid to invest to develop to UK society in the future. I shall give reasons to explain why some economy theories can be proved to satisfy these technological factors requirement to predict which can influence US and UK future economic growth. The economy theories include: welfare economic theory, Kaldor -Hicks efficiency theory, Pareto improvement theory, industrial production and pollution economy theory, natural rate of unemployment theory. Finally, I hope readers can attempt to analyze whether these economic theories can be applied to explain whether these technological factors can assist US and UK future economic growth.

This book second part concerns to be given my opinions to explain how artificial intelligent technology will impact our life and will influence economic development in the future as well as how to influence human job market change. In labor market part, I shall indicate how artificial intelligence technology influences future macro global economy change. Advances in artificial intelligence (AI) technology is for the progress in critical areas, such as health, education, energy, economy inclusion, social welfare and the environment. Which (AI) workers be instead of traditional human workers in these different new markets? In recent years, machines had been used to be human's tasks in the performance of certain tasks related to intelligence , such as aspects of image recognition. Experts also forecast that rapid progress in the field of specialized artificial intelligence will continue. Then, it also brings this question: Does (AI) exceed that of human performance on more and more tasks? If it is truth, will some of human jobs to be disappeared? (AI) will be instead of human some simple jobs, then unemployment rate to the low skillful and low educated workers will be increased. Whether (AI) will be raised either production or performance or unemployment to bring human job market more advantages or more disadvantages I shall give my opinions to indicate what reasons will cause (AI) artificial intelligent tools to be applied to social military defense weapon by human's intention. In my this books, I hope my readers can know what will cause human's immoral behaviors to bring our societies to bring more dangerous or risks or threats if human applied (AI) technology to achieve whose immoral or ambitious intention. Finally, I hope that human ought not apply (AI) technology to do any behavioral attack to satisfy ourselves interest or dominate global world ambition to avoid (AI) technological war occurrence in the future one day.

I shall explain what the (AI) functions are benefited to human and I shall indicate whether how it can impact human job nature as well as I shall explain whether (AI) can influence to change to human job market to be better or worse in future global labor market. I shall indicate how (AI) will influence global economic technology productivity change as well as I shall indicate whether (AI) technology can impact global service management industry development. I shall indicate whether (AI) technology will influence digital industry economic change as well as I shall indicate whether (AI) technology can influence developing countries' economic change and health and manufacturing change , even global economic change in our future life. In my this robot part, I shall explain whether (AI) will bring benefits or disadvantages to human job market. I shall give example to let my readers to think how to support my final view point.

In this book third part, I shall explain future robot development stages how may influence human societies to bring what advantages and disadvantages. What is future (AI) artificial intelligent products development trend and reasonable development stages? How to predict consumer behaviors to persuade who to feel (AI) products are more satisfactory to their needs? Why do consumers feel them to need to buy any (AI) products to use? Will it have other similar products to replace (AI) any products? How is the reasonable stages to achieve future (AI) development in success? Can AI development bring global economic growth or avoid recession? Can future non-manual driving public transport tool inventions bring global economic growth? AI can bring positive or negative to impact our society.

In this part, I shall give actual data to predict what the future (AI) products development trend is. Giving my opinions to predict how (AI) consumers' choices are more absolutely. In the (AI) past first stage, I shall concern travel, education, transportation, financial , hospital, administrative service etc. different job natures to indicate how to apply (AI) products to assist these industries more beneficial. In the (AI) nowadays second stage, I shall concentrate on how (AI) developing on education aspect. In the (AI) future third stage, I shall explain why (AI) will have possible to invent (AI) brain technology, even it will bring (AI) war occurence in possible.

In the first (AI) stage, it concerns to be given my opinions to explain how artificial intelligent technology will impact our life and will influence economic development in the future as well as how to influence human job market change. In (AI) labor market stage, I shall indicate how artificial intelligence technology influences future macro global economy change.

Advances in artificial intelligence (AI) technology is for the progress in critical areas, such as health, education, energy, economy inclusion, social welfare and the environment. Thus, it brings this question: Which (AI) workers be instead of traditional human workers in these different new markets? In recent years, machines had been used to be human's tasks in the performance of certain tasks related to intelligence , such as aspects of image recognition. Experts also forecast that rapid progress in the field of specialized artificial intelligence will continue. Then, it also brings this question: Does (AI) exceed that of human performance on more and more tasks? If it is truth, will some of human jobs to be disappeared? (AI) will be instead of human some simple jobs, then unemployment rate to the low skillful and low educated workers will be increased.

Whether (AI) will be raised either production or performance or unemployment to bring human job market more advantages or more disadvantages? In my this book, I shall explain whether (AI) will bring benefits or disadvantages to human job market. I shall give example to let my readers to think how to support my final view point.

What is future (AI) artificial intelligent products development trend? How to predict consumer behaviors to persuade who to feel (AI) products are more satisfactory to their needs? Why do consumers feel them to need to buy any (AI) products to use? Will it have other similar products to replace (AI) any products?

In (AI) second stage, I shall indicate how (AI) is developed on education aspect, artificial intelligent technology and online technology and online book stores are high technological intelligent product. Hence, human ourselves will have possible to cause artificial intelligent machine men to own human's mind to learn how to read books and/ or write books abilities. When artificial intelligent machine men can learn how to read books and/or write books. Consequently, it means that artificial intelligent machine men can own human mind to do any jobs.I shall assume when artificial intelligent machine men can learn how to write books and/or read books. Then, they will have human's mind ability in possible. Can future artificial intelligent machine men be invented to learn how to write books and/or read books ability? In my this book, I shall attempt to answer this question. Finally, I hope my readers can attempt to make judgement whether artificial intelligent machine men can really learn how to write and/or books. I shall apply online technology to answer this answer. Finally, I shall give my opinions what are the influences when AI machine men had invented to achieve owning human's mind and judgement ability to our future society.

In (AI) third stage development, artificial intelligence (AI) technology is popular to be applied to different industry aspects, such as medical, construction, transportation, hospital, education etc. Although, (AI) is a human invention new development. IN fact, it seems only beneficial to human's daily life. But, it will also have threats to influence human's safety in possible , if some scientists or self-interest mind people who aim to apply (AI) to earn more profit or apply (AI) tools to be weapon to attack other countries to achieve to dominate all human's ambitious intention. Thus, (AI) will bring negative influences to our society, instead of positive influences if we can not apply this kind of new technological tools immorally. In this stage, I shall give my opinions to indicate what reasons will cause (AI) artificial intelligent tools to be applied to social military defense weapon by human's intention. In my this books, I hope my readers can know what will cause human's immoral behaviors to bring our societies to bring more dangerous or risks or threats if human applied (AI) technology to achieve whose immoral or ambitious intention. Finally, I hope that human ought not apply (AI) technology to do any behavioral attack to satisfy ourselves interest or dominate global world ambition to avoid (AI) technological war occurrence in the future one day.

In (AI) fourth stage, I shall give some university lecturers' personal analytical mind to judge whether what will be occurred if artificial intelligence could be invented to match to own human brain's mind ability? What will be the advantages and/or disadvantages if (AI) robots would be invented to match to own human brain's mind ability? I shall follow current (AI) technological development to judge whether what the potential abilities are that (AI) will achieve to satisfy human's life needs when (AI) is invented to own human brain in future one day.

In AI fifth stage, we are experiencing technological development stages, human began to consider air pollution how brings our natural environment to cause worse. Our cars fuel emission can also pollute our air when we are often driving ourselves cars to go to anywhere for leisure or working aim. Instead of ourselves cars emission air pollution, public transport tools, such as buses, taxis, ferries, their fuel emission also influence our air to be polluted. Because passenger individual comfort need began raises. So, electric auto non-manual driving cars, battery charge energy cars are began to consider how to invent in order to reduce air pollution and raise driver individual comfortable need, instead of improving driving speed to be rapid, car inventors also hope to avoid air pollution to pollute our natural environment. When global fuel cars number continues increases.


Will it popular to accept to use any artificial intelligent vehicles? Is it possible to apply AI non-manual driving technology to AI non-manual driving transportation tools global transportation market? For example, in (AI) non-manual vehicle industry, driving automatic vehicle whether it will be accepted to drivers who have confidence to drive it on roads safely. Whether artificial (AI) intelligent non-manual driving systems are the improvement of traffic safety, reduction of energy consumption or improvement of the comfort of the driver. Whether will it be popular to accept to apply artificial intelligent non-manual driving technology from non-manual auto driving cars to be applied to any non-manual auto driving transportation tools transportation market development, such as train, tram, lorry, transportation air plane, passenger air plane, ferry, taxi, MTR. Etc. different kinds of transportation tools? If future human accepts to use any non-manual driving vehicles or non-manual driving transportation tools, what advantages and disadvantages will bring to influence our daily life.

How if (AI) non-manual auto driving technology stage is mature to achieve non-manual driving technology is safe driving. It is possible that (AI) non-manual driving cars can influence to change whole manual driving transportation tools to non-manual driving transportation tools. How it will influence (AI) autonomous cars change to influence global manual driving transportation industry development ? To achieve non-manual driving industry development success. (AI) non-manual driving vehicle manufacturers need to ensure (AI) driving system is more safe to drive to compare manual driving on the road. If they expect non-manual driving transportation market development success. So, self improving systems are a promising new approach to developing artificial intelligence. But will their behavior be predictable? Can will be sure that they will behave as we intended even after many generations of self improvement? This part presents a framework for answering any questions concern whether future non-manual driving transportation market whether it will be possible to bring global economic growth in success.

In AI final stage, I shall indicate whether AI can assist space tourism development in possible as well as what benefits it can bring to our societies. I write this book aims to let readers to feel whether AI can bring real benefits to our future societies in possible.

This book final part explains how robots may impact organizational and social behavioral change ? How social change influences human behavioral change ? Why human behavior may be influenced by social change? Our individual behavior whether can be influenced to bring negative or positive attitude by social change? I shall attempt to indicate cases to explain whether our individual behavior can be influenced to changed by social environment change. Readers can have more understand how and why social change may influence our behavior in possible.

# Prologue

High technology development can
reducing income inequality growth. p.91
Long term cheap medical cost trend factor
influences US economic growth. p.103
The impact of educational quality factor
influences US economic growth.
Bio-medical industry factor boosts US economy. p.110

Welfare economic theory, Kaldor-Hicks
efficiency theory and Pareto improvement
theory relationship explain US and UK technology development trend
- Natural rate of unemployment p .111
- Three Industrial production and
pollution economy p.116
- The future of job skills influence p.129
I. Introduction
II. What will be future human employment
trends?
III. How to change in ease of recruitment?

Can technology influence UK and US life of
standard to be fallen or grown
I. Has it close relationship between
climate change and technological process
to influence human living standard?
II. Why can climate change influence effect
on growth and inflation?
- Whether future technology can influence
human living standard p.149
I Can technology influence human habitable population dynamic change to impact living standard?
II Human's history technological development
III Can food technology production shift diets,
the type, combination and quantity of foods
people consume, contribute to a food
technological production life in future? p.155
Reference
Chapter 2 Robots how influences human social job market change
How Robots Influence Driven Automation Industry Development
1.1 (AI) - driven automation industry development how to influence work nature change
1.2 How (AI) influences labor market
Redefining management in the workforce of artificial intelligence
2.1 Change management
2.2 How (AI) influences organizational
change p.156-170
Future works change: Automation, employment and productivity
3.1 How (AI) influences employment
3.2 What occupations will be influenced
by (AI) technology

How can (AI) technology influence digital
economy?


What is the relationship between
(AI) and global digital economy development?
Could work activities in China be
automated making in the nation with the world's largest automation potential?
How does ( AI) technology influence
the future of employment change?
How can artificial intelligence impact
global economy growth?
Why will (AI) technology grow economic
development ?
How can (AI) technology impact to global
economic and social and psychological
changes?

Will (AI) technology influence digital
economy change to manufacturing
industry ?
What is artificial intelligence potential
benefits and ethical considerations?
How can (AI) technology influence to
global health care economy
development? P.191-216
Reference
Artificial intelligence and the future of defense
(AI) system immoral intention
(AI) soldier weapon ethical, social and
economic negative impact p.217-225
Chapter 3 Robots development stages brings what advantages and disadvantages to our societies
(AI) development first stage
Competitive influences between artificial
intelligence and human job
(AI) journalism, media publishing, digital
communication technologytrend
(AI) healthcare service industry
development
I Robot and internet things future machine
men invention

How artificial intelligence replaces human
job possibility p.226-240

(AI) development second stage
(AI) directions for future non-manual
control road vehicles market p.241-255

Reference

(AI) -driven automation
industry development
    5.1 How (AI) influences labor market
5.2 Redefining management in
the workforce of artificial intelligence
5.3 How (AI) influences organizational
change
5.4 Future works change:
Automation, employment
and productivity
5.5 What occupations will be influenced by (AI) technology. p.256-266


What does artificial intelligence(AI) mean?
● What (AI) function is?
● Can (AI) impact human job nature?
● How can human society job nature
to be changed to artificial intelligent society?
● Why does human need artificial intelligence machines?
● How does artificial intelligence influence future working changing in automation employment and productivity
aspects?
● Is artificial intelligence possible to replace labor ?
● Can (AI) technology replace human labour nature of work?
● Why can artificial intelligence satisfy human needs?
● Is artificial intelligence one good choice for human future technological benefit?
    What is relationship between
(AI) and economy growth?
● How can artificial intelligence technology influence economy?
● Can (AI) influence global economy growth?
● How can artificial intelligence impact global economy growth?
● What is the relationship between (AI) and (CRM)?
● What is the relationship between
(AI) and global digital economy development ?
    ● How can (AI) technology influence digital economy?
● Could work activities in China be automated
making in the nation with the world's largest automation potential?
● Will (AI) technology influence digital economy change to manufacturing industry ?
● What is artificial intelligence potential
benefits and ethical considerations?
● Can (AI) technology impact on customer relationship management (CRM) ?
● How can (AI) technology influence to global
health care economy development? p.267-283
    AI third stage development
Artificial intelligence education development
or the future of defense choice
8.1 Online technology and online book
technology influences artificial intelligence

defense system immoral intention
(AI) soldier weapon ethical, social and
economic negative impact

How does (AI) robots' brain invention
influence ourlives?
What is artificial intelligence
human brain invention?

How can artificial intelligent brain
satisfy to human beneficial and
natural needs?
Artificial intelligent brain future innovation and attribution p.310-325
16.1 (AI) brain invention successful factors
16.1.1 (AI) brain test experiments
16.1.2 (AI) brain invention successful factors

16.2 Artificial intelligence brain
invention opportunities
and challenges
16.2.1 (AI) brain legal remembering attribution
16.2.2 Whether human mind can create to (AI) brain mind
16.2.3 Brain-inspired intelligent
robotics

16.3 How can web intelligence need (AI) brain informatics?

16.4 Why model of sustainable development environment industry be (AI) robot attractive market
16.4.1 How to invent (AI) brain's environment protection ability?

16.5 (AI) Asia market
16.6 Is the brain a good model
for machine intelligence?
Reference
Does need to keep (AI) legal
4.1 Ethical and theological reflections
on artificial intelligence
4.2 Social choice ethics in artificial
intelligence p.326-333

● (AI) wearon moral or immoral
invention
Artificial intelligence and the future of defense

(AI) system immoral intention
(AI) soldier weapon ethical, social and
economic negative impact

Reference
Behavioral ethic in business
environment
1.1 Business ethic case
How scientifical ethic influences
business envirnment development
2.1 Scientist's social responsibility
2.2 What is scientific morality and scientists relationship?
    Rethinking morality in management
science
3.1 How common morality relates to business and professions?
Reference
    AI fifth stage development
Can non-manual driving public transport tools bring global economic growth
Why MTR underground train transportation needs to know passenger behaviour.
Why route choice can influence passenger behavioural choice.
Why trip time reliability and crowding factors can influence MTR passenger choice.
What is the crowding difference
between train and MTR underground
train.
How MTR can attract many
passengers. p.334-350
    What the psychological need differences between rail and bus and ferry passengers
    Reasons we need to improve public bus transport tool service quality
    What rail passengers really want rail innovation improvement
    What ferry passengers service improvement need
    ● How can ferry service be improved affordable, reliable, convenient, flexible and clean will get drivers out of
their cars ad onto environmentally responsible to passenger ferries?
    Future Human Transport Need Change

How future our transport need change? p.351-360
    What factors influence our future transport need change?
    How transport has changed from past to present?
    Future Non-Manual driving vehicle How
Influences Public Transport Tool Passenger Need

Why and how non-manual driving car owners need raise public transport quality on travel time and fare aspects
    How non human driving behavior can be influence
by non-manual driving cars
    Artificial Intelligent In Road Transportation Strategy

HOW DESIGNING UNDERGROUND MASS TRANSIT RAILWAY TO BRING PASSENGERS
    ● Designing transportation system advantages

Artificial intelligent public transport how influences passenger psychology
    How technology influence passenger psychology p.361-377

CAN NON-MANUAL DRIVING AIR PLANES EXCITE FUTURE AIRLINE AND AIRPORT TRANSPORT INDUSTRY DEVELOPMENT

Will airline industry's ticket price elasticity be influenced by demand and supply factor

● Boeing 747 manufacturing fuel cost strategy

● How airlines and airports implement successful netwpork strategies

● Performance measurement system strategy

● What factors influence cost-related management quality ? p.378-388

AI sixth stage development

How AI technology assists human to avoid energy and food waste

Factors influence householder energy using behavior at home

House quality influences householder
electricity energy consumption behavior

Environmental impacts of householder
greenhouse gas electricity energy

The effect of house space occupancy
and building characteristics on
householder electricity energy use

How to help low income household
earners to reduce not essential
electricity energy expenditure
spending at homes

Factors influence householder energy
efficient consumption behaviors
at homes

(AI) technology how impacts food
wastage behaviors

Why environmental factor can influence
food shortage

Whether environmental factor is the main
factorto influence food consumers to waste
food p.389-401

The failure of human education method
influences consumers to change food and
energy waste behaviours

AI seventh stage development

(AI) Technological Space tourism station development

1.1 Challenges to install (AI) automated system in space stations

How can artificial intelligence be
applied to adapt the nature of space
and satellite industry

What is (AI) unmanned aircraft system? Real-time data analytics

(AI) robotic detective system

The ethic challenges to (AI) space robotic learning system in space environment application

Why human network job behavior may influence economy

Robots take our jobs behavioral and economy influences
    Robot job behavior brings economy influences

Intellectual human economic behaviors
What does intellectual human economic behaviors
mean ?
    The relationship between social change and human
behavior
    How human productive behavior may influence economic development
● New Zealand farmer individual wine productive behavior
● America high technological productive behavior
● China share market investing behavior
Why has any individual country have many people invest share behavior which can influence the country's macro
consumption desire?
Can technology influence human shopping behavioral change?
    Why and how human behavior may influence the country's economic growth or recession?
Technology how impacts human behavior changing?
How and why employees behaviors may influence economy development?
Robots invention whether they can help organizations to raise efficiencies or inefficiencies? p.432-463

# CHAPTER I

# Technology how influences developed countries social change

UK Future Technology Development Trend

Online teaching technology

Future, online teaching method will be popular to be applied to teach to any university, even secondary and primary schools. Because internet service is free charge to any students in any countries. Many different age students who can know how to apply internet as well as internet studying is very convenient to any students who can to internet to learn or study in home or public library or school library conveniently. Teachers do not need spend much time to teach students in classroom. They can use internet to teach teachers by face to face seeing and talking to their individual student from every student's computer. So, students do not also often spend much time to go to school to learn. So, developing any fast speed and time saving and talking and listening online teaching methods will be popular needs to any UK primary and secondary and university students in the future. It will be one new technological teaching method to change the traditional classroom educational method in UK and schools. For example, when one UK student who had left UK and is living in another country long time. If any UK school did not provide online teaching service to any UK students. It means that the UK citizen can not choose study himself/herself any UK school if who still hope to study any UK course when who is living in another country. Even one foreign student who does not go to UK to study, if he/she can find any UK primary or secondary or university to study from online. Then, the UK school won't lose one foreign student, due to it does not provide online teaching method to any foreign students. So, online Technology educational learning method will be one popular learning method which is enhanced, supported, mediated or assessed by the use of electronic media. Technology also enhanced learning may involve the use of new or established technology and/or the creation of new learning material. It may be deployed both locally and at a distance ( i.e. a combination of traditional and e-learning approaches), to learning that is delivered entirely online. Online learning technology characteristics ( features) include identification of a project lead for each area of any learning strategy, identification of two " quick win" for example lecture capture, electronic submission and feedback.

How can online technology enhance learning at UK any schools? It will include these several aspects to analyze. On identifying, prioritizing and innovation hand, online technology is a process for resourcing, prioritizing, acquiring and evaluating school software and hardware for UK any school needs. On staff development learning plan and a student skills development plan hand, UK schools need to establish a base-line policy on the standard ( minimum) technology enhanced learning expectation for education each program and module and a mechanism for updating the schools' policies. On evaluation and research hand, a mechanism for engaging the owners of the technology enhanced learning strategy with best practice in the sector including contributing to and benefiting from pedagogical research and the evaluation of the student experience to UK any school.

Thus, UK schools can apply online technology to develop on educational aspect, such as digital literacies and appropriate technical skills that equip UK students for life-long learning, graduate level employment and professional practice, be empowered to learn how to learn with online teaching technology, using online technology to engage in interactive, creative and co-constructed learning with the potential for online learning in an interdisciplinary and international context, using online teaching technology to engage in learning with and from people from anywhere in the world, be supported on placement and in workplace learning through mobile applications and other supportive technologies that facilitate their online learning when away from the classroom, having access to innovative methods of online learning teaching and assessment that are the foundation of a research-lead academic environment, engaging with UK schools in developing , implementing and reviewing the technology enhanced learning strategy. Thus, in the future, it is important to build a capacity to the online education strategy to adopt future learning innovation and student individual online learning need (demand) to UK any school online teaching trend.

There are many examples where UK academics working in isolation or in small UK teaching organizations or classroom learning groups have developed teaching innovation that have a positive impact on UK students' academic experience , but these have remained isolated to particular modules or occasionally program. The aim of education researching online learning process is to identify the good online teaching innovation that is being developed and to prioritize those that have the potential to make a significant contribution to improving the academic student experience at UK any schools. This online teaching process will need any UK schools which can plan how to apply limited resources necessary to achieve online teaching. In addition, the online teaching research process would evaluate and prioritize large scale educational software and hardware requests for primary, secondary and university students' requests. An important part to this process will be to ensure the integration of online educational products and packages that school staff and students regular use to make routine working and access as seamless as possible.

Decisions about school administrative online technologies should not be taken in isolation before assessing the impact on UK teaching staff. In addition, a range of techniques such as, online expert facilitation, coaching and peer support will be used to support individuals, groups or longer academic units, who are learning on major technology enhanced online learning projects. Staff engagement may also facilitated through incorporating technology that is used in teaching staff research and/or professional activity that can be cooperated into their teaching.

Online learning technology can develop UK students skills, UK schools need to understand how UK students understand technology and learn with it, therefore the digital literacy strategy needs to be considered as part of the overall strategy as well as the relevant skills development in UK employability strategy. So, in the future, online learning strategy will make it clear that students will develop technical skills the appropriate level for graduate employability and professional practice. Also, in the future, the online technology can enhance learning working group to discuss external development, that are of educational strategic importance, understanding and evaluating current best practice and research and understanding and evaluating the online educational strategic contribution that pedagogical research and student feedback can have on online educational strategy, policy and practice. The e-learning unit is responsible for informing and educating. This could be done by, for example, providing a short digest of relevant information for each meeting and by setting aside a proportion of each school meeting to discuss a topic of particular online educational strategic interest to every school. Academics that have not got a specialist interest in online educational technology enhanced learning will need relevant information at an appropriate time. This could be provided at a school department or faculty level and this will have clear links to the staff online teaching development plan. Hence, online educational development strategy will influence any UK educational school technological improvement in the future.


Environmental protection technology

Why does environment technology valid to UK businessmen and government to develop? Nowadays, global air and water pollution is serious. Even, UK has many farming is polluted by the water and air pollution. It will influence UK farmers' income if whose farm land ( natural resource) is polluted by water or air ( natural resource). Even it will influence UK citizen will encounter food shortage if UK farmers can not grow any fresh and health food to provide the enough food numbers to eat every day. Moreover, air and water pollution will influence UK citizen drink the polluted water and breathe the dirty air to live every day. This natural resource ( air and water challenge) will influence UK citizen health to cause illness , even death very easily. So, UK government can not neglect the natural environment pollution challenge. The environmental protection technology will help the UK and development countries to solve the challenge of climate change to avoid or reduce farming, foods, or vegetable or fruits or rice, pork, livestock numbers loss threats, i.e. the development and deployment of low carbon energy technology, including technology for the efficient use of energy. The commercialization of low carbon energy and energy efficiency technologies in the UK, with a specific focus on the demonstration and deployment phases of bringing low carbon technologies to UK market.

The UK Government needs to deliver a low carbon economy and to meet UK ambitions emission reduction target. So, low carbon and environmental protection technology researching and development will reduce the carbon intensity of energy production as well as reduce energy demand, towards meeting the contributing UK's ambitions production as well as reduce energy demand, and renewable energy goals. The use of energy ( including transportation fuel) and the UK's targets on climate change, for example, by helping the UK make a step change in increasing deployment of renewable energy, improving UK energy efficiency and helping low carbon technologies reach the market. The development of low carbon technologies, and to realize the benefits of doing so in terms ensuring security of energy supply for the UK future economy development.

In UK, private sector investment in technology innovation in the low carbon energy sector will other sectors of the economy. So, in UK energy technologies are likely needed to be developed to avoid dangerous climate change, or an acceptable cost. So, in the future, UK government will need to consider to research environment protection and low carbon energy technology. The activities will reduce carbon emissions, or have the potential to reduce carbon emissions on the longer term, through the use of energy technology will accelerate development and deployment of low-carbon energy and energy efficiency technologies will capacity in the demonstration and deployment of low carbon technologies. Innovation in the energy sector is the only way to identify, develop and reduce the costs of new and improved technologies for the extractions, generation, distribution and use of energy. It has long been an important means of achieving the UK's energy policy aims of a secure and affordable energy supply, as well as to develop the environmentally friendly technologies that are required in UK response to climate change, i.e. nuclear, wind or water, sun energy technology, which is future new energy technology is suitable to research to create to apply instead of current electricity energy.

How global warming influences UK agriculture growth. Scientists have also been fighting the use of chlorine in municipal water systems to kill various strands of bacteria. Chlorine reduces by about 80% the number of alimentary tract diseases relative to polluted, unchlorinated water. A relatively new genetically modified agricultural products. They were partly successful in Europe, such as UK (some countries banned genetically modified products) in spite of the fact that neither history nor research supports their case. People began to modify plants as early as the beginning of the agricultural revolution (8000 to 10,000 years ago), when they started seed selection and who have continued ever since. The green revolution of the 1960 year brought about strains of grans and rice more resistant to a variety of local conditions. The effects have been that countries like India, which had suffered from recurrent famines over the millennia, became self-sufficient in food due to the resultant sharp increase in agricultural productivity. It was a real science and technology over the poverty dominating most of human history. But it is precisely the products of science and technology that ecologists are so deathly afraid of. In an interesting study in a quarter ( 28%) of clinically analyzed cases of obsessive compulsive disorder were cases resulting from the fear of global warming.

To destroy the modern, whether industrial or postindustrial, civilization, human have to destroy an important engine of economic growth, that is its energy sources. And this is what eco-warriors try to achieve under the banner of against global warming. Thus, UK government will have responsibility to attempt to research new technology to fight global warming challenge for itself farmer benefits and even global benefits both on the future.


Future Economist global warming technological protection economic influence opinion

Some future economists indicated reasons to explain why UK government and businessmen needed to consider how to develop natural environment protection technology to avoid global warming challenge to influence UK economy development. They indicated the anthropogenic (human-made) global warming resulting from the increase in "greenhouse gas". They offered their perspectives on the scientific valid of anthropogenic global warming phenomenon, its probability of occur and expected consequences and is dominated by technologists, economists and political scientists, who considered the need to make the horribly costly adjustments in energy generation and usage suggested by climate alarmists.

Many stress that global warming is primarily caused by other phenomena than human use of fossil fuels or human activities in general. They are looking at the activities of the sun and impact of the larger universe as the main source of global warming and stress that global warmings (plural) happen intermittently with global cooling. I shall explain why global climate warming will influence to the political and economics of the issue to UK country. For it is the latter, rather than the global warming itself, that will pose a challenge to the Western world, such as UK and the world at large in the future. Scientists concerned who should move forward with policy measures to avert the alleged disaster. They also apply manufacturing theories to support enough to frighten politicians into action and scare societies into acceptance of measures that would sharply reduce UK citizen their living standards. Otherwise, UK politicians had support that bureaucracies were established, money allocated and lobbies created dependent on the new kind of subsidies. In consequences, climate alarmism and resultant interventions in national economies and human activities have become the increasingly wide spread and increasingly cost reality. With the growing availability of money distributed, and even more promised, a range of benefit of the global warming machinery has been on the increase. So, if UK government did not concern how to innovate new weather protection technology to avoid climate change adverse (poor) influence. It is possible that billions of dollars of UK public money are needed to spend on research global warming challenge because global warming will influence UK agricultural industry. UK agricultural industry is one important export income source to raise UK GDP income every year. If global warming become very serious to influence UK weather to be bad to cause UK farmers who can not grow good taste food and vegetable to supply to domestic and overseas food consumers to eat. Then, UK will loss much GDP income from local agricultural export sale. It seems global warming and agricultural production which has direct relationship to influence UK economy development in the future.

The main problem with climatology is that it must be based as already stressed on very many variables affecting climate and too few hard data necessity. Differences apply not only with respect to the scale of changes obtained, but even to their direction (rising or declining temperature). Some weather scientists indicated to concern global warming challenge. In consequence, it would be impossible to discover if and where errors were made not only in estimating relationships between variables but also in the quality of data used. (Hauser, J. Tellis, G. J; Griffin, A. 2006) They were comparing average temperatures measured some 30, 40 or 50 years ago by, say, 90 % weather stations in the countryside and 10 % stations in the cities with contemporary average temperatures measured by weather stations located today on 50:50 basis in the countryside and cities. Then, one could obtain the increasing temperatures without any real world climate or even weather changes. Comparability would be ensured if the same number of countryside-located and city-located weather stations had been compared for different periods. The alarmists intentionally mix up " temperature growth" with the trend of temperature growth. To give an example, if in the first decade the temperature grew by 0.5 % degree, in second decade it grew by 0.3 % and in third decade it grew by 0.1%, what was registered was a growth in the temperature, but certainly not a trend of growing temperature. A fourth decade should, on the basis of the trend, bring about no change in the temperature.

To conclude, scientists believed that global warming was caused by human's bad behavior more than natural environment influence. So, it is human's responsibility needs to solve this challenge, due to who feel earning profit aim is more important to protect natural environment, e.g. air and water pollution , due to manufacturing process is the main factor. So, UK has responsibility to attempt to research how to solve global warming challenge , such as it has many famous scientists who can devote their scientific skills to cooperate to solve global warming challenge with other countries' scientists. Some weather scientists also hypothesized that human may be at the end of the present warming period. If they are right, it would be bad for humanity, as warmer periods have always been associated with better conditions for economic activity. To sum up, scientists believed that global warming will influence human economic activity to be bad.


The psychology and political economy of global warming success to UK farmers

Climate alarmists were able to convince a large part of the Western public and a majority of Western politicians of the cause of fighting against the global warming. It supposes itself in an instinctive preference for collectivist solutions in economic and social spheres, with negative to disastrous consequences when scientists are applied in practice, so UK government needs to concern global warming challenge, due to it is possible that it will influence UK natural environment weather to be poor to influence many UK farmers' agricultural and vegetable and fruit and rice wheat etc. food growth successfully. What is the global warming influence to cause disease? For example, ecological alarmists and activists ( eco-warriors ) never admit they are wrong, they long pursued their fear mongering campaign against chlorine. Their success in branding DDT a dangerous substance had a negative impact on the malaria eradication campaign in poorer parts of the world. Alternatives to be have been far less effective and the result has been the resurgence of malaria cases and the manifold increase in malaria -caused deaths to the largest extent in Africa.

Automation technology in manufacturing industry

Nowadays, UK computer and space explore technology had reached the mature stage. It means that UK government ought not need to continue spend much resource to research these two kind technologies. Otherwise, the automatic manufacturing technology, e.g. human intelligence new product. It has need to develop because human intelligence machines will bring beneficial to satisfy human everyday life need, e.g. hospital patients' activities need, if the patent who can not walk easily, but the human intelligence machine can assist the patient walk to anywhere conveniently. So, he/she does not need to sit on wheel chair and apply the human intelligence machine man to help him/her to drive on the intelligence automatic driving vehicle to go to anywhere conveniently.

Otherwise, increased automation in low wage countries, e.g. China, Korea, Africa, Hong Kong etc. which have traditionally manufacturing firms, could use automatic technological manufacturing to bring lose cost advantage and potentially lose their ability of achieving rapid economy growth by shifting workers to factory jobs. So, UK government and businessmen needs to consider automation technology development, i.e. 3D printing manufacturing industry will encourage UK companies to move manufacturing process, closer to gain the biggest advantage from this 3D automation technology development.

A growing concern of premature de-industrialization in energy and developing countries could require new models and a need un-skillful the UK workforce. In the future, the best way toward for UK cities will reduce their exposure to automation is to boost their technological dynamic and attract more UK skilled workers. Automation technology progress can give UK manufacturers' employee benefits, such as long term healthy productivity improvement, raising productivity efficiency and product quality, macroeconomic and microeconomic effects of automation technological change, it's change will be beneficial to UK society, i.e. automation active labor market policies, which could help UK job seekers find jobs from training to incentive to support self-employment to create high technological job employment chance in UK society. So, raising science, technology, engineering and math subjects update skills level are needed to UK any universities, which can be increasingly important in UK society, these factors could complicate the ability of UK high automation technology education to adopt to the UK automation manufacturing technological change. A talent mismatch already exists in UK, with many well UK educated workers can find employment in lower-skilled jobs. To combat this, greater coordination will be needed between the education, training and employment sectors in UK society.

Why are high automatic technology product development models needed to research to UK any manufacturers? UK government and manufacturers need to consider how to achieve high technology product development models. According to Hauser et al. (2006) indicated the high technology (high tech.) development process, is influenced by the innovative process, bringing products on exception value which stimulate product market demand. Innovation provides products the specific basis for which world economies compete with each other on the global market. Able to find new solutions, innovations generate significant changes in existing markets, destroy them, or create new marketing ( Hauser et al. 2006). So, UK manufacturers need to concern on any manufacturing high technology product development process because which can influence any new products development to manufacture to sell to any overseas or domestic both markets successfully.

What is high tech. product meaning? Mohr et al. (2010) argues that there are two reasons why it is important to clarify and specific high technology : (1) due to the impact of technologies on the economy, attempts are made to classify economic production and incomes ; (2) due to the impact of high tech. on the environment. Standard marketing strategies are being modified and adopted , therefore, it is necessary to know the products to focus on. Why UK manufacturers need to consider high technological product process. Nowadays, high tech. products are complex, advanced, requiring specific technical knowledge, which is technologically not discontinued and being produced at the companies which have twice as many technical personnel and invest twice as many in scientific research and development than other companies. Moreover, these products are time-sensitive as scientists are continuously searching for new approaches for invention of more advanced technologies which make all preceding ones lower-ranking. The most important, nowadays global consumers will adopt the particular technology. It means that global customers may delay adopting new high-tech. products and in order to mitigate the prolonged uncertainty require a high degree of education and information about the product and need post-purchase reassurance.

Anyway, nowadays customer individual needs in high tech. environments are characterized by sudden changes related to unpredictable fashion. Even, consumers concern about how to preserve new product' competitive technological standard is completely incompatible with technological uncertainty. The most important factor is the prevalence rate of any new products development process, which is influenced by slower than of traditional products. In many cases high-tech. automatic product market are being materialized slower than which are expected. The technological uncertainty challenges will exist in development process, such as uncertainty related to the timetable for development of the question whether the new product will be function as promised. In automatic high-tech. industries, the time requires for product development is difficult to predict as , commonly, it takes longer than expected , uncertainty related to unanticipated consequences and uncertainty about the product life cycle related to competition products. In conclusion, these factors will influence new automatic technology product development process unsuccessful, so UK manufacturers will need to concern on any high technological automatic product's manufacturing process.

Future economists predict automatic technology how to influence future UK economy

Before, all over the world presented picture of demonstrate in London on the occasion of the meeting of the G20. Some economists indicated disastrous economy consequences will occur to any one of Western country , such as UK, so if any one of Western country did not consider automatic technology development to itself country. They indicated one example, such as material incentives to produce disappeared throughout Russia and, when Society leadership called off the experiment, the country faced industrial output reduced to 10% of what had been registered in 1914 and agricultural output reduced to such low levels as to cause widespread famine.

Why would UK encounter disastrous economy consequences if UK government did not encourage manufacturers spend money to invest to innovate automatic technology industry? According to a variety of anthropological studies, a collectivity is unable to operate efficiently with everybody giving talent workers have chance to devote whose best effort to manufacture any high technological products, e.g. human intelligence vehicle or airplane. Hence, economic incentives are needed to UK manufacturers to invest high technological automatic industry development. Because the economists predict UK will have many talent worker numbers, their number will be more than a certain number of normal effort workers, due to UK technological education level is very excellent to provide to train many young technological manufacturing students to find this kind of high technological manufacturing job. So, the high technological manufacturing job seekers will increase and it won't decrease to UK job market in the future.

Assuming that UK high technological automatic manufacturing workers who would desire only to introduce changes in the workings of the international economic order and policies of countries participating in the present economic order rather than change the order itself, what will be UK manufacturers their specific economic preferences in the future? It implies tnat either concentrate on spending more investment to automatic high technological development, e.g. human intelligence automatic high technological products or still concentrate on spending more investment to common traditional technological products.

However, UK was a developed Western country which had had strong automatic high technological development effort very long time. Otherwise, it compared to some developing countries, such as Asian China, Hong Kong, Korea etc. Asian countries their future economic growth rate will show un- surprising , different patterns, so the Asian countries has weak effort to invest high automatic technological product development, such as human intelligence technological development. The catching-up process suggests low economic growth rate in the high automatic technological product development to the Asian developing countries in the future.

Hence, the future economists predict that it views as probable successors of the Western world economic leadership if any Western country , such as UK manufacturers who prefer to invest to any high automatic technological products development , e.g. developing on human intelligence automatic technological products more than traditional common technological products development. On the one side, but it seems important to stress that two very poor countries among the challengers-China and India-are examples of countries that changed their institutions and economic policies from no or little economic freedom to more economic freedom. Because there two countries whose governments prefer to lend loans to encourage their country manufacturers prefer to invest high automatic technological products manufacturing. On the other side, attitudes toward foreign direct investment (FDI) have undergone change since the 1960 s and a large majority of less developed countries, e.g. China and India are now competing strongly among themselves and with developed market economies for direct investment from multinational companies. So, UK will face China and India high automatic technological product competitors in the future. And in fact, all countries that joined Western developed economies did that without much (if any) external inflow of public resources. It is right time that UK government needs to lend loans to encourage domestic manufacturers to invest high automatic technological products to raise whose international high technological products sale effort to win its future competitors. So, machine resources will be increased demand to o UK manufacturers if who chose to spend machine resources to innovate to manufacture any new and high technological automatic products to raise human daily life needs in the future. It means that it is right time UK manufacturers need buy much machines to prepare to manufacture many future high technological automatic products when these machine prices are low. Because the future global machine prices will possible be raised if many China and India manufacturers will also buy many machines in the future. For example, USA government had provided much financial support to assist sugar cane producers to develop their businesses. And they are dependent to a much larger extent than sugar cane producers and sugar processors in the USA on government. Without very high subsidies to renewable energy generation, they would not have survived at all. So, USA government had been the first country which could lent much financial assistance to encourage domestic renewable energy generation manufacturers to develop high technological energy manufacturing business. So, UK government needs follow USA to lend financial assistance to encourage domestic high technological automatic industry development.

Future economists also predict China and India will be competitors for future leadership in the global economy, special high technological products. China has been the media and analyst's favorite for quite some time. Quantitative projections have seemingly supported such expectation. Such as China and India had manufactured many high technological new space rockets products, ocean war large ships etc. Moreover, China has become one of the major world trade players in the early twenty-first century.

Many long-term forecasts, assuming similarly high economic growth rates in the decades ahead, predict that China will surpass the USA in terms of aggregate GDP somewhere between 2020 and 2030 or later, say between 2030 and 2050 year. The future economists conclude on the basis of these predictions that China will not only pass the USA in aggregate product (GDP), but its economy and economic policies will influence the rest of the world to a similar extent that the USA does at present.

I stressed a very important point, namely that the UK future high technological automatic product competitor China and India, namely that economies not only grow, but in the process change their structure. China and India have been industry very rapidly (the first transition) and building the physical infrastructure that accompanies industrialization changes to technology in the future. However, at a certain per capita GNP level the two countries, such as China and India will face another structural shift when which technological development will reach the mature stage in the future. China and India had been primarily historical pattern of economic development because the shift in the role of engine of growth from industry to services is to a much greater extent a qualitative shift. Both higher and different skills are required. And, even more importantly, interactions generating ideas driving the highly human-capital-intensive service economy require a much freer environment, not only in the economic area. Chinese exports have been heavily labor-intensive. This being the case, they contributed to the expansion of industrial employment, offering for the first time in the history of China a taste of (very modest) prosperity to more than 100 million new industrial workers and their families. This is the major component of the success accomplished by Chinese economic growth. Richer trade partners create room for more trade, so the Chinese should hope that intra-South trade, that is, trade between the emerging economies of Asia, the Middle East, Africa and Latin America, will open up new and growing opportunities. I presume that if Western economy , such as UK did not developed high technological automatic industry to stable their social welfare, so thoroughly slowed down their economic growth.

Will it allow China to accomplish the transition to a mature, innovation, service-sector-based market economy? It has allowed the economy to industrialize much more successfully, even if the labor shift from agriculture to industry has not yet been completed. But it is a long way off the next major test: the second high technological industry transition of the economic structure to China. Bear in mind that Russia attempted it twice and failed at both attempts.

But even, assuming that China at some point in the future does succeed in accomplishing the second transition, will it be able to supersede the USA, for example, as the main global high automatic technological innovation center if it wants to become the No.1 global high technological industry economy? Given the nature of the centralized state and its stability to collect financial resources , China's ability to increase research and development expenditure to high automatic technological products and to hire a mass of researchers, engineers, technicians and other specialists should not be doubted. This process in already taking place.

But , again, Soviet Russia already exceed the USA in the R&D/GDP ratio in the 1970s, long before the communist collapse, with no effects on its innovativeness. Inputs matter less than outputs, quantity in the innovation process mean much less than quality. The latter characteristics depends importantly on economic, civic and even political institutions. Otherwise, independent India had three options open to it in 1946s. It could pursue spontaneous economic development, with some state intervention to be sure, along the lines of basically free market capitalism; it could turn the clock back and try to recreate the rural-agricultural and handicraft based. The dominant way of thinking was Society -style priority to industrialization and , within industrialization , priority to heavy industry. In other words, not textiles and clothing, which has been developing well in India since the mid- nine teen century, but production of sewing machines and , even better, production of machines the produce sewing machines.

The results were only to be expected. The heavy stress on the expansion of capital-intensive heavy industries in a very poor country quickly strained the ability of the Indian economy to generate adequate savings. Moreover, some of these industries were above the level of industrial competence of an underdeveloped economy. Thus, the amount of required resources (capital, skilled labor) was usually larger per unit of output than in the same industries in more mature, richer industries economies. In another view point, India will develop light industries, just as any other poor country with a great deal of unskilled labor, had a comparative advantage and no less importantly, an economy in which, due to their low capital/labor ratio, light industries could employ many more people, spreading prosperity more widely in a poor country. So, it explain that why China will have more effort to develop heavy high technological industry in the future. Thus, India got less economic efficiency, less employment than in a spontaneously developing economy, less ability to compete internationally in light industries suitable for an underdeveloped economy and finally got heavy industry unable to compete even on the domestic market and, therefore requiring no less heavy a dose of protection. Overall India got an underperforming economy, in particular in its relations with the rest of the world.

To conclude by comparing the performance of the traditional sectors of the Indian economy and the performance of its modern, human -capital-intensive subsector of manufacturing and skill intensive service sector. The latter both employ workers with high-and medium -high skillful level ( in branches ranging from computer software and biotechnology and pharmaceutical high technological light industry). India is ahead of China in terms of the output and export of such products and services. Thus, it implies that UK ought concentrate on developing high automatic heavy high technological industry, e.g. human intelligence technological products because these industry is not better development to other many countries' strong effort , such China and India large population countries.

US future Technology Development Trend

 Developing countries cities technological

competitive investment factor

 Some economists predict developing countries every city in the global 750 is projected to have a larger future technological economy growth. But the diversity of developing countries' economic performance is large. Developing economy cities, such as China, Japan, Hong Kong, Korea cities can grow rapidly by acquiring capital and technological know-how and putting them to use by their rapidly growing urban labor forces. Even, these developing countries cities' rapid technological development can impact to US labor market supply.

Due to future rapid technological development to these developing countries, the result will cause developing countries cities, such as Asia China, Hong Kong, India cities economy growth will rapidly. Otherwise, developed western countries cities, such as US, UK, Canada, Australia, Span, Germany etc. countries lie close to the technological frontier have stable urban populations and more limited investment and job creation opportunities. It will influence these developed countries' economic growth is slow than the developing countries. Due to developing Asia countries cities will prefer to invest more technological development to compare to developed countries, such as US, UK, Canada, Australia, Span, Germany etc. countries. Therefore developed countries, such as US, UK, Canada, Australia, Span, Germany etc. countries tend to grow more slowly. It seems developing Asia countries, such as Hong Kong, China, Korea etc. countries urban and central cities economic performance within developing countries will be better than developed countries, such as UK, US urban and central cities within five years.

Thus, I suppose that future developed countries, such as US the speed of technological development will be slower to compare developing countries, such as Hong Kong , China, Korea, India etc. countries. Then, it will bring this question: How to solve developed countries, such as US, Washington, New York etc. cities future economic slow growth challenge? For future developed country cities, such as New York, Washington etc. large cities' technological investment and location decision, it will need understand that diversity is essential. Various factors can have an impact on US country intra-national urban cities performance, including sector structures, agglomeration benefits, infrastructure quality. For example, US country central government needs have tolerance of diverse performance, land supply and US city governance plan. It aims to raise US central and urban cities' future economic performance. Moreover, US country will need to concern above issues to arrange how to improve central and urban city economic development . Specially, it will need to concern technology innovation to encourage many overseas technology investors to anticipate different kinds of technology products innovation, e.g. human intelligence machine, mobile, computer etc. high technology products. It aims to absorb different countries' new technological knowledge of different kinds of high technological products to excite US domestic or foreign countries consumers' desires to choose to buy many different kinds of high technological products to raise US GDP growth in US technological product sector industry in the future.

What are the trends impact on rural America's future economy?

Some economists indicate that there are five trends reshape to impact rural America' future economy. They include that digital economy will shift future America rural economy. US quality of life will change a lot, the US rural economy will stay uneven, US commodities will compete in global markets and will give less benefit to US rural economy and US new products will revolutionize US agriculture economy.

The first aspect impacts to US agriculture economy, the US future rural economy stays uneven. Growth will concentrate in 4 out of 10 rural places and they have scenery, a retail hub, or one next to a city in US. The impact on rural America includes some rural places will try to manage growth , but many places on a quest for new economic engines. Thus, it will bring these questions to US rural economy impact, such as : Who will be US businessmen clients? The struggling farmers? The struggling farm-dependent country? The booming mountain area? The rural area transforming into city?

The second aspect impacts to US agriculture economy, due to US commodities will compete in global markets if it will bring a smaller benefit in US rural economy. Then, US farm scale will cut costs and competition fewer US farms and places will depend on farm income. What is the impact of commoditization on rural America? There are more farms depend on area jobs and it creates a new imperative to add value. What is the impact of commodity on US clients? The need for competitive commodities remains, but the payoff for added value is rising and community impacts are important.

The third aspect impact to US agriculture economy, new products will revolutionize agriculture. Major, shift from commodities to products, spurred by biotech, means two agriculture in the future and two rural America . Hence, US future agriculture determination the rural economy will be declined. A future new US agriculture supply chain integrator will be caused from the traditional farming supply chain procedure the change to outsourced contractor farming supply chain procedure, such as: In beginning, from farmer will outsource supply chain contract to processor, then contract to distributor and contract to food retailer final step.

The fourth aspect impacts to US agriculture economy, what will be two agricultures impact to the US future economy? The first US agriculture impact will be US commodity agriculture. It focuses on production capabilities, farming foods production will be thin margins maintained with technology and big sale. The second US agriculture impact will be product agriculture. It focuses on consumer needs, farming foods production margins will be protected by capturing value and building business relationship. Thus, it will bring these questions concern what the impact of product agriculture which can influence to US economy. How to apply biotechnological agricultural techniques and farming product application to raise US farming productivity? How to build world class agricultural ( farm) producing chains that benefit to US farming produces production method? How can US faming foods producers participation or lead in product chains?

In the fifth aspect impacts to US agricultural economy, it is digital economy impacts US rural agriculture. Future diversity economic base can encourage US farming product agriculture to enter rural digital agricultural service consumption sector to change US agricultural consumer individual shopping habit. What will impact to US farmers? Digital agricultural economy will bring these questions: Who will lead in broadband digital agricultural technology? How to launch e-markets and businesses and how to cooperate with US rural farmers? What will impact on US agricultural on natural resource management aspect and on how to US commodity to build amenity strategies? However, rural America's future will be shaped by policies that will encourage agricultural technology adoption, it will enhance farming worker skills, it will improve rural quality of life and it will ensure access to capital to bring US agricultural economy growth.

High level education and high
birth rate factor influence US technological economic growth

According to John G. (2016) indicated that " a point forecast is that GDP per capita will rise well under 1% per year in the longer run, with overall GDP growth of a little over 1 to 12%. The main drivers of slow growth are educational attainment and demographics. First, rising educational attainment will add less to productivity growth than it did historically. Second because of the aging and retirement of baby boomers, employment will rise more slowly than population which in turn, is projected to rise slowly relative to history." Thus, it seems that some economists believe education and born rate will influence future US employment method. They assume that the employment and growth ratio will rise and unemployment ratio will deadline of US will increase birth ratio and future there are many young people ( students) can have chance to accept high education degree to raise whose education level to prepare to enter different kinds of high educational jobs market to work in the future.

Considering a more growth accounting perceptive on productivity growth. According to the analysis in Fernald (2015) which uses a multi-sector growth model for the projections. It indicated that although, the details differ and if bases projections on TFP data since 2004, it also implies a preferred point estimate of 1.6% per year in US over this short period. The " fundamentals" of labor-productivity growth (namely, growth in total factor productivity , TFP) have exceeded the actual realization for reasons that reflect the unwinding of dynamic of labor quality and capital depending associated with the great recession. Thus, how to raise labor quality which concerns how to raise educational level to young people in any country will be one labor economic challenge in US in the future.

Fernald (2015) also explained " the shortfall" in productivity growth relative to fundamentals reflects, at least in part, the unwinding of two dynamics associated with the great recession. He indicated that first, at the end of the great recession business had a lot of capital relative to labor, which has attended the need of add capacity to meet demand in recent years. In contrast to the outsized growth in capital deepening from 2007 to 2010 year, we have seen capital "shallowing" since 2010 year. Second, businesses fired low-skilled workers during the recession, which raised labor quality in 2007 year to 2010 year period. As these potential workers have been rehired, the growth rate of labor quality has added less.

Thus, on the one hand, when one country encounters economic recession, many employers will like to employ high educational level and high skilled labor to raise productivity. If US encountered economic recession, but it had none (lacked) enough high educational level and high skilled knowledgeable workers to be supplied to the labor market in US. Then, US will encounter long term economic recession period, it can't shorten economic recession period. Thus, US will need to let many US young people have effort to study to prepare enough high educational level knowledge to do different kinds of professional or high knowledgeable or high technological jobs in future US labor market.

However, US high knowledgeable labor can also assist US businessmen to raise productivity growth. Due to labor quality could be grown by high educational and increasing number of knowledgeable workers. On the other hand, moreover, if there are many US young married parents who will either choose not born any babies for next generation or choose born only one to four children. Then, future US will have a small numbers of young people to be supplied to primary schools, high schools, even universities to study. It also means the university level graduate student numbers will decrease also. So, low birth rate factor can also influence the high knowledgeable labor numbers to be supplied to future US labor market.

So, I recommend that US government needs to encourage many young parents choose born next generation and provides student loans to lend to poor families to support whose children can have more chance to attempt to go to high schools or universities to study in the future.

Socio-economic and political factor
impact future US strategic
policy change?

US development practitioners are increasingly aware of the role that US social and political structures play in future shaping US's development paths and results. In this context, US macro social analysis needs to understand the ways in which power relations act to circumscribe the opportunities available to poor US people to improve their situation. For example, US donor organizations need to understand of relevant US social structures, such as informal institutions or other relevant US social practices in US.

This provides an entry point for understanding the broader US political environment or challenges in a particular sector or process. Furthermore, any US donor organizations need to place greater emphasis on the analysis of livelihoods and economic opportunities and their relationship to reduce the unequal of gap between rich and poor US citizen in US societies . Thus, US government needs to encourage donor organization participants choose to do the reasonable donor behavioral to aim to reduce the unfair donor spending to the unneeded donor assistance beneficiaries. To let us socio-economic has more balance chance to every US poor citizen in US society. So, it can also raise average every US poor citizen feels better quality of life in US society.

How does the rise of US exports to East Asia factor influence US economy change? Export have become an increasing important source of revenue for both national and regional forms in the United States. How does the primary growth market for US exports to influence US economy growth? I recommend the developing nations in East Asia will the developing nations in East Asia will soon rival today's industrial nations as the most important US trading partner . In view of the rapid growth of US exports and their geographic shift toward developing nations.

Some economists indicates that two tends in recent US export performance are particularly notable. First, US exports have increased rapidly relative to total US output, development with important implications for the entire economy. Approximately 130,000 US firms employing over 10 million domestic workers, export their products ( US Bureau of the census,1993). Second, the geographic distribution of US exports has been shifting dramatically. The industrialized nations of the organization for economic cooperation and development ( OCED) account for about 57% of all US exports . However, in recent years of the start of US exports to major trading partners in East Asia. America has been rising and now accounts for 26% of the total. Thus, economic policymakers can no longer make trade related decisions without considering their effect on US exports to East Asia market.

From 1981 to 1987 year, two factors temporarily stemmed the rising tide of 1980 year. First, the US dollar began to appreciate significantly in 1981 year as the United States enacted a restrictive monetary policy and an expansion fiscal policy. From 1981 to 1985 years, the dollar rose almost 50%, thus making US exports more expensive relative to foreign products ( Hakki & Whittakersand J, Gregg Whittaker, 1985. " The US dollar recent developments, outlook, and policy options". Federal reserve bank of Kansas city, economic review, Sept./Oct., pp.3-15.) Thus, future US will need to make good international export and import trade relationship with Asia countries, e.g. China, Hong Kong, Japan etc . countries to raise GDP income.

Education factor influences US future
technology labor market change?

In the past, there fundamental forces have shaped US work labor market, includes an increase in the returns to education. General education upgrading and the large numbers of female need to work. How educational attainment, demographics and human capital will be predicted to influence US future labor market. Some economists believe education is useful to influence US future labor market. They indicate the social returns to education policies today depend on the relative prices that labor of different educational levels will command in future US labor market; current US labor market trends appear to leave a large group behind; less educated males. The rationale for social policies target specifically to this population is strengthened if predicted future outcomes in US labor market will lead this less educated male group numbers to fall down.

How to supply educational level components to influence future changes in US labor supplied? It can change in the size of the US working age population, it can change in hours worked conditional on being of working age, and it can change in the skills (effective human capital units) to US workers of different education levels, gender and age . In US , the labor is supplied by both highly educated men and women increased substantially relative to the supply of labor by the less educated. Among US males this is largely , due to an increase in educational attainment; highly educated US males did not differentially increase their supplied compared to less educated males nor did their experience differential increases in their human capital policy in US. For US females, some economists found large increases in the labor supplied by US working age women that are due to both large increases in hours worked and increase in educational attainment. So, I assume that future US will have many high educational level female labors to work in US society.

Today, human history is at the beginning of a growth industrial revolution. Developments in genetics, artificial intelligence, robotics, nanotechnology, 3D printing and biotechnology will be popular to influence US future job market change. For example, smart systems, product-homes, factories, farms or cities with help to solve problems ranging from supply chain management to climate change. The rise of the sharing economy will allow US human to monetize everything from their empty house to their car.

Due to the future patterns of consumption will change to trend high technology life enjoyment, it will cause production and employment will also change to employ high technological production labor. As entire industries adjust, most US occupations are changed to high technological manufacturing industries. When some US jobs are threatened by others grow through a change in the skill sets required to do them. a key element in understanding how the benefits and burdens of the growth.

Nowadays, the current technological changes of humans and machines , but rather an opportunity for work to truly become a channel though which human recognize full potential. As US is a high technological developed country. I assume many US employers will choose to act to be the first high technological and invention manufacturing leader to encourage which labors need to learn high technological production methods to prepare manufacture any new technological products to sell in the future US domestic or foreign both markets. Thus, it is possible that future high technological manufacturing labor numbers will be shared large go e.g. 50 to 60% future total US labor market.

How can future US labor market affect job creation and productivity growth? US future economic growth requires factor reallocation across US firms and continuous replacement of technologies. US labor market influences US economy dynamism by their impact on the supply of a key factor, skilled workers to US new and expanding firms, US growth -favoring labor market includes portable pension plans and health insurance united to the current US employers, individualized wage-setting and US public income insurance systems that encourage mobility and risk taking.

US future economic growth arises as production shifts from less to more successful firms though the reallocation of factors of production . US labor market can advance restructuring. Overly regulations tend to create a system in which a large share of economic activity occurs in US small firms without the ability to grow. US labor market should be organized to promote potential high growth US firms, especially through decentralized and individualized wage setting, portable jobs.

In the future, US will have many key importance of high growth firms. US capitalism entails a process of creative destination. New ideas continuously challenges act structures, giving rise to structural transformation as successful innovations and new products firms and industries will arise and obsolete ones will decline in US society.

Martin , J. P. (2012) studies pointed to high-growth forms ( sometimes known as gazelles) as the main drivers of this process. In the US , an estimated 1% of firms creation 40% of all new jobs and 5% create almost 70% of new jobs. A review of the studies of US firms growth reveals some common findings . US high-growth firms are crucial to net job growth, generating a large share of all net jobs . This is particularly pronounced in recessions, when US high-growth firms continue to grow when other US firms deadline. US small firms are over-represented among high-growth firms, but these US firms come in all sizes . A small subgroup of large high growth firms are major job creators. Such as US high -growth firms are younger on average, US young and small high high-growth firms grows, not through mergers and acquisition and make a larger contribution to net employment growth than do US larger and older higher growth firms, high growth firms are present in all industries. Through they are slightly overrepresented in service industries in US.

Some economists predict that future US will be a flexible labor market, the marginal product of labor and the average wage in an industry should tend toward equally across US firms. Taking advantage of a legislative change to raise cost to US employers a study measured the gap between the marginal product of labor and the average wage in an industry before and after the reform. The gap increased after the legislation, which suggests that the legislation reduced allocative efficiency. Their studies have suggested that total factor productivity could increase by as much as 30% in China and India of they were to attain the US level of allocative efficiency across firms within individual industries. The result implies that plants with low total factor productivity are too large and plant with high total factor productivity are too small relative to the US benchmark of allocative efficiency.

What is allocative efficiency meaning? It occurs when the mix of product produced matches consumer preferences ( where marginal benefit equals marginal cost). There products and services are the most profitable, thereby promoting economic growth other research also indicate a strong quantitative effects of strict employment protection legislation on the rate of reallocation in US industries experiments. By relaxing employment protection rules to US developed countries, such as UK, UK etc. with the strictest legislation could increases their reallocation rate by an estimated 50% in the most dynamic sectors, those that benefit most from flexibility.

The effect appears to be particularly strong on the entry-exist margin, which is arguably, especially importation for creative destruction. In future US, if manufacturing industry can have high technological production method to achieve reallocation rate to efficiency to every manufacturing industry. It can have benefit to economic growth, due manufacturing process can be move efficient to avoid cost. Also high technological manufacturing reallocation rate innovation method can influence the future US number of jobs lost in contracting or existing US firms as well as the number of jobs gained in new or expanding US firms in future a certain period divided by the average number of existing jobs. So, US high technological manufacturing reallocation rate innovation method will bring disadvantages to cause many contracting or existing firms job lost, but it will also increase the number of jobs gained in new or expanding firms in US. Thus, if future US manufacturing industry can have high technological reallocation rate innovated successfully. It will raise many new job chances in new or expanding firms in US, but it also have chance to cause many contracting or existing firms job lost or the same time.

What will expected to impact of future computerization on US labor market outcomes? Some economists estimated, about 47% of total US employment is at risk, due to that US wages and educational attainment will exhibit a strong negative relationship with an occupation's probability of computerization. They indicated that the poor performance of global labor markets across advanced economies has intensified the debate about technological unemployment among economists more recently. Indeed, over the past, decades, computers have substituted for a number of jobs, including the functions of bookkeepers, cashier and telephone operators ( Bresnahan, 1999, MGI, 2013).

Although the extent of these developments remains to be seen, estimates by MGI (2013) suggests that sophisticated algorithms could substitute for approximately 140 million full time knowledge workers world wise. Hence, when technological process throughout economic history has largely been confined to the machine of manual tasks, requiring physical labor, technological progress in the twenty-first century can be expected to contribute to a wide range of cognitive tasks, which until have largely remained a human domain. Of course, may occupations being affected by these developments act still far from fully computerization, meaning that the computerization of some tasks will simply free-up time for human labor to perform other tasks. Nonetheless, the trend is clear: computers increasingly challenge human labor in a wide range cognitive tasks ( Brynjolfsson and Mc Afee, 2011).

Thus, it seems that future US computerization job trend will influence some US traditional labor hand made manufacturing industry to divide some part job duty to let computerization work. However, it will not cause many US manufacturing labor to be dismissed. It can still keep some US manufacturing labors to US employers' needs in the future.

Popular science, technology , engineering and mathematics kind of labor supply will be increased demand in US.

Science, technology , engineering and mathematics workers will drive US innovation and competitiveness by new ideas, new companies and new industries in US. However, US employers frequently concern the supply and availability of this kind of workers. Over the past 10 years, growth in this kind of jobs was three times as fast as growth to general jobs in US. This kind of US workers are also less likely to experience joblessness than other kinds of workers. In the future, science, technology, engineering and mathematics workers will play a key role to grow and raise stability of the US economy.

Bureau of labor statistics, ESA calculation 2010 and 2018 year. indicated these kinds of technology, science and engineering workers had been increasing 7.9% from 2000 to 2010 year growth as well as these kinds of workers had been increasing 17% from 2008 to 2018 year growth. Otherwise, other kinds of workers had been increasing 2.6% from 2000 to 2020 year growth as well as these kinds of workers had been increasing 9.8% from 2008 to 2018 year growth. Hence, these kinds of science, technology and engineering workers increasing rate and increasing level are more than other kinds of workers both twenty years.

The other occupations include positions, such as educators, managers, technicians, health-care professionals or social scientists. The science occupation divides four categories: computer and mathematics, engineering and surveying, physical and life science four categories. The reasons why these kinds of jobs will be trend popularly. We define these kinds of degree holders as persons whose primary or secondary undergraduate major was in a science, engineering or technological field. To using similar logic to what we used in our occupation selection, we exclude business, healthcare, and social science majors.

The US department of commerce, economics and statistics administration analysis showed that a science, technology, engineering and math. (STEM) degree is the typical path to a job related to those kinds of degrees or more than two-thirds of the 4.7 million (STEM) workers with a college degree has an undergraduate (STEM) degree. However, this does not necessarily mean that (STEM) field or their jobs . For example, only 35 per cent of college educated computer and mathematic workers have a degree in computer science or math. Thus, these past data explained that it is possible US science, technology ,engineering and mathematics employers will increase needs and why many US students choose to study these subjects in US. Thus, future US economic growth will depend on developing these kinds of science, technology ,engineering and mathematics industries.

Increase development in genetics, human intelligence, robotics, nanotechnology, 3D printing and biotechnology technological industry

In US future, these kinds of jobs will be needed to increase development in genetics, human intelligence, robotics, nanotechnology, 3D printing and biotechnology. For example, smart systems homes, factories, farms grids or cities will help tackle problems ranging from supply chain management to climate change. The rise of US economy growth will allow US people to monetize everything from their empty house to their car in US. These new technological products development will change US patterns of consumption, production and employment adaption are also be changed by US corporations, US government and individuals.

Why will the technological revolution be broader socio-economic, geopolitical and demographic drivers of change to influence future US social economic and consumption pattern change? Future US most occupations will also be changed. When some traditional old jobs are threatened by redundancy and other new technological jobs will grow rapidly, existing jobs are also changed in the skill sets required to do them. The debate is between some economists foresee limitless new job opportunities and foresee massive dislocation of US jobs. In fact, the reality is highly specific to future US high technological production industry, region and high technological occupation in question as well as how US production workers can be raised themselves ability to actions the upgrade level of high technological production ability from various stakeholders to manage high technological production method change.

Overall, this is a modestly positive outlook of US high technological production employment across future most high technological production industries with jobs growth expected in several sectors. However, it is also clear that this need for more talent in certain job categories is accompanied by high skills instability across all job categories. Combined together, future US net job growth and skills instability result in most US businesses with face major recruitment challenges and talent shortages, a pattern already evident in the result and set to get worse over next five years in possible.

The question is how US businesses, government and individuals will react to these new technological job changes, due to talent shortage, mass unemployment and growing inequality challenges will encounter in future US society.

The current technological revolution does not need become a race between humans and machines , but rather an opportunity for work to truly become a channel through which US people recognize their potential. So, if US traditional low manufacturing skillful workers lack talent to learn new skills to prepare to do future new technological manufacturing jobs, such as 3 D printing, robotics, nanotechnology, biotechnological high technological products manufacturing jobs. Then, it will cause increasing of unemployment rate to some not talent US low manufacturing skillful workers. So, US government or high technological product industry employers need to consider this future unemployment challenge will be caused by high technological products manufacturing changing influences. It seems high technological development will cause these low manufacturing skillful workers unemployed rising numbers as well as high manufacturing skillful workers human capital shortage global challenges will exist.

In the future, the driver of changes to influence US demographic and socio-economic growth. They may include: changing work environments and flexible working arrangements. It means new technologies are enabling workplace innovations , such as remote working, co-working spaces and teleconferencing. Rising of the middle class in Asia markets. It means the world's economic center is shifting towards the Asia developing countries.

Some economists predict that Asia will be projected to account for 66% of the global middle class and for 59% of middle class consumption by 2030 year. In addition, climate change, natural resource will be constraints to a greener economy. It means that climate change is a major driver of innovation as organizations search for measures to help adjust to its effects. As global economic growth consumers are needed to lead to demand for natural resources and raw materials, over explanation implies higher extraction most and degradation ecosystem and these challenges will also impact US employment changes needs. All US government also needs to concern future global economic change influence.

US Entrepreneurship Innovation How Influence US Future Geography
Economic Growth.


Some economists believed the interplay between entrepreneur , innovation and economic geography growth in the United states of America have close relationship . Because future innovation is the driving force of growth in the knowledge economy to US . They assumed that if future US entrepreneurships chose to act as new firm formation, which will prefer to accept any new technological innovation to manufacture any new technological products and reallocate any new resources to apply to concentrate on manufacturing any new technological products within firms. For example, Bernard, et al. (2006) find that one third of the net increase in real U.S. manufacturing output between 1972 and 1997 is due to the net adding and dropping of products by surviving firms, a contribution that dwarfs that of net firm entry and exist. Clearly new firm formation is only one dimension of innovation and existing firms account for a large share of total research and development (R&D), in many industries such as pharmaceuticals. It seems US any industries will need spin off of existing operations and the acquisition of independent start-ups are now important dimensions of new process and product development. Thus, innovation is an important factor to influence US economic growth in the future.

However, it will have challenges to US different industries when the causal connections between entrepreneurship, innovation and growth arises. For example, US employment growth of large, middle and small sizes of entrepreneurships will be strongly positively correlated with US any new firms formation, but this doesn't necessarily imply that entrepreneurship causes growth . There may be one factor that causes US employment growth and US firms formation to co-vary, and it is hard to find instructions that affect US firm formation, but have no independent effect on US employment growth.

The interpretation of the correlation between US employment growth and US firm formation relates to old debates in US economic geography about whether US workers follow US firms, or US firms follow workers or there are mutually reinforcing feedbacks between US firms' and workers' locations decisions. So, some economists believed that, US any domestic geography economic growth will possible be caused by these factor , such as either US geography firms innovation factor or US geography high technological workers high level productivity factor or US geography firms innovation or US geography high technological workers high level productivity both combining factor. But the challenge will cause, it assumes that future many US entrepreneurships choose to innovate to manufacture whose products. Although, many high technological manufacturing workers employment chance will be raised, but it also cause many low technological manufacturing workers employment chance will be reduced as the same time. So, it is difficult to keep high technological and low technological both workers who have same fair employment chance as the same time in the future US society if US planned to achieve the future knowledgeable innovation economic society.

However, to achieve this innovation dream, the economists give opinions to US government which will need to encourage the foundations of US entrepreneurial policy and distinguish four broad actors: (a) individual agents who identify business opportunities and choose to exploit them. (b) new formed businesses which innovate using new knowledge and other resources, (C) the economy including all institutions that influence economic growth, and (d) US society is as the collection of all agents who are the ultimate beneficiaries of wealth creation. Within this organizing framework, US entrepreneurships will be easily shaped or changes the overall business climate to achieve knowledge economic society in the future.


The role of intangible assets influences the regional economic growth in US.

What are the four big factors of intangible assets? Some economists indicated they include knowledgeable capital, human capital, social capital and entrepreneurs capital. Nowadays, globalization and increased competition will cause new types of pressure to influence US economic growth. So, US companies need have flexibility, the ability to immediately adapt to market developments, and pro-activism in creating future markets. The relative importance of physical growth: However, soft production factors, that is those related to personal knowledge are becoming more important to influence US future economic growth, which regards to human capital and knowledge as driving factor of economic growth in industries developed countries, such as US.

All these soft production factors can be grouped in what is known intangible assets. These assets can be defined as non-material factors that contribute to enterprise performance in the production of products or the provision of services, or that are expected to generate future economic benefits to the entities or individuals that control their deployment ( Akerlof & Kranton , 2000).

Whether it has a close relationship between these intangible assets and US regional economic growth? Some economists had researched to have more reliable quantitative statistical information to do report in above four big factors how to effect on US regional economical growth influence. Their report indicated that returns from human and social capital are taken as homogeneous for US all regions. They also indicated these two sets of questions in their report. The first one, dealing with knowledge accumulation, addresses, among others the following issues:

(a) How does innovation and knowledge accumulation occur within firms and hoe does it impact on economic performance?

(b) What is the role of universities in regional, national and global knowledge accumulation processes?

The second set of questions addresses the key knowledge diffuses over space and how this diffusion impacts on economic performances.

In particular:

(c) To what extent knowledge diffusion is conditioned by spatial proximity?

(d) What is the impact of knowledge accumulation and diffusion on economic performance?

They indicated to answer these questions, having reliable measures of how innovation process occurs, of the actors that take part in it and the mechanisms that are in place is important. They pointed out that the systemic nature of US regional intangible asset demands indicators that grasp two kinds of capabilities: network capabilities , i.e. connectivity, both intra-and-inter-regional and organizational capabilities, as well as their dynamic in US. These indicators have been applied to the study of linkages and relationships between US firms and between US firms and US universities, highlighting the mechanisms through which those actors contribute to the processes of knowledge accumulation, generation and diffusion.

They also found results: How does knowledge accumulation occur within US firms and how does it impact on US economic performance? They included collaborations with competitors are most commonly undertaken at early stages of the development projects, those with buyers and suppliers are more likely to result in the introduction of new products and processes: At the same, it has been highlighted that, regardless of the industry, the most innovative firms, i.e. those more able to absorb knowledge are more likely to participate in collaborative networks.

Thus, US future entrepreneurships need to offer knowledge accumulation opportunities for US regional firms and research institutions to benefit and contribute to global network of US expects to achieve rapid regional economic growth. US's entrepreneurships also needs to explore organizational innovation, those US firms whose structure enables learning by doing, using and interacting by relying, among other things on parallel development teams. Semi-autonomous work teams and reduced management layers are more likely to introduce new products to the future US domestic and foreign both markets.

What is the role of US universities in US regional, national and global knowledge accumulation processes? US universities can heavily influence US regional, national and global knowledge accumulation processes. For instance, when collaborating with multi-national enterprises, they may affect simultaneously the three level. Their report had highlighted interesting results on the local impact of universities: when it has been found that top ranked departments are significantly associated with partnerships involving spatially close industry partners, it has also emerge geographical proximity, is nt the main driver of collaboration choices. These are found to depend largely on US firms' networks and universities' specific characteristics. Among other things, the cultural traditional of academic institution has been shown to influence the ability to collaborate with industry and commerce research.

Networks characteristics have emerged as crucial in determining academic knowledge transfer: the better the access to international networks, the higher the patenting activity, have the knowledge transfer to US industry. This implies that the set of tools of knowledge based economic development should include not only research and development, but also clever ways of supporting academic research network.

To conclude, if US expected research development can assist economic growth. Then, US academics universities and companies both need to co-operate to carrying on experimenting any kinds of research and prepare to assist any US product innovation more easily. So, the intangible asset, not as scientists, technology, US universities' and firms' research fund which is an conditional factor to influence US regional research and development economic growth in the future.

Social factor impacts US technological development growth.

Has it relationship between social influence and economic environment in future US? For example, the social factors that are positively correlated with the economic growth ( i.e. the expected years of schooling and the life expectancy) and respectively, the factors that are negatively correlated with US future economic growth ( i.e. the US population or risk of poverty and the unemployment rate).

The improvement of the US future economic environment will be an objective of the macroeconomic policy on short, medium and also long term. The importance of social factors upon US future economic growth, considering that the future used macroeconomic indicator, GDP per capita, is not most proper measure for the future US nation welfare. Due to GDP per capital fails to take into consideration some specific sectors of the US social economy, such as the black market.

Until recently, some economists rely on culture is as a possible determinant of economic phenomena. However, in current years, better techniques and more date made it possible to identify systematic differences in people's preferences and beliefs and to relate them to various measures of cultural environment suggest an approach to introduce cultural-based explanation that can be tested and are able to substantially understand economic phenomena.

The increased importance of social factors relies on a basic concept. Some theory is measured to economic growth which has wrong assumption. For example, the fiscal and monetary policies focused on increasing the national income, which lead consequently to economic growth. The reason most of economic opinions have been argued because whose opinions are based on a wrong hypothesis, according to which the nation welfare is based only on the level of income.

Can social factors influence US future economic growth? Human development history, global life expectancy has been experiencing these stages: from the industrialization process, the technologic progress, the medical evolution, the scientific research, these stages were also related to internal causes, specific to some developed countries, e.g. US developed country. Thus, the differences are significant and are linked both to US life expectancy level and the GDP /capita. Such as US population is less than China population too much. Although, US land area is near to China area. It seems US will encounter life expectancy level need to prepare its technological development to raise economy growth of opportunity. For example, Africa and Asia are still facing major economic and social issues. The access to a health life and medical services are still long terms objectives for countries with low life expectancy.

According to Harrison & Huntington (2000), the analysis of social factors helps understanding the human behavior with respect to consumption, savings, investment system, expectations and attitudes towards the economic circumstances, which also have a major impact on the economic growth. The evolutions of economic and social environment are needed for US future development. In order to eliminate the gap of living standard, outside resources and support US needs have good social indicators study plan to concern econometric model to rise poor people living standard in future US society between rich and poor people who are living in US. However, I believe the social factors include demographic and culture, population's structure factors which are one important social indicator to influence the distribution of the US social public income.

Barro & Sala-i-Martin (1996) defined culture is as the sum of symbols, meanings, habits values, behaviors and social artifacts which characterize a distinctive and specific human population group. For decades, economists and social thinkers debated the influence of population change on economic growth. Bloom et. al (2001) defined three alternative hypotheses: that population growth restrict, promotes or is independent of economic growth . Each hypothesis was sustained with strong arguments, and all the arguments mostly focused on population size and growth. The debates revealed other important issues, such as the age structure of the population, the way in which the population is distributed across different age groups.

The economists indicated that people's economic behavior varies at different stages of life, changes in a country's age structure can have significant effects on its economic performance. So, such as US, if it had a high proportion of children are likely to devote a high proportion of resources to their care, which would tend to depress the pace of economic growth. By contrast, if US's population falls within the working wages, the added productivity of this group can produce an increase in the economic growth. This is how the combined effect of this large working age population and health, family, labor, financial and human capital policies can create cycles of wealth creation. On the other hand, if US a large proportion consists of the elderly , the effects can be similar to those of a very young population; a large share of resources is needed by a relatively less productive segment of the population , which likewise can inhibit economic growth in future US society.

Further Bloom et al. (2001) analyzed the three main mechanisms of population's structure for determining economic growth ( labor supply, savings and human capital ) and their dependence of policy environment. They indicated that a growing number of adults will only be productive of there is sufficient flexibility in the labor market to allow its expansion, and macroeconomic policies that permit and encourage investment, people will only save if who have access to adequate saving mechanisms and have confidence in domestic financial markets and the demographic transition creates conditions where people will tend to invest in their health and education, offering great economic benefits, special in the modern world's increasing sophisticated economies.

It seems future US labor market will need a growing number of adults to supply , a stable domestic financial market to encourage US people have more confidence to save money in banks, a stable health and education sector industry can encourage US people have confidence to invest to do this kind of health and education service industry. Instead of population of working age factor will influence US economic growth. However, US people living culture will also influence US economic growth. For example, if it has a trend that there are many US young people like often go to shopping in relax time, because who feel shopping is their young group entertainment in the year. Then, the year US GDP growth will be possible raised, due to there are many US young people prefer to spend for shopping expenditure suddenly in the year. In conclude, in social environment of cultural and population of working age size will influence US future economic growth.


Online tourism industry influences US
future online industry economy growth.

Today, US online tourism sale industry has always been one of America's great home growth industries. Today, more than 8 million Americans are employed in travel and tourism. For example, US domestic South Carolina, hospitality and tourism has been as the largest local industry, providing tens of thousands of families directly or indirectly with jobs. Predicting global tourism consumption needs work be increased. US will experience in a changing economy, tourism industry will provide job security to many Americans as well as the service -oriented nature of many travel positions and these jobs are difficult, if it is impossible to outsource.

Many people accept to buy air tickets from online sale channel. Will future America government play an important role to influence future US online air ticket tourism sale method to raise online travel consumption behavior? However, some have questioned in recent years whether US policies have harmed the ability of the tourism industry to expand. One of the most frequently discussed concerns involves American visa policy, when European Union allows tourists from 26 European countries and almost all of the America and Australia to visit without a visa. America policies require some travelers from friendly countries, such as Brazil, to travel thousands of miles from home to attend the in-person interview needed to secure a tourist visa. Although, the US White House has taken steps to reduce rise wait times in recent years, in some parts of the world, like Vietnam and Turkey, the US tourist visa application process can be a multi-month process ( US Travel Association , 2014).

Will the visa waiver program impact driving increases in US impact driving increases in US tourist volumes? Some economist analysis found that when a typical country joins the visa waiver program, it sees a notable increase in the number of countries who chose to visit the US in the immediate years. That follow over the course of its first five years in the US visa wavier program, the number of tourists arriving from a participating country rises by 16.4%. Also their analysis indicated to expand the visa waiver program to some countries would result in $7.66 billion in additional tourist spending within a five year period. It would also create at least 50,000 American tourism jobs within 5 years. Purchases of products and services by visitors contribute foreign significantly to job creation and economic growth in US international travelers to US purchased more than $180 billion amount . In addition, tourism generates a trade surplus as foreign visitors spend more in the US. So, US enjoyed a $57 billion trade maintained a trade surplus in tourism every year. Some economists indicated that India is likely to contribute to the growth in foreign travel to the US . If India government can ensure which have enough numbers of foreign travelers are encouraged to choose to travel to US. In the future, India will be one important tourism cooperation partner to US. So, US ought achieve online travel sale strategy to promote to Indian to know what online travel advantages can give to them in the future, e.g. reducing time to buy air ticket, convenience, cheap e-ticket price, avoiding to lose air seat supply, more airlines choice providing. These are online e-ticket sale attractive characteristics to India online e-ticket sale market. Thus, US ought concentrate on promoting this kind of online travel sale service to India in order to raise US many airline e-tickets sale numbers.

The effects of population growth influence US economy growth.

Nowadays, China and India population growth rate are rapidly. Population growth can have several disadvantage effects on the economic expansion and performance to China and India. Does it influence cultural and sociological links between the following per capita income, rates, technological advancement , education aspects? China with 1.32 billion people and India with 1.1 billion people. Though, they each enjoy a large labor force advantage, several key economic factors have contributed to how the Chinese and Indian populations have grown and what differing effects that growth has had on their developing economies. Although, there two countries have high population growth rates, but the poor people has low standard of living.

So, when these two countries population sizes are large, but the standard of living will be low, and population will be reduced by either the preventive check ( international reduction of fertility) or by the positive check ( malnutrition, disease, and famine). It seems china and India governments think overpopulation challenge will cause social crime increasing, education competition, raising the job applicants number will increase, but the jobs supply number won't increase faster than job applicants number, and it will cause unemployment etc. social challenge in the future long term. So, why China and India governments will choose to control birth rates every year. However, in these developing countries, children are often as an economic asset, a tool to help increase agricultural production. Demand or need for large number of children is driven down by improvements in living standards and child survival and by the modern of economies. The movement away from agricultural production and into more modern industries in the manufacturing and services sectors need the necessity to have numbers of children to prepare to supply to China and India future manufacturing and services job markets.

However, US is a developed countries. The country main GDP growth source is high technological industry. Hence future children birth rate is not necessary to prepare to supply for manufacturing or service industries needs mainly. It seems US population birth rate is stable. So, overpopulation challenge won't be caused easily. Can population growth influence US economic growth? US is different to China and India. US's population growth rate is less. Hence, if US population growth rate can rise. It won't have overpopulation challenge. US needs have many young people number who can have high technological knowledge to prepare to have enough supply to do the future high technological different kinds of positions to build social economic growth in the future. Hence, US ought encourage young people choose to marry and to born next generation children to supply future different kinds of high technological jobs of needs to raise technological productivity and they ( high technological manufacturing workers) will be the main production of factor to influence future US economic growth in the high technological manufacturing sector.

Thus, I believe the effect future US of population growth will influence US future economic growth as the same time. Also, it is the right time, US government needs to assist universities have enough afford resources, e.g. university fund, technological educational courses, lectures etc. to provide future US technological education students' needs in the future. So, it brings this question: What can determine US economic growth?

In global macro-economic influence, since 1973 year, per captia income growth in the US and other advanced countries has slowed to 2.2 % a year or almost half the 3.9% annual rate of the preceding quarter century . If the US had maintained the level of growth experiences in the 1950 year and 1960 year, real per capita income today would be about 71% greater than it actually is. In contrast, it has been estimated that eliminating the variability in US consumption since world war II would be equivalent to boosting current real consumption by only about 4.8%. If the choice is between long-term growth policies and further short-term stabilization policies long term clearly have the potential for higher benefits. Hence, what factors determine US economic growth ( David M & Roy J, 1993).

According to traditional growth theory, the main determinants of long run economic growth are not influenced by economic incentives. Perhaps the reason why economists have neglected long time, to profession relied on a theory that offered litter scope for policy to influence important sources of growth.


High technology development can
reducing income inequality growth.

Some economists found one key is that education and anti-discrimination policies well designed labor market and large and/or progressive tax and transfer system can all reduce income inequality . In many OECD developed western countries, income inequality has increased in past decades. In some countries, top earners have captured a large share of overall income gains, when for other income has risen only a little. Some see poverty as the relevant concern with the type of growth enhancing policy reforms advocated for each OECD developed countries and economic growth might have positive or negative side effects on income inequality.

OECD (2011), it first highlights differences in some income inequality across the OECD and the factors driving them, such as cross-country differences in wage and non-wage income inequality as well as in hour worked and inactivity. OECD developed western countries can be divided into five groups to their pattern of inequality. For example, in five English-speaking countries ( Australia, Canada, Ireland, New Zealand, the United Kingdom) and the Netherlands wages are rather dispersed and the share of part-time employment is high, driving inequality in labor earnings above the OECD average means-tested public cash transfer and progressive tax.

It seems income inequality will influence the developed countries, such as UK, Canada, Australia, New Zealand , US etc. economic growth. Although, technology change and globalization have played a role to influence the distribution of labor income. Some economists believe that any countries' policies will also influence income inequality. These policies factors include: technological education policies can increase different technology graduation rates from upper secondary and tertiary education and that also promote equal access to a well-designed different sector technology labor market policy can reduce inequality.

A relatively high minimum wage narrows the distribution if labor income, but if set too high, it may reduce employment of inequality reducing effect. It tends to reduce labor earning inequality by ensuring a more equal distribution of earnings. Job protection reforms that make permanent and temporary contracts more even in their provisions low income wage dispersion of earnings is rather mixed, removing product market regulations can reduce labor income inequality by boosting employment, policies the faster the immigrants and fight all forms of discrimination reduce inequality, progressive tax policy play a key role in lowering overall income across the OECD developed countries.

However, the redistributive fair income level between low level income labors and middle level income labors and high level income labors impact of developed countries, e.g. consumption taxes and real estate taxes tend to be regressive tax policy. Hence, it seems that reducing income inequality can cause the more fair income distribution between the high income level and the middle income level and the low income level labors. Then, it will let the developing countries or developed countries , such as US citizens feel that who can get real social welfare fairly, due to high technology development can boost economic growth in US society.


Long term cheap medical cost trend factor influences US economic growth.

Some economists predict to make long term forecasts to reduce medical cost trends how will influence US economy. Also, they indicate short run cheap medical cost forecasts for first 1 to 5 years to reflect the particulars of specific groups, benefit packages, regional markets or cheap medical cost network providers and use their local cheap medical choice information and actuarial skills to improve accuracy and reasonability. They also oversight group review on the consistency of baseline assumptions with factor influencing future US patients' medical choice taste, medical technology trend component and historical annual percentage increase in medical costs, premiums, income and excess cost growth are found to influence future US patients consumption behavior. However, their research predicting expensive medical cost is the main factor to influence to reduce US patient numbers among of ward room sleeping environment, doctor loyalty, hospital entertainment and service facilities provision hospital cheap car parking charge service, cheap or free telephone call provision, hospital meal quality provision etc. different factors .

They suggest that how to reduce medical cost. They indicate many other factors, e.g. patient individual illness aging, physician supply, medical insurance plan benefit will affect the cost of a particular patient, group, organization or plan in a specific locality or over the short run. Also they indicated that additional analysis is still needed to determine if reliable estimates of separate medical insurance premiums, medical care payments, pharmaceuticals or other cost categories are possible to influence medical cost to be increase.

Hence, for long run US cheap medical cost will encourage or attract many patients prefer to pay medical cost to see doctors in US any hospitals or medical centers. Then, US hospitals or medical centers income will increase and US GDP income will also increase in medical service sectors in the future. It seems that long run cheap medical cost will influence US economic growth.


Long term cheap medical cost trend factor

influences US economic growth.

The next generation of US talent management practices and solutions will largely be driven by US economic evolution, demographic changes and technology advancements. These factors are dramatically influencing the way US people work, the way US companies are organized and the way talent is managed.

Some economists explained the key economic factors driving changes in talent management include: the knowledge economy is value to companies, talent is now a required strategic asset. Key changes in the future include a line between inside and outside talent that will result expansion of the talent management, globalization, such as European expansion is well-known top expansion prospect for global companies now include China, Russia and Eastern Europe and America and the rest of Asia. So , US government needs to know globalization can give what opportunities to US future .

Although, knowledge management or knowledge economy will bring advantages to US. However, it also bring disadvantages to this aspect, such as skill gap and structural unemployment will also influence US organizations structural unemployment and skill gap challenges issues will be caused between low skillful and high skillful workers, generational geographies changing will occur in US.

Although, baby boomer retirement has been top of mind for many years in the US, even more significant demographic changes are happening outside the US, where population growth rates and aging population will influence as local economies. Thus, the ability for organizations success will need global talent or effectively more talent from areas of abundance to scarcity is becoming a strategic issue to any US companies development , increased health and longevity mean that seniors are working longer enabling US organizations to keep experienced team members into their retirement years. But it also raise US workforce planning and generational challenges, workplace and diversity is increasing, a more diverse pool of talent affords new opportunity . Such as hiring workers raise productivity of needs, digitization of candidate or employee profiles can meet US business employment needs because employee talent data has been digitized and integrated into comprehensive talent profiles.

Techniques , such as attribute matching and recommendation technologies can be applied in talent management can be applied in talent management to find and match the most right candidate to do any positions in short time efficiently, with the increasing need of telecommunication in the US market penetration of smartphones and tablet devices a significant portion of the world's human potential will have access to rich web and application experiences from anywhere. Thus enables US organizations to source and collaborate on knowledge work with any part of the world into a global talent pool.

How can intellectual assets impact to US to change to knowledge economy? The information age moved the basis of economic value from products to intellectual assets, information and the talent that develops them. It is now widely acknowledged that intangible assets, with largely consist of know-how, unique intelligent property, and patent right, drive move than 80% of the valuations of publishing US publicly traded companies.

In future US knowledge economy, US leading edge organizations will have efforts to use into the talent and intellectual capital of not only employees, but also clients, partners and the public at large in an effort to create an extended electronic community. For example, client support portals where clients can answer each other's questions, which will reduce costs by expensive client support calls for US companies in future US knowledge economy environment.

Also, knowledge economy will assist US companies to be partner and customer innovation efforts easily. Such as P&G daily product company's ability to source more than 50% its new product ideas to external innovation, driving industry-leading standards for new product launch rates. In future knowledge economy environment, it is clear that US organizations will be increasingly deriving value from talent that is outside the company.

In future US knowledge economy environment, the link between employees within the US organization and those outside, it is also driving talent management changes, particularly around sourcing, strategic workforce planning and employees engagement . Given these changes, executives in leading US companies will be increasingly focused on talent management issues, recognizing that talent, wherever knowledge economy will give knowledge management method to any US organizations to build more competitive advantages in the future. However, knowledge economy will also bring structural unemployment to US society. Unfortunately, today's global talent market is largely inefficient and characterized by a high degree if competition relative to the redistribution of talent.

One result of this inefficiency is long period of skill gaps structured unemployment is a particularly different scenario in which section skills are no longer required, not just within a US particular company, but within US an entire sector. Structural unemployment may be the future result of cyclical boom and bust cycles, offshoring of partial. In conclusion, whatever the causes of future knowledge economy to US , the result is an uneven distribution of talent relative to the available jobs in US. In part of structured unemployment in US will be caused from knowledge economy influence. However, knowledge economy will also influence US future businesses speed increases, bandwidth increases and growth in mobile internet access during US encounters the knowledge economy influence in society.

The impact of educational quality factor influences US economic growth.

Schools can train skills of an individual and human capital for future businessmen need. It is not the only factor. Schools nonetheless have a special place, not only because education and skill creation are among that prime explicit objectives, but also because which the factor(s) most directly are affected by public policies. So, US is well established that the distribution personal incomes in US society is strongly related to the amount of education US people have had. Thus, any noticeable effects of the current quality of US schooling and the distribution of US people skills and income will become apparent some years in the future, those US students now in US school become a significant part of the future US labor force. Then, the question arises as: To whether are these skills correlated with US student's subsequent performance in the future US labor market and will the US economy's ability to grow ?

The quality of human resources are measured by scores is directly related to individual earnings , productivity and economic growth. A range of research results from the United States shows that earnings advantages , due to higher achievement on standardized tests are quite substantial . These studies typically find that measured achievement has a clear impact on earnings, after allowing for a differences in the quantity of schooling, age or work experience, and for other factors that might influencing US people earnings. In other words, for those leaving school at a given grad, higher quality school outcomes ( represented by test scores) are closely related to subsequent earnings differences. So, it seems quality of education and individual income has direct relationship to influence US economic growth.

According, educational program that deliver those skills will bring higher individual economic benefits. Obviously, students who don't better in schools as evidenced by either examination grades or scores on standardized achievement tests tend to go further in school or university. By the same way, the net costs of improvement in school quality if reduction in rates of student repetition study. Then, it brings this question: Will education quality bring future economic return to US?

Early some economists' research found that personal and behavioral traits, such as perseverance and leadership qualities had a significant influence upon labor market success, including earnings. To answer above question, we need to know if US schools which generally have good student performance, it will influence socio-economic advantage improved performance, changes in school climate, teacher morale and commitment, school autonomy , teacher-pupil relations and disciplinary had some compensatory influence towards greater equity. In American, pupil socio-economic background and classroom climate appeared to be the most important predictors of achievement. So, it seems human development will influence US economic growth in the future. So, US development organizations seem better will then non-specialized educational organizations to design and deliver effective combinations of livelihoods and literacy.

Whether people who attain literacy actually make much use of it is subject to debate. On balance, however, literacy seems more used where economic development is better established. This supports the argument that same degree of economic and political improvement is necessary to sustain literacy. Then, US people will use their literacy skills where conditions make it useful or desirable for them to do so. However, schooling is a social process, and improvements in resources, technology and quality of student and teaching inputs should in principle to able to be enhance its overall quality. In a good number of countries, large increases in average real expenditure per student and other measures of school resources in primary and secondary schools over the last four or five decades have not remotely been matched by a comparable increase in average test scores.

What can be influenced from quality of education to US? Some economists suggest the earning can provide an important incentive mechanism, which can influence both the quality and motivation of teachers. If teachers' real average earnings had kept the same level with other professional groups over the period, the productivity impact of their earnings growth would literacy have been small. The first view point, in fact, however, in many countries teachers' earnings have increased considerably less sharply than those other groups. Teachers may feel worse off, because of their decline in status relative to other professional groups. This circumstance could well explain part perhaps an important part of the apparent lack of impact on learning outcomes of increase in real per student spending over time. One to teachers' general salary is low to compare other professional in the global society.

Then, it brings this question. Can quantity of education influence US economic growth? In this modern economic approach to investigating the determinants of educational outcomes has borrowed well established technique from other economic applications. The idea that there is a determinate relationship between ( teacher number) inputs to a production process ( teaching process) and the outputs( student examination results and learning performance ) that subsequently bring has being been important in micro-economic analysis. If US schooling teaching possibilities are governed by certain teachers' quality between factors of production, e.g. teachers; teaching performance , teachers' supply numbers to every schools, teachers' educational qualification. So, the product function describes the maximum feasible output ( student examination results and learning combinations ) obtained from alternative combinations of these inputs ( teachers' quality teaching performance, supply numbers, education qualification).

Production factors are powerful analytic tools, which have been applied to the analysis of most forms of economic production. Since the mid-1960 year, they have also been widely used in the economic analysis of education. Thus, in schooling educational organizations, teacher salaries will influence teachers' performance ( production of factor) to be good or bad. If teachers' educational performance is good, then production of output ( student learning abilities and students examination results) will be good usually. Otherwise, if their educational performance is bad, then production of output ( student learning abilities and students examination results) will be bad usually .

In US society, if many poor learning performance students have been produced. Then, there will not many talent young people who can contribute to serve US society in different professional aspects. It means that they can not contribute US economic growth easily. Thus, it seems that US quality of education can impact US economic growth for long term and US universities, secondary schools and US government ought need to find methods how to raise quality of education for next generation of labor supply market.


Bio-medical industry factor boosts US economy.

In the future, the perspective of bio-medical industry will be an alternative growth scenario. So, US government will need to modest improvements in key policy areas to adopt bio-medical industry needs, e.g. more favorable coverage and payment policies for medical innovation, improvements in regulatory policy to create efficiencies in research and development process and improvements in policy to incentive R & D ( research and development) create efficiencies in the R & D process.

As an industry rooted in science and advanced manufacturing, the innovation bio-medical industry is uniquely position to help maintain US leadership in new technologies and scientific to continue to create high quality, high wage R&D and manufacturing jobs and enhance America's global competitiveness in the future. Today, the US bio-medical industry supports a total of 3.4 million jobs across the US economy , including over 810,000 direct jobs: contributes $189 billion in economic output and is responsible for about one in five dollars spent on domestic R&D by US businesses. Hence, innovation bio-medical industry will be one high value knowledge-based industries as a driver of US economic growth.

The reasons include: The first reason is such as , the US history, it has been the world leader in bio-medical research and development of new medicine over the past 30 years. It has been one leader of world class life science ecosystem and innovation among different countries in medicine development history. So, bio-medical industry will be a high technological innovation related activity of industry leader, as is measured by R&D investment, e.g. generation venture capital and share a total R&D employment in manufacturing industry. Next reason is that US bio-medical industry employs a total of 813, 523 worker amount. These workers lead a wide range of occupations that offer high wage, high quality employment. The total economic impacts of the industry, it is estimated (via the generally accepted methodology of input/output analysis) that supports nearly 3.4 million total jobs and generates nearly $ 789 billion in US economic output ( National Science Board, 2014). Finally, reason is that US people concern life health issue, In addition to improving individual health and lengthening life ages, medical advances have contributed to substantial societal health gains, such as reducing disability and improving productivity.

According to two Universities of Chicago economists indicated the estimated economic gains from declining morality alone in the US from 1970 year to 2000 year had a value to US society of more than 3 trillion a year. Hence, US medical health need of patient numbers will increase, due to who have large needs for life health. Hence, quality of education can influence bio-medical industry development to impact US economic growth in the future.

Welfare economic theory, Kaldor-Hicks
efficiency theory and Pareto improvement
theory relationship explain US and UK technology development trend

Economists explain that welfare economic theory, Kaldor-Hicks efficiency theory and Pareto improvement theory which have close relationship. Welfare economic theory means general social equilibrium. They indicate the demand of commodities must same to the supply of commodities when the commodity market stays in the stable level. So , when production is increasing, the manufacturers also need to keep their products supply and demand keep to the stable level and consumers' demand of their products must also need to increase as well as bank interest rate must be fallen in order to raise consumption desires. Also, the low bank interest can also excite investors' investment desires. Such as to explain why the developing countries cities technological competitive investment factor can impact US and UK economic growth. Welfare economic theory means general social equilibrium. The economists indicate the demand of commodities must same to the supply of commodities when the commodity market stays in the stable level. Due to the future developing countries investors will invest much on developing new technological products research. So, these countries' economy growth will be possible faster than US and UK both developed countries because it is possible that the technological product investor number will decrease. So, US and UK government will need to attract and encourage many foreign developing countries' investor prefer to invest their technological products research to assist US and UK both developed countries to develop its economy. As Welfare economic theory means general social equilibrium. If US and UK government can give benefits to these developing countries' technological product investors, then these developing countries' technological product investors will prefer to invest to developed country US urban cities, such as small cities or farming location to build factories to manufacture their new technological products. Because , London, New York, Washington etc. large cities' technological investment and location where land supply is limited and offices and factories rents are also expensive. Otherwise, UK and US, farming location or small cities location where have much land supply to build large offices and facilities and the rent is much cheaper than US and UK main cities, such as London, New York, Washington etc. Even, if UK and US government and banks can lend low interest rate of bank loans or government loans to encourage foreign investors to do technological product research in its country. So, UK and US bank low bank interest rate can encourage US consumers prefer to buy any future new technological products, due to low bank interest rate can not attract who choose to save more money in US and UK any banks as well as US and UK bank low bank interest rate can also encourage foreign developing countries' high technological product investors to choose to invest in UK and UK , due to who can pay low bank or government loan interest to compare themselves countries or other countries.

As Kaldor-Hicks efficiency theory indicates the policy is suitable to be implemented if the policy beneficiaries' welfares can compensate to the policy benefactress. So, US and UK bank and government low loan interest rate policy is valuable to implement if benefactress, such as US and UK government which can raise economic growth as well as the US benefactress, such as the high technological product businessmen can sell many different kinds of high technological products from foreign developing countries' investors' assistance in US and UK both developed countries.

As Pareto improvement theory indicates that it can not carry on continue, due to the country has shortage of natural resources after the country improved its welfares to reach the maximum standard. The result will be caused, such as the country will reduce this group of citizen's welfares, if the country decides to continue to improve another group of citizen's welfares. So, future US and UK will have shortage of natural resources to supply to technological product manufacturers. So, the number of technological product production will be fallen , due to the shortage of natural resources are supplied to US and UK high technological product manufacturers. The only solvable choice is encouraging foreign high technological product investors to attract them to supply themselves high technological products any resources to manufacture high technological products and provide to US and UK high technological product sellers to help them to sell in US and UK domestic market or export to overseas market to earn income. Then, US and UK will earn raise GDP income from high technological product sale industry in the future.

Natural rate of unemployment

Economists believe why natural rate of unemployment will occur in any countries. They indicate that during economy condition keeps at the balance situation ( condition), anyway any country government adopts any policy. However, the natural rate of unemployment won't reach to zero level as well as there are some people will still unemployed or it is possible that some people will be alternative employment between any time. Hence, it means that although US and UK is an developed country. It can not guarantee there are not any people unemployed , so the natural rate of unemployment will not reach to zero level.

So, it seems that US and UK current economy condition had kept at the balance situation ( condition). Thus, even future science, technology , engineering and mathematics workers will drive US innovation and competitiveness by new ideas, new companies and new industries in US. However, US and UK employers frequently concern the supply and availability of this kind of workers. Over the past 10 years, growth in this kind of jobs was three times as fast as growth to general jobs in US. In the future the Natural rate of unemployment of US and UK science, technology , engineering and mathematics worker number must not reach zero level.

Anyway, in the future, the perspective of bio-medical industry will be an alternative growth scenario. So, US and UK government will need to modest improvements in key policy areas to adopt bio-medical industry needs, e.g. more favorable coverage and payment policies for medical innovation, improvements in regulatory policy to create efficiencies in research and development process and improvements in policy to incentive R & D ( research and development) create efficiencies in the R & D process.

As an industry rooted in science and advanced manufacturing, the innovation bio-medical industry is uniquely position to help maintain US leadership in new technologies and scientific to continue to create high quality, high wage R&D and manufacturing jobs and enhance America's global competitiveness in the future. In the future the Natural rate of unemployment of US and UK of bio-medical industry worker number must not reach zero level.

Thus, if future US had enough job number which could supply to these high technological and bio medical industries of labors to do. However, the natural rate of unemployment must not reach at zero level and it does not represent its economy is poor because one developed country , such as US had been experiencing economy condition keeps at the balance situation( condition), so it's natural rate of unemployment must not reach zero level . Otherwise, the developing countries' economy condition does not keep at the balance situation ( condition), so their natural rate of unemployment have more possible to reach to close zero level.

Industrial production and pollution economy

Economists indicate that it has close relationship between industrial production and pollution. When traditional products are manufactured, the air and water pollution will be caused in the economic activity. So, human needs to protect environment to keep health, the demand of investment of money reduces pollution and labor number will increase to achieve to reduce air and water pollution in natural environment. Thus, product manufacturers need to analyze which economic stage(s) will encounter shortage and find methods to solve challenge.

This Industrial production and pollution economy theory can explain why US and UK needs to plan achieve digital economy to shift future America rural economy. Some economists indicate that there are five trends reshape to impact rural America' and England future economy. They include that digital economy will shift future America rural economy. US and UK quality of life will change a lot, the US and UK rural economy will stay uneven, US commodities will compete in global markets and will give less benefit to US and UK rural economy and US and UK new products will revolutionize US and UK agriculture economy.

US and UK needs to plan digital economy to future agricultural production of the reasons include: The first aspect impacts to US agriculture economy, the US future rural economy stays uneven. Growth will concentrate in 4 out of 10 rural places and they have scenery, a retail hub, or one next to a city in US and UK. The impact on rural America includes some rural places will try to manage growth , but many places on a quest for new economic engines. Thus, it will bring these questions to US and UK rural economy impact, such as : Who will be US and UK businessmen clients? The struggling farmers? The struggling farm-dependent country? The booming mountain area? The rural area transforming into city? The second aspect impacts to US and UK agriculture economy, due to US and UK commodities will compete in global markets if it will bring a smaller benefit in US and UK rural economy. Then, US and UK farm scale will cut costs and competition fewer US and UK farms and places will depend on farm income. The third aspect impact to US and UK agriculture economy, new products will revolutionize agriculture. Major, shift from commodities to products, spurred by biotech, means two agriculture in the future and two rural America . Hence, US and UK future agriculture determination the rural economy will be declined. A future new US and UK agriculture supply chain integrator will be caused from the traditional farming supply chain procedure the change to outsourced contractor farming supply chain procedure, such as: In beginning, from farmer will outsource supply chain contract to processor, then contract to distributor and contract to food retailer final step. The fourth aspect impacts to US and UK agriculture economy, what will be two agricultures impact to the US and UK future economy? The first US and UK agriculture impact will be US and UK commodity agriculture. It focuses on production capabilities, farming foods production will be thin margins maintained with technology and big sale. The second US and UK agriculture impact will be product agriculture. It focuses on consumer needs, farming foods production margins will be protected by capturing value and building business relationship. In the fifth aspect impacts to US and UK agricultural economy, it is digital economy impacts US and UK rural agriculture. Future diversity economic base can encourage US and UK farming product agriculture to enter rural digital agricultural service consumption sector to change US and UK agricultural consumer individual shopping habit.

Thus, future US and UK agricultural production needs to concern how to reduce air and water pollution to influence natural environment clean quality as well as it needs to find what economic stage of agricultural production will be innovated in the future.

Reference

Akerlof, G.A. & Kranton, R.E. (2000): Economics & Identity, Quarterly Journal Of Economics, 105(3) , 715-753.

Barro, R., Sala-i-Martin, X., ( 1996). The classical approach to convergence analysis. The economic journal, vol. 106, no. 437.

Bernard, A., Redding, S & Schott, P. (2006). " Multi- product firms and product switching, " NBER working paper, 12293.

Bloom, D., Canning, D., Sevilla, J., 2001. Economic growth and demographic transition. National Bureau Of Economic Research.

Brynjolfsson and Mc Afee, (2011). Race against the machine: how the digital revolution is accelerating innovation and driving productivity , US.

Bresnahan, T.T. (1999). computerisation and wage dispersion: an analytical reinterpretation. The economic journal, vol. 109, no. 456 pp. 390-415.

David. M Gould & Roy. J. Ruffin. What determines long run economy growth? Economic review, second quarter, 1993.

ESA ( 2010 & 2018 year) calculations using current population survey public use micro date and estimates from the employment projections program of those Bureau of labor statistics.

Fernald, John (2015). " Productivity snd potential output before during, and after the great recession" In Jonathan A Parker & Michael

Hakkio, Craig S. 1992. " Is purchasing power parity a useful guide to the dollar?" Federal reserve bank of Kansas city, Economic review, third quarter, pp. 37-51.

Harrison, Huntington, S . (2000). Culture matters: how values shape human progess. Basic books.

John, G. 2016 " Reassessing longer run US growth. How low? " Federal reserve bank of San Francisco working paper 2016-18. http://www.frbsf.org/economic-research/publications/working-papers/wp2016-18/pdf.

Martin, J.P. & S. Scarpetta. " Setting it right. employment protection, labor reallocation and productivity." De Economist , 60: 2 ( 2012): 89-116. Online at: http://ideas. repec.org/p/iza/izapps/pp.27 html (3).

MGI (2013). Disruptive technologies: Advances that will transform life business and the global economy. Tech. Rep: Mckinsey Global Institute.

National Science Board, 2014 , Science and Engineering Indicators, USP TO Patent applications and grants by industry.

OECD (2011), Divided We Stand: Why Inequality Keeps Rising, OECD publishing.

US Travel Association " US Travel Employment Reaches An Time high" ( press release ) , Nov. 7, 2014 Accessed Dec. 8 , 2014. available here : http://www.ustravel.org/news/press-releases/travel-industry-employment-reaches-all-time-high.

US Bureau of the census, 1993. Statistical abstract of the United States, 113 the ed. Washington.

Technology influences future of job skills need change

Technology innovation will change businss models to impact on the employment. Many of the major drivers affecting global industries are expected to have significant impact on jobs, ranging from significant job creation to job replacement and from heightened labor productivity to widening skills gaps. This global technology innovation brings this question: How will technology change impact on human job skills?

Economists predict that there are many industries and countries the most in-demand occupations or specialists did not exist ten or even five years ago and the pace of change is not to accelerate. They also predict that by one popular estimate 65% of children entering primary school today will altimately end up working in completely new job types that don't yet exist.

In such a rapidly employment needs change, the ability to anticipate and prepare for future skills requirements, job content and the aggregate effect on employment is increasingly critical for businesses, governments and individuals in order to fully seize the opportunities presented by these trends.

The World Economic Forum's Future Of Jobs Report seeks to understand the current and future impact key disruptions on employment levels, skill sets and recruitment patterns in different industries and countries. It indicates that human is today at the beginning of a fourth industrial revolution. Technological development will be in these fields , such as artificial intelligence and machine learning, robotics, nanotechnology, 3D printing and genetics and biotechnology. For building technology example, smart systems, homes, factories, farms, entire cities will help tackle problems ranging from supply chain management to climate change. Concurrent to this technological innovation is a set of broader socio-economic, geopolitical and demographic developments.

II. What will be future human employment trends?

The global workforce is expected to experience significant change between job families and functions. Across the countries are covered by the report, current trends could lead to a net employment impact of more than 5.1 million jobs lost to disruptive labor market changes over the period 2015 to 2020 year, with a total loss of 7.1 million jobs, two thirds of which are concentrated in routine white collar office.

Future of jobs survey, World Economic Forum indicated that the drivers of change , industries overall, share of respondents rating driver as top trend percentage. Such as demographic and socio-economic: changing nature of work, flexible work 44%, middle class in emerging markets 23%, climate change, nature resources 23%, geopolitical volatility 21% , consumer ethics, privacy issues 16%, longevity , ageing societies 14%, young demographics in emerging markets 13%, women's economic power, aspirations 12%, rapid urbanization 8%. Technological industry includes mobile internet, cloud technology 34%, processive power, big data 26%, new energy supplies advance technologies 22%, internet of things 14%, sharing economy, croud sourcing 12%, robotics, autonomous transport 9%, artificial intelligence 7%, advance manufacturing , 3 D printing 6%, advanced materials, biotechnology 6%.

Some economists also predict how the timeframe to impact industries, business models: The impact is felt already included: rising geopolitical volatility, popular mobile internet and cloud technology, advances in computing power and big data, crowdsouring, the sharing economy and peer-to-peer platforms, rising of the middle class in emerging markets, young demographics in emerging markets, rapiding urbanization, changing working environments and flexible working arrangements, e.g. online working style, climate change, natural resource constraints and the transition to a greener economy. Then, from 2015 to 2017 year period, our life will be impacted by these influences. Such as new energy supplies and technology development, internet , advanced manufacturing and 3D printing, new consumer concerns about ethical and privacy issues, longevity and ageing societies, women's rising aspirations and economic power.

Further, in the future period from 2018 to 2020 year. Our life will be impacted by these influences. Such as advanced robotics and autonomous transport, artificial intelligence and machine learning and advanced materials , biotechnology and genomics will be mature developed. Thus, these technological changes will bring these impacts of this human job question, such as: How to change in ease of recruitment?

III. How to change in ease of recuitment?

Given the overall, disruption industries are experiencing, it isn't surprising that with current trends, competition for talent in-demand job families, such as computer and mathematical and architecture and engineering and other strategic and specialist roles will be fierce, and finding efficient ways of securing a talent technological development for every industry. Thus a tangible impact to further adequacy of employees' existing skill sets can already felt in a wide range of jobs and industries today. However, if skills demand is evolving rapidly at an aggregate industry level, the degree of changing skills requirement. Within individual job families and occupations is even more pronounced. Even, those jobs that are less directly affected by technological change and how a largely stable employment outlook. How can technological development influence occupations or jobs future changes? For example, the mobility industries expect employment growth is accompanies by situation, where nearly 40% of the skills are required by key jobs in the industry are not yet part of the core skill set of these functions today.

As the same time, workers in lower skilled roles, particularly in the office and administrative and manufacturing and production job families, many find themselves caught up in a cycle where low skills stability means who could face redundancy without significant reskilling and upskilling even when disruptive change many erode employers' incentives and the business case for investing in such reskilling.

Future of jobs survey World Economic Forum also indicated skills stability from 2015 to 2050 year industries overall unstable and stable percentage such as below:

Media, entertainment and information industry skills stability 65% and skills unstability 35%,

Consumer industry skills unstability 30% and skills stability 71%,

Healthcase industry skills unstability 29% and skills stability 71%,

Energy industry skills unstability 30% and skills stability 70%,

Professional service industry skills unstability 33% and skills stability 67%,

Information and communication technology industry skills unstability 35% and skills stability 65%,

Mobility industry skills unstability 39% and skills stability 61%,

Basic and infrastructure industry skills unstability 42% and skills stability 58%,

Financial services and investors industry skills unstability 43% and skills stability 57%.

Thus, economists predicted the future skills serious shortage industries included basic and infrastructure industry unstable skills was 42% as well as financial services and investors industry unstable skills was 43%. Both were the highest percentage of the unstable skills among of them. It implied that university students could choose these both kinds of subjects to learn because these kinds of both industries will be shortage of graduated student number of supply. Otherwise, the other industries will have higher stable skills supply to employment market. So, it brings this question: How can stable and unstable skills influence to different industries and global employment market and living quality? For example: efforts to place unemployed youth in apprenticeships in certain job categories through targeted skills training may be self-defeating of kills requirements in that job category are likely to be different in just a few year's time. Indeed is some causes such efforts may be more successful if businesses models were on future expectations. Thus, businesses will need to put talent development and future workforce strategy front and centre to their growth.

It is therefore critical that long term changes to basic and lifelong education systems are needed to complemented with specific, urgent and forced reskilling efforts in each industry. This entails several major changes in how business views and managers talent both immediately and in the longer term. Thus, if university students can choose to study these any one subject among of them. Their living standard will be raised in possible, due to it will have many these kinds of job skills requirement in the future. Finally, it is also possible that unemployed young students will find jobs more easily and it will reduce young people unemployment number and it will raise life standard of future young people if nowadays any country universities began to encourage young students to choose to study any one of these above subjects to prepare to study . Thus, these industries' future unstable skills shortage will be decreased to influence any businesses development to be better in the future.

Can technology influence UK and US life of standard to be fallen or grown

Some economists predicted long term global economic growth will be remain till to 2050 year at least. There are based on a model that takes amount of projected trends in demographics, capital investment and technological process.

Will the shift in global economic power be continued by technology process factor? They give these reasons to explain why technology process can influence global economic growth as below:

The first reason, they feel future consumer online transaction market will trend mature that are becoming increasingly mature, sophisiticated and digitally technological process online transaction in popular. The second reason, emerging markets vary greatly in their institutional strengths and weaknesses and need to be assessed in a technological transaction way. These could also be major differences in institutional strengths between industry sectors within countries. Deep local knowledge that is updated in real time is critical here to manage businesses successfully in an emerging market environment, e.g. shorten online shopping transaction time and online product choice process. Having the right local partners to navigate online business through local political, legal and regulatory systems is also critical. Identifying and promoting local talent who understand local online business and social cultures better than any outside will also be an increasingly source of comparative advantage ( As discussed in more detail in a recent PWC Growth Markets Centre Report, " Presence to prosperity" ). The third reason, for larger western companies making strategic investments in emerging markets, part of their contribution could be to try to improve the local institutional framework. This could involve offering, appropriate technical assessment and advice to local governments in areas like corporate governance, fiscal policy and intellectual property rights protection. It could also involve investing in social and economic infrastructure ( e.g. schools, roads, railways, power and water networks) where there are critical to a company's long term success in a region. Finally, there are existing markets in North America and Europe. There will remain very significant players in the global economy for decades to come. PWC report analysis shows average income levels will remain much higher than in even the best performing emerging markets for the foreseeable future. Advanced economics will also generally speaking, still be easier and lower risk places to do business given their political and institutional strengths. Thus, it seems that technological process will be the main factor to influence global GDP growth and human living standard in our future societies. It also means that the economists predict the technological process will be one main factor remain to influence global economic growth and rasing human living standard.

Developing countries will face stronger headwinds in the decades and because technological changes are rendering manufacturing more capital and skill intensive. So, it brings this question: How past, present and future of human living standard is influenced by technological process development? To answer this question , we need to assume that economic growth is a precondition for the improvement of living standards and lifetime possibilities for the " average" citizen of the developing world. Optimists would point to improvements in governance and macroeconomic policy in developing countries and to the still not fully exploited political of global economic technological globalization to foster new industries in the poor regions of the world by outsourcing and technology transfer

Otherwise, pressimists would feel rich countries control world economy, threaten to technological globalization and obstacles that late industrializers have to surmount given competition from China and/or America owned high technolgical development etc. countries. Thus, it also brings this question: What will be changed to influence past, present and future economic growth and human living standard by technological development?

It is a global technology structural transformation change issue, it means that the birth and expansion of new ( higher technological productivity change) industries and the transfer of labor from traditional or lower : productivity activities to modern ones. Thus, this history change will occur from past industrialization, then it will change to present technological process and it will change to future artificial technological industry in the future.

I. Has it close relationship between climate change and technological process to influence human living standard? Some economists point out global growth will be hindered by rising operational costs as global temperatues rise, with studies suggesting that a worst-case impact of a 1% reduction in GDP growth per year could be realized. Research also suggests that the impact will be disproportionately damaging to developing economies and only through a collective effort to enact strict carbon emissions policies can be the long term financial challenge of climate change potentially. If it was truth, it will bring this question: Can climate change influence to reduce human living standard to be fallen? To answer this question: We need to know why climate change can raise economic cost to global societies. Generally, we know global warming is likely to impact global activity. Such as, the effect on growth and inflation , climate damage functions: quantifying the impact on activity, regional poor effects. Accessing the impact of cimate change is complex with uncertainty about both the degree of future global warming and the subsequent impact an global activity. There are clearly some benefits as well as costs as the planet warms. There is also the unknown of how technological process will respond and potentially alter the path of global warming.

II. Why can climate change influence effect on growth and inflation? Climate change at varying levels of warming, the impact of rising temperatures will be widespread, in part due to the financial, political and economic integration. Global warming will primarily influence economic growth through damage to property and lost productivity, mass migration and security threats. For Hurricane Sandy example, which flooded much of New York in 2012 year, it caused economic damage such extreme weather events can cause. Rising sea levels will also likely harm economic output and businsses become impaired and people suffer damage to their homes.

In economy view point, using a production function, we can demonstrate the likely effort climate change will have on output. If we assume less capital stock is available , due to the damage inflicted from climate change, we would see a fall in the productive capacity of the world economy. Then, the world products: function as each unit of labor produces less output, lower labor productivity may not just occur , due to a lower level of capital stock. However, higher global temperatures may affect food security, promote the spread of infectious diseases and impair those working outdoor. Such factors are likely to cause greater incapacity and social unrest and as a result will reduce both the effectiveness ( productivity) and the amount of labor available to produce output. The argument whereby the world's population is seen not to respond to climate change. It is possible that, preventive measures , such as flood defenses are put in pace in order to avoid the costs of climate change. It is a short-term economic cost to this action as resources are directed from more productive uses. It can include two aspects influence , such as food inflation and energy cost increasing , the reasons are as below:

On the one hand, inflation is likely to rise as shortage, particularly in agriculture. Climate change will cause agriculture supply shortage and food demand rising. Due to a reduction in output, but an increase in the general price level as a result of global warming. This leads human onto the possible, an inflationary effects of global warming on the agriculture world economy. Because agricultural yields are sensitive to weather conditions and as our climate becomes ever more extreme, more frequent droughts may reduce crop yields in areas where food production is vital. Higher global food prices will likely thus causes the pressure to global consumers. Then, it will lead global food consumers' living of standard to be fallen down because we need to save more money to buy food. So our entertainment and daily essential spending expenditure will be also decreased, due to saving more money to buy inflationary food.

On the other hand, energy costs will increase, due to climate change influence natural resource supply number of energy production to be shortage. Thus, it will bring higher energy costs are also likely to boost inflation. As our climate become more extreme, we are likely to demand greater energy to both our working and living environments during the summer and heat , then when we experience cold winters. Thus, if human lacks high technological process development to prepare to win climate change challenge influence, e.g. invention of renewable energy or new food planting or growing method which can adopt to grow or plant in bad climate in the future easily. Then, we shall face our living standard to be fallen , due to lacking technological development to solve climate change negative influence challenge in our future.

To conclude, I believe that human living standard will be fallen, due to we lack high technological process to prepare to solve climate change challenge, such as global warming, sea level rising, flooding, etc. natural crisis to cause food nature resource and energy nature resource supply shortage. Then, it will cause inflation of our essential product or food price rising seriously. Finally, as an effect, human living standard will be fallen.


Whether future technology can influence human living standard

I Can technology influence human habitable population dynamic change to impact living standard?

To explain above question, we also need to consider these questions, such as: What factors influence human population growth trends most strongly? How does population growth or decline impact the environment? Does urbanization threaten our quality of life or offer a pathway to better living condition?

Nowadays, human population trends are centrally important to environmental science because they help to determine the environmental impact of human activities. Rising population put increasing demands on natural resources, such as land, water and energy supplies. At the same time, pollution challenge will cause, due to human needs , spend or waste much natural resource to raise living standards, such as air and water pollution and greenhouse gas emissions, along with increasing qualities of water. Then, it will bring this question: Will technology influence environmental impact to human living standard?

Paul R. Ehrlich & John P. Holdren (1974) indicated one widely-cited formula is " I=PAT", which concerns population interacts with several other factors to determine a society's environmental impact as below:

Environmental impact = " population" x "affluence" ( or
consumption) x " technology"

In fact, the question depends on assumptions about human preference. What standard of living is seen as acceptable , and what levels of risk and variability in living conditions will people tolerate? Many of these issues are not just matters of what humans want, rather who intersect with physical limits, such as total arable land or the amount of energy available to do work. In such instances nature sets of bounds on human choices ( Joel E. Cohen, 1995).

To research whether technology can influence human habitable population dynamic change to impact living standard. We need to answer this question. How did industrialization alter population growth rates so sharply? One central factor was the mechanization of agriculture, with enabled societies to produce more food from available in inputs.

In our past agricultural society, as food supplies expanded, average levels of nourishment rose and diseases declined over succeeding generations. Improvements in medical care and public health services, which took place more in uraban than in rural areas, also helped people to live longer. So , death rates face in our nowadays technological society. After several decades of lower mortality, people realized that they did not have. So, many families decided to born many children to achieve their desired family size as well as birth rates began to fall as well because human was educated popularly in our technological society. Thus, in our technological society, it will cause expanded work forces, it can help nations increase their economic output, raising living standards for everyone. They also can strain available resources and services. Anyway, it also may cause shortages and economic discuption.

However, I believe pollutants, medical services, mortality and fertility factors can influence human living standard to be better or poor, instead of technology factor. But, I also believe technology will be one important factor to influence human living standard. I shall indicate these reasons to explain why technology can influence human living standard to be better or worsen and it has effort to reduce these bad factors causing to influence our future living standard.

The first reason, exposure to pollutants is a major factor contributing to infant motality and lower life expectancy in developing countries, e.g. China, India, Hong Kong, Korea. Thus, if any these developing countries lack enough technological environmental investments, such as providing cleaner energy sources and upgrading sewage treatment systems which can significantly improve public health. So, these development countries' population living standard will have negative impact if they lack technological environment investments to reduce pollution serious level to influence citizens' unhealth.

In second reason, increasing life expectancy is creating a public health infrastructure that can identify and repond quickly to disease outbreaks, families and other threats. For example, when severe alute respiratory syndrome ( SARS) emerged as a disease that might cause an international epidemic. The health centers for disease control and prevention launched an emergency response program that required health departments to report suspect cases for evaluation, developed tests to identify the SARS virus, and kept health care providers and the public informed about the status of the outbreak. The United States and many other countries also reported their SARS cases to the World Health Organization. These types of close surveillance and preventive steps to control infection. Thus, it seems nowadays, countries need to invest or medical technology on attempt to invent more unique medical machines to avoid any unknown diseases existing.

The third reason, that drives population trends migration when includes geographic population shifts within nations and across boards. Migration in less predictable over long periods than fertility or morality since it can happen in sudden waves, for example, when refugees brings a war. So, slowly over many years, immigration often changes host nations or regions ethics mixes and strain society. Immigration often changes host 's or regions' ethic mixes and social services.

On the position side, it can provide needed labor ( both skilled and unskilled). In my view point, if the country has good technology resources to provide skilled immigration labor to attribute their technological reource is one important factor ( production of factor) to let immigration skilled laboe to contribute their effort to serve its country to raise productivity to develop its economy growth for long term. For technological process source countries, however, technology can assist immigration to attribute valuable talent. Specially, since educated and motivated people ( labors ) are most likely to migrate in search of opportunities.

II Human's history technological development
In human history development, through the early decades of the industrial revoluation, life expectancies were how in Weatern, Europe and the United States. Due to lack high technological medical invention. Many people died from infactious diseases, such as typhoid and cholera, which spread rapidly in the crowded, bad environment conditions that were common in early factory towns and major cities or were weakened by poor nutrition, due to lack high technological environment science resources to provide environment protection professionals to produce any new technological environment protection product to reduce air and water pollution. But from about 1850 year through 1950 year , a cascade of health and safety advances radically improved living conditions in industrialized nations. Due to technological environment industry protection development. Some this technological environment protection industry development brings these benefits to human, includes: improving urban sanitation and waste removal, improving the quality of the water supply and expanding access to it, forming public health boards to detect illnesses and quarantine the sick, researching causes and means of transmission of infectious diseases, developing vaccines and antibiotics, adopting workplace safety laws and limits on child labor, and promoting nutrition through steps, such as fortifying milk, breads and cereals with vitamins.

By mid-20 th century, most industrialized nations had passed though the demographic transition, as health technologies mere transferred to developing nations, many of these countries entered the mortality transition and their population. The world's population growth rate peaked in the late 1960 year at just over 2% per year 12.5% in developing countries. Thus, it seems health death chance from diseases and environment pollution importantly. It implies our environment pollution to influence our health will be serious, so we need continue to develop my technology on health and environment protection aspects to raise human living standard to avoid or reduce death chance increasing.

III Can food technology production shift diets, the type, combination and quantity of foods people consume, contribute to a sustainable food future?

Building on the United Nations Food And Agriculture Organization's food demand projections, it estimates that the world needs to close a 70% "food gap" between the crop calories available in 2006 year and expected calorie demand in 2050 year. It indicates the food gap stems primarily from population growth and changing diets. The global population is projected to grow to nearly 10 billion people by 2050 year, with two thirds of those people by 2050 year, with two-thirds of those people projected to live in cities. In addition , at least 3 billion people are expected to join the global middle class by 2030 year. As nations urbanize and citizens became wealthier, people generally increase their calorie intake and the share of resource-intensive foods, such as meats and dairy in their diets. As the same time, technological advance, business and economic changes and government policies are transforming entire food chains from farm to fork.

In human future, the demand of increasing technological agricultural production will increase to efforts to reduce the food shorten to close the food gap. However, relying solely on increased production to close the gap would exert pressure to clear additon natural ecosystems. For example, to increase food production by 70% , when avoiding further expansion of harvested area, crop yields would need to grow one-third more quickly than who did during the Green Revolution. In short, yield increases alone with likely be insufficient to close the gap. In the fuure, farming technology development will be applied to solve these challenges. Such as human plan use the globagrimodel farming technology to quantify the land use and greenhouse gas consequences of different foods and then analyze the per person and global effects of the three diet shifts on agricultural land needs and greenhouse gas emissions. Moreover, scienists estimate to apply farming technology to these diet shifts, if implemented at a large scale, it can close the food gap by up to 30%. When substantially reducing agriculture's resource use and environmental impacts. Thus, it seems farming technology industry will be increased demand to applied to solve future food shortage challenges to raise human living standard in the future.

In conslusion, due to human's history development, it proved in our past, we had been encountering agricultural production to industrial production . Then in the present, we are encounting technological production. In the future , we will encounter human intelligence technological production in possible. So, it implies " future technology change will influence human living standard to be better if human continue consider to how to innovate medical health technology and agricultural growth technology and human intelligence technology and climate environmental protection technology and space technology research. Otherwise, human living standard can not be raised , even fallen if human is lazy and we do not continue to spend time and money investment to research these kinds of technological improvement. Thus, I think human ought continue to research these kinds of technological improvement to achieve human living standard to be better in the future.

Reference

Future of jobs survey, World Economic Forum.

Hauser, J. Tellis, G. J; Griffin, A. 2006. Research On Innovation: A Review And Agenda For Marketing Science. 25(6): 687-717.

J.P. Holdren & P.R. Enrlich, " Human population and the global environment", American Scientist, vol. 62 (1974), pp.282-92.

Joel E. Cohen, How many people can the earth support? ( New York: Norton, 1995), pp. 212-36, 261-62.

Mohr, G. J. Griffin, A. 2010. Research On Innovation : A Review And Agenda For Marketing Science. 25 (6): 687-717. Names of drivers have abbreviated to ensure legibility. Future of jobs survey, World Economic Forum.

" Presence to prosperity", PWC Growth Markets Centre Report: http://www.pwc.com/gx/en/growth-markets-centre/presence-to-profitability.jhtml

# CHAPTER II

# Robots how influences human social job market change

How Robots Influence Driven Automation Industry Development

1.1 (AI) - driven automation industry development how to influence work nature change

(AI) -driven automation industry will create wealth and expand economy growth to any countries, but it will be accompanied by changed in the skills that workers need to learn. One of main ways that technology increases productivity is by decreasing the number of labor hours needed to create a unit of output. It implies (AI) technology will influence low educated and low skillful labor number to be decreased ( reduction employment number).

In contrast, technological change tended to work in a different direction throughout the nowadays. The advance of computer and the internet raised the relative productivity of higher skilled workers. So, routine-intensive occupations that focused on predictable tasks disappearance, such as switch board, operators, filming checkers, travel agents and assembling line workers etc. were particularly replaced by new technologies.

However, today, it may be challenging to predict exactly which jobs will be most immediately affected by (AI) driven-automation. The reason is because (AI) is not a single technology, but rather a collection of technologies that are felt unevenly through the economy to influence job changing both negatively and positively. In positively view point, (AI) driven-automation will make many workers more productive and increase demand for certain skills. Consequently, new jobs are likely to be directly create in areas , such as the development and supervision of (AI) as well as indirectly created in a range of areas throughout the economy as higher incomes lead to expanded demand. Otherwise, in negatively view point, many traditional human needed ( demand) skillful jobs will be threatened by automation are highly concentrated among lower-paid, lower-skilled and less -educated workers. It means automation will cause pressure on demand for this group, pressure and employment, if (AI) can replace the low skilled and less educated workers' jobs. Thus, (AI) will have negative influence to impact on the labor market.

(AI) capabilities will enable automation of some tasks that have long required human labor. Why can (AI) replace some simple human jobs? For example, advances in robotics are expanding machines' abilities to interact with and sharp the physical world. Combined , (AI) and robotics will give rise to smarter machines that can perform more sophisticated functions than ever before and brings more advantages that humans have exercised. This will permit automation of many tasks now performed by human workers and could change the shape of the labor market and human activity.

1.2 How (AI) influences labor market

Today, it may be challenging to predict exactly which jobs will be most immediately affected by (AI)-driven automation. Because (AI) is not a single technology, but rather a collection of technologies that are applied to specific tasks.

Some specific predictions are possible based on the current (AI) technology. For example, driving jobs and house cleaning jobs, bank counter service jobs, telephone enquiry service operators. Restaurant cooking jobs, simple accounting record service jobs etc. that require relatively less education to perform. Advancements in computer vision and related technologies have made the feasibility of fully appear more likely, potentially displacing some workers in driving-dominant professions. Seemingly similar robot, for which the operational tasks is less specific of navigating to a specific destination when following a set of given rules and preserving safety.

In the future, the effects of (AI) on the labor market in the decade ahead will continue the trend toward skill-biased change that computerization and communication innovations have driven in recent decades. Thus, some human driving occupation will be disappeared or replaced by (AI) automation driven. For example, bus drivers, light truck or delivery services drivers, heavy and tractor-trailer truck drivers, school drivers, tax drivers, travel bus drivers. However, (AI) technology could enable some workers to focus time on other job responsibilities, boosting their productivity, and actually raised wage growth among those still holding the reshaped jobs. For example, salespeople, who currently spend a considerable amount of time driving could find themselves able to do other work when a car drives them from place to place, or inspectors and appraisers could fill out paperwork, when their car drives itself. This (AI) -driven technology should make these workers more productive, with (AI) -driven technology serving as a complement, not a substitute. New jobs will also likely be created, both in existing occupations cheaper transportation costs with lower prices and increase demand for products and all the related occupations, such as service and fulfillment, and in new occupations not currently foreseeable.

What kind of jobs will be created by (AI) technology? Predicting future job growth is extremely difficult, due to it depends on technologies or substitute for existing today as well as they may complement or substitute for existing human skills and jobs. However, (AI) will also lead to substantial indirect job creation to the degree it raises productivity and wages, it may also lead to higher consumption that would support additional jobs from high-end draft production to restaurant and retail. The future(AI) " augmented intelligence", the technology's role is as assisting and expanding the productivity of individuals rather than replacing human work. Thus, based on the biased-technical change framework, demand for labor will likely increase the most in the areas where humans complement (AI) automation technologies. For example, (AI) technology , such as IBM's Watson may improve early detection of some cancers or other illnesses, but a human healthcare professional is needed to work with patients to understand and translate patients' symptoms, inform patients of treatment options, and guide patients through treatment plans. Shipping companies may also partner workers who pick up and deliver products over the last feet with (AI) enabled autonomous vehicles that move workers efficiently from site to site. In such cases, (AI) augments what a human is able to do and allows individuals to either be move effective in their specially task or to operate on a larger scale. Thus, it seems (AI) technology will also create new jobs, raise productivities and workers' efficiencies.

Redefining management in the workforce of artificial intelligence

2.1 Change management

In the future, due to artificial intelligence influences to some kind of human jobs nature. So, the kind of human jobs of management methods will also need to change to adapt the artificial intelligence technology input to their organizations. It will cause challenges for every executive and manager if who won't have effort to manage their teams how to apply artificial intelligence technology to work efficiently and easily. For example, division of labor will change among humans and machines will increase. Thus, companies will have to adapt their training performance and talent strategies how to emphasize on work that how to make human judgment and skills and experimentation. Thus, (IA)'s greatest impact will be on administrative coordination and control tasks, such as scheduling , resource allocation.

In fact, mangers will encounter this challenges: How to apply human experience and expertise to judge critical business decisions and practices when the information available is insufficient to suggest a successful course of action? Due to this kind of work will require new skills and mindsets. I shall indicate these change management methods to adapt (AI) technology. Such as: administration and routine tasks, scheduling , allocation of resources and reporting will fall within the intelligence machines, responsibilities that have long been reserved for humans. For example, a typical store manager or a lead nurse at a nursing home most constantly arrange shift schedules, accounting for staff members' absences owing to illness, vacation time or sudden departures.

Thus, the managers need to learn how to arrange new division of labor within the organizations after (AI) technology had been implemented to the organization. Artificial intelligence is currently influencing into once considered exclusive to humans: assessing and acting on human emotions and personality traits. The influences to managers need to change their strategies to adapt (AI) technology implements include such as below:

Firstly, managers need to spend the bulk of their time on coordination and control tasks from intelligent system implements. Their time spending on these major three aspects from impact of intelligent system: coordinate and control, solve problems and collaborate and people and community , strategy and innovation three aspects. Thus (AI) will influence managers need to change their judgment method to teach whose teams how to adapt the (AI) system operations in any organizations.

Secondly, (AI) will influence top, middle and low level management needs to change to adapt the (AI) technology operations to any owned (AI) technology organizations in the future. Intelligent machines must be trained in context. Just like humans , on-the-job training is a requirement for such machines because they typically arrive with only very general capabilities. To get the most from (AI), managers at all levels must participate in the instructional experience and in the learning process and provides managers' familiarity with such systems on these aspects, e.g. How the system works and generate advice, how the system has a proven track record , how the system provides convincing explanations , how the system can make simple rule- based decisions.

Thirdly, managers need to learn how to make judgment more accurate (AI) systems assistance. Although (AI) will invariably take on more routine work and even augment human decision-making, it won't judgment work, the application of human experience and expertise to critical business decisions when the information available is insufficient to suggest a successful course of action or reliable enough to suggest an obvious course of action. For a sense of the nature of judgment work, consider big data marketing and sales analytics. Such analytics often provide insights that can inform promotional campaigns, including predicting which promotions will generate desired sales brand further into the future, marketing executives need use judgment, combining analytics with their own and others' insight and experience.

The application of experience and expertise to critical business decisions and practice represents the real value of human judgment. But, when artificial intelligent machines are implemented to any organizations to assist the low, middle and top level management to make any business judgment. These forms of judgment work that managers can gather data interpretation, idea development more absolute from (AI) machine assistance. Thus, why these level management executives need to learn how to apply (AI) machines to help them to make any business judgment more accurate.

2.2 How (AI) influences organizational change

Consequently creative and social intelligence will be in even greater demand as (AI) makes in management and the workforce. This development will represent a long term trend in labor markets , one characterized by intensifying demand and reward for social skills with a growing desire for creative capabilities, managers will seek to fashion of ideas and hypotheses from inside and outside of the enterprise to shape solutions to their most pressing business problems. Thus, (AI) will influence overall organizational team members who have chance to participate any decision to make more accurate business judgment.

Many managers mistakenly view judgment work as only an individual discipline, failing to appreciate that it can also involve decide interpersonal and organizational practices. In more complex settings, judgment is typically a collective outcome of individuals' and teams' diverse perspectives, insights and experiences. And often , the resulting choices are better informed than decisions that an individual would have arrived at on his or her own.

Thus, when any organizations apply (AI) technology to assist managers to gather data and ideas to make any judgment. In these cases, organizations can create the conditions for effective collective judgment by establishing structures , such as " shadow advisory boards" that prompt managers and employees to source and synthesize multiple perspectives. Thus, a traditional organization (firm) might freshen its thinking is t put together a shadow advisory board, comprised of young, digital people who can apply (AI) machine assistance to make judgment work more accurate whether related to people development, problem-solving or strategizing and innovating for considerable degrees of creative and social intelligence.

Thus, on the one hand, (AI) technology machine augmentation and automation can give these advantages to human (organization managers) , e.g. developing people and community, solving problems and collaborating, coordinating and controlling work, shaping strategy and leading innovation. Besides, on the other hand, the next generation managers need have these individual attitude to treat intelligent machines to be as colleagues.

When, judgment is a human skill, intelligent machines can accelerate human learning that supports it, assisting in data -driven simulations, scenarios and search and discovery activities. Focuses on judgment work, some decisions require insight beyond what data can tell them. This is the sweet sport for human judgment, the application of experience and expertise to critical business decisions and practices. Thus, managers will also need to find ways to learn how to use digital (AI) technologies to tap into the knowledge and judgment of partners, customer external stakeholders and role models in other industries after the (AI) machine had been implemented to the organization.

Future works change: Automation, employment and productivity
3.1 How (AI) influences employment

Human future " micro to macro" industry trends will be affected business strategy and public policy by (AI) technology. In the future (AI) technology will influence those six themes: productivity and growth, natural resources, labor markets, the evolution of global financial markets, the economic impact of technology and innovation and urbanization. However, (AI) technology will bring economic benefits of tackling gender inequality, a new global competition, Chinese innovation and digital globalization.

Nowadays, advances in robotics artificial intelligence, and machine learning are in a new age of automation, as machines match or outperform human performance in a development to any countries. For example, automation of activities can enable businesses to improve performance by reducing errors and improving quality and speed, and in some cases achieving outcomes that go beyond human capabilities. For example, some research indicated automation could raise productivity growth globally by 0.8 to 1.4 % annually; more than 2,000 work activities across 800 occupations. When less than 5% of all occupations can be automated using demonstrated technologies about 60% of all occupations have at least 30% of constituent activities that could be automated. Many occupations will change that will be automated away: Activities most susceptible to automation involve physical activities, in highly structured and predictable environments, as well as the collection and processing of data. They are most prevalent in manufacturing , accommodation and food service and retail trade and include some middle-skill jobs. For example, such as natural language processing is a key factor. Beyond technical feasibility, the cost of technology competition with labor including skills and supply and demand dynamics, performance benefits including and beyond labor cost savings, and social and regulatory acceptance will be affected by (AI) automation technology. Thus, (AI) automation will impact to influence global employment in those aspects as below:

Firstly, assuming that people are displaced by automation will find other employment. The anticipated shift in the activities in the labor force is of a similar order as the long-term shift away from agriculture and decreases in manufacturing share of employment. Both of manufacturing and agriculture industries which would be accompanied by the creation of new types of work not foreseen at the time.

Secondly, for business, the performance benefits of automation are relatively clear. Thus, the businessmen have opportunities for their micro economies to benefits from the productivity growth potential and macro economies to benefit to encourage continued progress and innovation , investment and market incentives. At the same time, employers must innovate policies to help workers and institutions adapt to the impact on employment.

This will likely include rethinking education and training, income support and safety nets , as well as support for those dislocated, when employees need to leave themselves homes to move to other cities to learn new (AI) automation works. Thus, individuals in the workplace will need to engage move comprehensively with machines as part of their everyday activities, and acquire new skills that will be in demand in the new automation age. Consequently , the scale of shifts in the labor force over many decades that automation technologies can be a similar order to the long -term technology -enables shifts in the developed countries' workforces away from agriculture in the 21 th century. Those shifts did not result in long-term mass unemployment because they were accompanied by the creation of new types of work not foreseen at the time. However, human will still be needed in the workforce when the total productivity gains are caused by (AI) technology.

3.2 What occupations will be influenced by (AI) technology.

In the future, scientists predict that these occupations will be influenced by (AI) technology mostly. They include : retail salespeople, food and beverage service workers, language or translation teachers, health practitioners. Since these work activities have a more relevant occupations are made up of a range of activities with different potential for (AI) automation . For example, a retail salesperson will spend more time interacting with customers, stocking shelves , or ringing up sales. Each of these activities is distinct and requires different capabilities to perform successfully.

Thus, these job activities have similar simple control characteristics. Simple activities include greet customers, answer questions about products and services, clean and maintain work areas, demonstrate product feature process sales and transactions. All these activities can have similar simple activities in order to (AI) machines can be learn how to do these activities from (AI) technology . For example, the capability perception includes sensory perception, cognitive capabilities, such as retrieving automation, recognizing known patterns( supervised learning), logical reasoning problem solving.

Thus, (AI) machine is such human, which has feeling and emotion, such as social and emotional sensing, judgement reasoning methods, natural language understanding and physical capabilities, such as mobility , navigation, gross motor skill, fine motor skills. It seems that the future, (AI) human invents machines which will have these human characteristics to do human similar behavioral job duties more easily and efficiently. It implies these above human occupations will be replaced by (AI) human invention machines in the future. Due to (AI) creation, it is possible to cause unemployment number of these above workers will increase because (AI) machines can do their similar job behavioral activities.

Consequently, employers won't need to employ many of these skillful labor. Otherwise, they can buy less number (AI) machines to attempt to do whose job activities more easily and efficiently. So, it seems (AI) machines will have more high work performance to replace these occupation workers' work performance. Finally, these occupation worker unemployment number will only increase when the (AI) machines had been invented to achieve to do their work behavioral activities absolutely success in the future.


3.3 Whether (A) technology machine labor
will replace human worker more or assist
human worker more


There is no single agreed definition of a robot how outcome of a task that is completed without human intervention. When some definitions require the task to be completed by a physical machine moves and respond to its environment, other definitions use the term robot in connection with tasks completed by software , without physical embodiment.

However, to answer the question : Whether (AI) technology machine labor will replace human worker more or assist human worker more. I shall indicate some examples to let readers to judge whether (AI) technology can create new jobs or reduce old jobs.

Firstly, I shall explain what (AI) function is. (AI) is a service robot that performs useful tasks for humans or equipment excluding industrial automation application . Thus, the classification of a robot into industrial robot or service robot is done according to its intended application. It is also a personal service robot or a service robot for personal used for a non commercial task, usually by lay persons . Examples are domestic servant robot, and pet exercising robot. It is also a professional service robot or a service robot for professional used for a commercial task, usually operated by a properly trained operator. Examples, are cleaning robot for public places, delivery robot in offices or hospitals, fire-fighting robot, rehabilitation robot and surgery robot in hospitals. Thus, these functions will be future (AI) application to our daily life necessaries or business necessaries.

However, some authors agree (AI) will bring negative outcomes of automation, due to raise competiveness, reduce human job nature. Otherwise, other authors argue (AI) will bring positive outcomes of automation, due to raise productivities, job creation, assist humans work.

On the positive outcome hand, robots can increase productivity . This is particularly important for small-to medium sized businesses both are in developed and developing countries economies. It also enables large companies to increase their competitiveness through faster product development and delivery. Increased use of robot is also enabling companies in high cost countries to re shore, or bring back to their domestic base parts of the supply chain that will have previously outsourced to sources of cheaper labor. Currently , the greater threat to employment is not a automation, but an inability to remain competitive. Automation has led overall to an increase in labor demand and positive impact on wages. The reason is that the middle-income/middle-skilled jobs have reduced as a proportion of overall contribution to employment and earnings leading to fears of increasing income inequality, the skills range within the middle income bracket is large. Thus, robots are driving an increase in demand for workers at the higher

-skilled and with a positive impact on wages. This issue is how to enable middle-income earners in the lower-income range to unskilled or retain. Finally, the (AI) positive impact supporter who argue the future will be robots and humans can work together.

However, on the negative outcome hand, robots can substitute labor activities, but don't replace jobs. They believe that less than 10% of jobs are fully automatable. Increasingly , robots are used to complement and augment labor activities, the net impact on jobs and the quality of work is positive. Automation can provide the opportunity for humans to focus on higher-skilled, higher-quality and higher-paid tasks. Robots can improve productivity when they are applied to tasks that which perform more efficiently and to a higher and more consistent level of quality than humans. For example, increased productivity is enabling some firms, such as Whirlpool, Caterpillar and Ford Motors company in the US restructure their supply chains, bringing back parts of the manufacturing process to the country of origin. Thus, productivity gains due to robotics and automation are important not just at the company level, but also for build industry and nation competitiveness.

I suppose that productivity can be raised. What are the impacts of robots on employment? Firstly, the main focus of development has been on personal entertainment, which does not drive worker productivity ( manufacturing production). When the internet ( information and communication technology (ICT)) innovation. This is borne and by findings that manufacturing productivity, which has been driven by innovations in automation rather than consumer technologies, has government strongly than productivity in the services sectors of the economy in most nature economies. It seems (AI) automation will create many jobs in internet communication entertainment game industry. For example, many young people like to use internet to play any electronic games from computer or mobile at home or outside home conveniently. Thus, (AI) automation will increase demand to be invented to any new entertainment game from internet channel. It will need to employ many (AI) entertainment game inventors to create many automation entertainment games. Thus, (AI) automation in internet entertainment game industry will need human (AI) entertainment game inventors to invent the knowledge-based capital of (AI) automation entertainment games. The (AI) entertainment game inventors will need own research and development skills, form specific skills, organizational know-how skills, databased knowledge, design and various forms of intellectual property to do these (AI) automation entertainment game invention occupations in the future.

International Federation Of Robotics(2016) indicated that China will be as a major robotics manufacturer and user of robots, benefiting from jobs created by robot manufacturing and productivity gains from robot use. Chins had sold of robots to any one single market every year since 2017 year. The Chinese government has included a focus on robotics in its 10 year strategy. In order to achieve its target of a robot density of 150 units per 10, 000 workers by 2020 year. Thus, Chinese companies will have to install around 650,000 new industrial robots between 2016 to 2020 year, 2.5 times more than installed globally in 2015 year.

Hence, China (AI) manufacturing industry will need to employ many workers . It implies (AI) manufacturing industry will create many new occupations in China. Also, ministry of economy, trade and industry (2015) also showed that Japan currently has the largest stock of industrial robots in operations, primarily in the automation industry. Driven by a rapidly aging population and low productivity rates, the Japanese government has sights on a 20-fold increase in the use of robots in the non-manufacturing sector and a three-fold growth rate of labor productivity in the service sector both by 2020 year. Thus, it also implies Japan will need many robots to be provide to service industry. Due to robots will provide to serve any businessmen's clients. Thus, it is possible that the service workers won't be dismissed as well as it is depended on the serving job nature to decide whether Japan's service workers can still serve to their employer when the service (AI) robots are applied to whose employers.

Consequently, it seems that (AI) can create employment, Ministry of economy, trade and industry (2015) showed that such as China will develop the major (AI) automation manufacturing industry. The (AI) employers will need to employ many workers to manufacture any these different kinds of (AI) robots to satisfy China or overseas individual or business buyers needs. But, (AI) can also cause unemployment to the low skillful service workers. Such as if Japan some service businesses choose to buy any (AI) service robots to replace their service staffs to serve their clients. It is possible that the service staffs will be dismissed, due to (AI) robots can do such as their same service job duties to achieve better service performance.

Thus, today, it is increasingly common for people to use robots in various situations at home and in retail stores, hotels and hospitals these service industries. Robots are classified into server types based on their functionality ( service and utility robots or those designed to communicate with humans) and appearance ( humanoid robots or mechanical robots). The type of robot, to which each country allocated particular importance in the advance of robotics, reflects the sense of values and preferences of its population. Thus, if the country has high population needs to use robots, then they will influence either more new jobs creation or more old job loss in the country's (AI) manufacturing or (AI) service industries both. For example, Japan respondents often associate the term " robot " with humanoid robots that can communicate with human and they have a high level of familiarity with robot. The US has the highest level of robot utilization at home and in retail stores with its people being the most enthusiastic about the future use of robots. Germany shows a strong tendency to consider robots for industrial purposes and its people feel strong effort to the presence of robots in their households.

In conclusion, to judge whether how (AI) will influence the country's employment to be better or worse. It will depend on the country home buyers (users) or business buyers (users) how to use (AI) for their daily needs. If the country , such as US retail stores need to use (AI) , it will have possible to reduce some or many retail service workers. Even, if the country , such as Japan has many home users need to use (AI) , it will not influence the employment market. Otherwise, it will raise (AI) salespeople numbers. Even, if the country, such as Germany and China will have many (AI) manufacturers, then it will create many (AI) manufacturing occupations for these (AI) manufactory workers.

Consequently, (AI) robots manufacturing and service needs will have positive or negative impact to any country's employment. It will depend on the (AI) service provision and service workers' job nature as well as the manufacturing workers of (AI) knowledge level to decide their employment chance in their country's employment market.


How Robots Influence Our Societies

What does artificial intelligence(AI) mean?

● What (AI) function is?

Some scientists explain that artificial intelligence means which is an expert system, computer software that embodies a portion of the specialized knowledge of a human portion in a specific, narrow domain, owns decision making ability of human expert. The (AI)technology is based on the premise that what makes a person an expert is years of experience that enables who recognizes certain patterns in a problem as being similar to pattern. For example, in the future artificial intelligence system can be applied to control air traffic, design to computer configuration, medical diagnosis, instruction/training, speech/interpretation, monitoring to ( nuclear plant), planning to mission, factory scheduling, prediction weather, repairing telephone, automatic driving etc. different industries.

Artificial intelligence characteristics include: creative, adaptive , common sense, fact processing, quick replication, broad focus permanent and consistent skill. Otherwise, traditional computer expert system disadvantage includes perishable, unpredictable, slow reproduction, expensive, slow reproduction, slow processing lacks inspiration, needs instruction, narrow focus only machine knowledge. So, artificial intelligence is a branch of computer science devoted to creating computer to influence software and hardware to attempt to create human intelligence or human intelligent behavior. It is learning from experience, responds flexibility in situation that are, new or not anticipated.

Thus, (AI) can be learnt programmed knowledge to solve problems, using reasoning in solving problem, understanding and inferring facts and rules, recognizing the relative importance of different elements in a situation. In summary, artificial intelligence is concerned with two basic ideas mainly: The first idea, it involves studying the thought processes of humans to understand what intelligence is; the second idea, it deals with representing thought processes using companies to create artificially intelligent entities for testing the theories of intelligence.

● Can (AI) impact human job nature?

Human need concern this question: Will artificial intelligence (AI) reduce some human jobs in order to instead of replacing machines to do? Due to artificial intelligence is the ability of machines to do thing, that people would require intelligence. For example, artificial intelligence machine man driving( self-driver), it (AI) machine man driving research is an attempt to discover and describe aspects of human intelligence that can be simulated by driving machine functions. Alternatively, (AI) mathematical research may be another viewed as an attempt to develop a mathematical theory function to describe the abilities and actions of things ( natural or man-made) exhibiting intelligent behavior and server as a design of intelligent calculation machine function.

Why do humans need artificial intelligence machines to instead of traditional human service job? For example, can artificial intelligence machine man (self-driving) driver drive to replace human driver? I shall compare the differences between humans and computers : The characteristics of humans are good at recognizing various things, either seen before or not, recognizing the relationship patterns between things. Human thinking is common sense reasoning, combining all types of sensory input, acting appropriately in novel situations, learning new things and changing behavior patterns, making decisions , even when given incomplete information, working with noisy, incomplete information gathering behaviors . However, characteristics of computers are good at: The tasks humans do naturally are extremely difficult for a computer program as intelligent, which must be able to do the same kind of tack as humans do naturally.

Hence, (AI) is an combination of many different success and technologies: Linguistics - computational and socio, philosophy-logic, philosophy of mind and of language, electronical engineering -image and speech processing, pattern recognition, robotics, machine learning, neural networks, optimization scheduling, management information system and decision making. So, it is possible that (AI) can impact human job nature to instead of human working behavior in the future.


How can (AI) influence labor market?
● How can human society job nature
to be changed to artificial intelligent society?

From the first intelligent perspective reason view point, artificial intelligence is making machines " intelligent" acting as humans expect people to act. Artificial intelligence has ability to distinguish computer responses from human responses, it owns knowledge to solve expert problem. From another research perspective reason view point, artificial intelligence is the study of how to make computers do things which, at the moment, people do better ( Rich & Knight, 1991, p.3).

(AI) researchers are native in a variety of domains, e.g. formal tasks ( mathematics, games), tasks ( perception, robotics, natural language, common sense reasoning), expert tasks ( financial analysis, medical diagnostics, engineering, scientific analysis and other areas).

From the second business perspective reason view point, (AI) is a set of many powerful tools, and methodologies for using those tools to solve business problems. From a programming perspective reason view point, (AI) includes the study of symbolic programming problem solving and search .

From the third human technological perspective reason view point, today's computer can do many well-defined tasks, for example, arithmetic operations, are much faster and more accurate than human beings. However, the computers' interaction with their environment is not very sophisticated yet. How can human test whether a computer has reached the general intelligence level of a human being? Can a computer convince a human interrogator that it is a human? But before thinking of such advanced kinds of machines, human will start developing our own extremely simple " intelligent" machines.

So, it is possible that human society job nature will to be changed to artificial intelligent society when (AI) technology is developed to the mature stage in the future.

● Why does human need artificial intelligence machines?

One of major division in (AI) is between humans who think (AI) is the only serious way of finding out how we ( human) work and human who want companies to do very smart things, independently of how we ( human) work. This is the important distinction between cognitive scientists vs engineers. One of another major division in (AI) is between symbolic (AI), which represents information through symbols and their relationships. Specific Algorithms are used to process these symbols to solve problems or deduce new knowledge and connectionist. So ( AI) , which represents information in network. Biological processes underlying learning, task performance and problem solving are imitated from human mind behaviors.

Thus, it is possible that artificial intelligence machines can do the better judgicious behavior to compare human.

● How does artificial intelligence influence future working changing in automation employment and productivity aspects?

In the automation changing influence aspect, as companies increasingly use robots on production lines or algorithms to optimize their logistics manage inventory, any carry out other core business functions. Technological advances are creating a new automation age in which ever-smarter and more flexible machines will be deployed on an ever larger scale in the marketplace. However, researching artificial intelligence with how influences human working nature. We need to answer these questions: How will automation transform the workplace? What will the implications for employment? And what is likely to be its impact both on productivity in the global economy and on employment?

Advances in robotics, artificial intelligence, and machine learning are growing in a new age of automation as machines match or outperform human performance in a range of work activities, including ones requiring cognitive capabilities. What factors are determined the changing in workplace adoption by artificial intelligence innovation? What advantages are automation? Automation of activities can be enabled businesses to improve performance by reducing errors and improving quality and speed, and achieving outcomes that go beyond human capabilities.

Some scientists indicated based on their scenario modeling. They estimated automation could raise producing growth globally by 0.8 to 1.4 percent annually. Almost, the activities people are paid almost $16 trillion in wages to do in global economy have the potential to be automated by adopting currently demonstrated technology. According to their analysis of more than 2,000 work activities across 800 occupations. When less than 5% of all occupations have of least 30% of activities that could be automated. They also indicated that technical economic and social factors will determine automation. Continued technical progress, for example, in areas such as natural language processing is a key factor beyond technical feasibility , the cost of technology, competition with labor including skills, and supply and demand dynamics, performance benefits including and beyond labor cost savings and social and regulatory acceptance will affect ( alter) the scope of automation.

Other some scientists also indicate U.S. country for example, the anticipate shift in the activities in labor force of a similar order of magnitude as the long term sight away from agriculture and decreases in manufacturing. Share of employment in the United States both which were achieved. So, those factors can influence why artificial intelligence technology needs. So, it is possible that future agriculture and manufacturing both industries will apply (AI) technology manufacturer-kind of job nature to raise productivity instead of farmers, fruit picking workers, farming transportation labours as well as factory manufacturing workers and supervisors etc. human-kind of job nature.

● Is artificial intelligence possible to replace labor ?

Not just intelligence, but also debating, if machines are capable of having a conscious minds. Artificial intelligence has those characteristics as below:

On functionalism aspect, artificial intelligence inputs mental states, sensory inputs, ( beliefs, desires being in pain feeling) and behavioral outputs. Since mental states are identified by a functional role, which are thoughts to be manifested in various systems. Even, perhaps computers which are physical devices with electronic substrate that inform computations on inputs to give outputs similar to brains which are artificial intelligence composed of part any intrinsic relationship to each other. Thus, artificial intelligence activities is not the whole itself, but into parts or on external influence on the parts.

On dualism aspect, artificial intelligence is a set of views about the relationship between mind are matter. On materialism aspect, it builds the only thing that exists is matter, including consciousness.

On biological naturalism aspect, it is similar a human brain than feels pains makes mental situation. So, artificial intelligence is similar biologist which might to be excited to human labor work.

Hence, it seems artificial intelligence can change ( alter) or replace human labor work of nature in possible in the future.

● Can (AI) technology replace human labour nature of work?

On technological innovation reason view point, the history development of artificial intelligence studying the intelligence is one of most ancient scientific discipline. The history development of artificial intelligence what aims to achieve human use to sense, learn remember and think, logic probability, decision making and calculation develop from mathematics, instead of replacement human labor functions.

Artificial intelligence history development aim is the scientific analysis of skills in connection and practice with the appearance of computers from 1950 year beginning. The artificial intelligence (AI) can deal with the ultimate challenges. How can ( either biological or electronic) mind sense, understand and manipulate a world that is much simple and more complex than itself? And what if would human like to construct something with such capabilities? The general-purpose software of the early period of (AI) were only able to solve simple tasks effectively and failed when which should be used in a wider range or an more difficult tasks. One of the sources of difficulty was that early software had very few or mix knowledge about the problems which handled, and activities successes by simply syntactic manipulation. Moreover, the other difficulty was that many problems that were tried to solve by the (AI) were untreatable.

The early (AI) software whether trying step sequences based on the basic facts about the problem that should be solved, experimented with different combinations till which found a solution. From the end the 1960 year, developing the so-called expert systems were emphasized. These systems had ( sue-based) knowledge base about the field which handled. Till to the beginning of the 1970 year, ( Prolog) the logical programming language was born, which was built in the computation realization of a version of the resolution calculus. ( Prolog) is a remarkably prevalent tool in developing expert systems ( on medical, judiciary and other scopes), but natural language parsers were implemented in this language. Then, in 1981 s, the Japanese announced the fifth generation computer system project, a 10 years plan to build an intelligent computer system that use the ( Prolog) language as a machine code. Nowadays, (AI) can be applied any industries, such as car manufacturing industry can use (AI) technological machine-men manufacture car, instead of replacing human labors in factory. Even, in the future, using (AI) machine-men drivers can drive any private cars or public transportation tools, instead of replacing human drivers, e.g. bus, train, tram, ferry etc. Also in the future, machine-men can replace housewives to serve families to do housekeeping clean job , e.g. cleaning toilets, bathrooms, kitchens, even cooking functions at home. So (AI) machine-man can reduce housewives works at home. Moreover, (AI) machine man can take care old people , when who are living at homes or elder care centers.

So, it seems artificial intelligence (AI) will be possible developed to manufacture a new generation machine-man to assist ( serve) families to do any simply cleaning or cooking jobs at homes. Moreover, the overall demand of ( AI) general social needs will also rise, such as security, driving transportation tools, restaurant cleaning, elder centers care service etc. So, it seems that individual or families or social needs of (AI) will be increase in the future. Thus, it will influence macro economy growth (GDP) if there are large house family consumer group and hotel or bus or taxis or ferry etc. different business consumer group demand any artificial intelligence machine numbers increasing. Then, the artificial intelligence products and material manufacturers must need to buy many artificaial intelligence materials to produce any kinds of artificial intelligence machines to prepare to satisfy consumer individual needs. Consequently, macro economy will grow to the owned artificial intelligence development countries, e.g. US, China, UK.

● Why can artificial intelligence satisfy human needs?

First, On machine-man satisfactory demand aspect view point, it makes computers that think, it is the automation of activities. We associate with human thinking: like decision making, learning. It is the act of creating machine that perform function that require intelligence when performed by people. It is the study of mental faculties through the use of computational models. It is the study of computations that make it possible to perceive, reason and act. It is a branch of computer science that is concerned with the automation of intelligent behavior. It is anything in computing service that human don't yet know how to do property.

Second, on thought aspect artificial intelligence means systems thank think like humans, systems that think rationally.

Third, on behavioral aspect, artificial intelligence systems that act like human and that systems act rationally. However, the basic objective of (AI) is to represent human's thought processes in computation . These machines are supposed to exhibit behavior that. It is performed by a human being, would be considered intelligent. However, some authors feel (AI) has disadvantages, such as it is not creative, it is excited in the use of sensory devices, it can't make use of a very wide context of experiences and it does not use common sense.

For speech recognition and understanding function needs example, (AI) can be applied in speech recognition and understanding function, which (AI) speech or voice recognition is a data input method. For example, the computer recognizes and understands one ( or a few) word commands. Speech understanding on the other hand is the computer's ability to understanding a spoken language. That is , the computer understands the meaning of sentences, an paragraphs through (AI).

So, (AI) can be attempted to learn human language how to speak. It is similar to translate human language skill, instead of actual human speaking skill. Also, (AI) can assist handicap learning or language student how to listen different languages by machine-man sounds from computers more accurately.

So, it seems that it (AI) can replace human language teachers speaking function and can change teaching language nature of job in language speaking and listening education industry.

- Is artificial intelligence one good choice for human future technological benefit?

Nowadays, new technology development is popular. However, artificial intelligence is one kind of new technology choice among different technologies innovation. So it brings this question: Is artificial intelligence technology value to invest? To answer this question. I shall indicate some other new technology developments to compare (AI) technology development to judge which has urgent needs to achieve human expectation nowadays.

For example, why is green peace interested in new technologies? New technologies features prominently in our ongoing campaigns against genetic modified crops and number power. However, which are also an integral part of our solutions to environmental challenges, including renewable energy technologies, such as solar, wind and wave ( water) power energy as well as waste treatment technologies, such as mechanical, biological treatment.

It seems humans need concern how to apply (AI) technology to solve environment pollution challenges in our future. So, environment protective, agriculture, natural energy technology will be popular demand to attempt to apply (AI) technology to solve their challenges or apply (AI) to assist to develop their industry.

How Robots Influence Economy Development

- How can artificial intelligence technology influence economy?

Advances in artificial intelligence (AI) technology and related fields have opened up new markets and new opportunities progress in critical areas, such as health, education, energy, economic development, social welfare and the environment pollution.

(AI) automation will continue to create wealth and expand the global economy development in the future. However, when many will benefits that growth won't be costless and will be accompanied by changes in the skills, that workers need to increase productivity in the economy and structural changes in the economy. So, in the skills that workers need to succeed in the economy and structural changes.

I shall indicate why aggressive policy action will be needed to help Americans who are disadvantaged by these changes , due to (AI) technology is caused. For automation industry change example, artificial intelligence (AI) capabilities will enable automation of some tasks that have long required human labor. These artificial intelligence technology introduction can increase new opportunities for individuals. The economy and society, but (AI) has also the potential to disrupt be current livelihoods of many Americans. However, (AI) leads to unemployment and increase in inequality over the long run depends not only on the (AI) technology itself, but also on the institutions and policies that are changed.

Thus, it is possible that (AI) technology will raise some countries unemployment number if the employer apply (AI) technology workers to work instead of human labor in their factories, but it can also raise productivities for these employers.

- Can (AI) influence global economy growth?

Technological progress is main driver of growth of GDP per capita, allowing output to increase faster than labor and capital . However, technology can increase productivity, but also decrease the number of labor hours needed to create a unit of output. So (AI) causes unequal to labor wage decreases, even reduces the number of labor to manufacture, e.g. artificial intelligence technology of automation car manufacturing industry; clothing manufacturing industry; plane manufacturing etc. high technology of artificial intelligence manufacturing method. But (AI) should be potential environment benefit, although it raises unemployment ratio. Moreover, it can rise production , due to many skilled craft were replaced by the combination of machines and lower-skilled labor. The result of (AI) technology introduction , it causes output per hour risen when inequality declined, driving up average living standards, but the labor of some high-skill workers was no longer as valuable in the market. Otherwise, if (AI) technology is continue developed to be success. Some routine intensive occupations will be loss, which focused on predictable, e.g. easily programmable tasks, such as switchboard operators, filing clerks, travel agents, and assembly line workers would be particularly replaced by new (AI) technology. However, at the same time, (AI) technology development will bring these benefits: improvement in education ( training (AI) technology scientists) , due to (AI) manufacturing technology needs are raising to businesses and institutional changes, such as the reduction in unionization and raising in the minimum wage to the (AI) manufacturing technology skilled labor in factories.

Because (AI) technology is not a single technology, but rather a collection of technologies that are applied to specific tasks, the effects of (AI) will be felt unevenly though the economy. It will bring some tasks will be most easily automated than others , and some jobs will be affected more than others, both negatively and positively. Finally, new jobs are likely to be directly created in areas , such as the development and supervision of (AI) as well as indirectly created in a range areas though out the economy as higher incomes lead to expanded demand.

However, if (AI) technology could dominate global labor markets. If labor productivity increases, do not influence into wage increases, then the large economic gains brought about by (AI) technology could be increased wealth inequality, due to employers can reduce production cost, but workers ( labors) wages will not be increased, even will be decreased. Hence, it seems the (AI) technology will bring disadvantages to labor market to cause unemployment or reduce wages in possible, although it can reduce employer individual salary (wage) expenditure and it can raise productivity.

- How can artificial intelligence impact global economy growth?

Artificial intelligence (AI) technology is a branch of computer science that aims to create intelligent machines that work and react like humans. So, (AI) is a technology that appears to impact ( influence) human preference by learning, understanding complex contents, enhancing humans in executing both routine and non-routine tasks. In the future, (AI) technology that can be virtual personal assistant, as well as it may exist, such as robots with human-like processing capabilities.

How can (AI) technology impact global economy growth over the next 10 years? During this time period, (AI) technology is predicted to have wide-ranging applications including: Machine learning that automates analytical model building by using algorithms that allow machines to operate without human assistance.

In global education aspect, potential applications include predicting cause-and-effect relationships from biological data, identifying new drugs, self-driving cars, and protecting against fraud, improved natural language processing that allows computers to continue to better analysis, understand and generate language to interface with humans using natural human languages. For example, transcribing notes dictated by physicians, automatically drafting articles and translating text and speech. So (AI) technology can be applied to education aspect to improve humans' knowledge level.

In visual art aspect, (AI) machine vision that allows computers to identify objects, scenes and activities in images. Current applications of (AI) machine vision include providing objective descriptions for the blind seeing( visual) needs.

We except the economic effects of (AI) technology to include both direct GDP growth from sectors that develop or manufacture. (AI) technology and indirect GDP growth through increased productivity in existing sectors that employ some form of (AI). If (AI) technology is an increasingly critical component of more products, it will become an integral part of many people's lives. Thus, (AI)'s ability to influence economic activity, rather than the economic or development status of the region. (AI) has the potential to impact income classes and to bring significant gains to both developed and developing countries. For example, (AI) has the potential to optimize good production around the world by analyzing agricultural regions and identifying what is necessary to improve crop yields.

In estimating the future economic effects by (AI) technology innovation, it is important to note that it is challenging to accurately predict which applications of (AI) will ultimately be commercially successful. In micro level economic influence, we need to apply methodologies to estimate the economic effects of investment in firms developing (AI) technology since investment levels in a technology are a telling sign of the future potential of that (AI) technology.

● How can (AI) influence GDP of high income countries in the next ten years?

How (AI)'s development may affect the global economy over the next ten years. In fact, (AI) technology has the potential to affect business across the global in a wide range of industries in ways only a number of technologies have done in the parts. For example, (AI) technology's expected to be a useful tool for enhancing human capabilities and in some instances replacing functions, such as driving a car, adoption of broadband internet, mobile telephone, industrial robotic automation have served to enhance human capabilities.

However, significant public debate has focused on projections of (AI) technology's effect on the labor force. However, large companies prefer to invest in (AI) technological industry. For example, face book's (AI) research lab., google machine intelligence lab. and micro soft machine learning and artificial intelligence research division are all making advances in (AI) technology and investing in the industry's top talent. Additionally, between 2010 year and 2015 year, nearly $5 billion in venture capital funding invested in firms across the global developing and employing (AI) technology ( Facebook (AI) Research).

● How can artificial intelligence impact on workplace?

Modern information technologies and the labor economy growth of machines is powered by artificial intelligence have already strongly influenced the world of work in the 21 ST century. Computers, algorithms and software simplify every tasks and it is impossible to image how most of our life could be managed without them. How can be the information economy characterized by exponential growth replaces the most production industry based on economy of scales? What will the future world of work look like and how long will it take to get? Will the future world of work be a world where humans spend less time earning their livelihood? Alternatively, are mass unemployment, mass poverty and social distortions also possible scenario for the future, where robots, artificial intelligence systems play an increasingly central role? These questions concern how artificial intelligence further development . Can influence labor economy growth on workplace ? When the labor market has widespread impact on intelligence property, information technology, product liability, competition and labor and employment laws.

How (AI) technology impacts on labor workplace.

The future influence any organizations how labor economies use of (AI) can be analyzed, such as deep machine learning is based on a set of model high level data. Unlike human workers, the machines are connected the whole time in workplace. If one machine makes a mistake, all autonomous systems will keep this in mind and will avoid the same mistake the next time.

Over the long run intelligent machines will win against every human expert. Production robots have been replacing employees because of the (AI) technology. They work more precisely than humans and cost loss. Creative solutions like 3D printers and the self learning ability of these production robots will replace human workers, the automatic data recording and data processing, traditional back office activities are no longer in demand. Autonomous software will collect necessary information and will send it to the employee who needs it. Additionally, dematerialization leads to the phenomenon that traditional physical products are becoming software. For example, CD or DVDs are being replaced by streaming services. The replacement of traditional event ticket, e-travel ticket service products or hard cash will be the next step, due to the possibility of payment by smartphone. So, (AI) technology will impact human's daily life consumption behaviors in the future. For another example, transportation tools, such as boats and ferries and private vehicles will use sensors and navigating without human input. Taxi and truck drivers will become obsolete, the stock store applies to stock managers and postal carriers of the delivery is distributed by (AI) machine delivery method.

What is the relationship between (AI) and (CRM)?

● Can (AI) technology impact on customer relationship management (CRM) ?

Nowadays , (AI) is a technology almost as old as the computer industry itself, it is similar with the advent of personal assistants function to businesses and personal promotion channel, such as ( Amazon's Alexa, Apple's Siri, Google's Assistant) image recognition ( face book), personalized recommendations ( Netflix , Amazon). Those innovations have been driven by a increase in processing power, lower cost hardware, and the exploding creation and availability of data. It seems, (AI) technology can impact global customer service management method.

How to forecast economic impact modeling to (AI) will affect global economy? Can human forecast business revenue growth and job creation ( or destruction) based on (AI) applied to customer relationship management (CRM) activities? In addition to the economic impact on (AI) or (CRM) which can include an estimate of the economic impact attributable to sales forces customer base. What can economic benefits be brought to (CRM) from (AI) technology?

Artificial intelligence(AI) comprises a set of technologies that use natural language processing, machine learning, knowledge graphs, and other tools to answer questions, discover insights and provide recommendations. Computer systems can use (AI) hypothesize and formulate possible answers based on available evidence can be trained through the ingestion of vast amounts of content, and automatically adapt and learn from (AI) self mistakes and failures.

So, any business organizations (customer service departments) can provide efficient and effective customer relationship management of excellent customer service quality if which applied (AI) technology system. The different type of (AI) systems include: (AI) system platforms, machine learning (AI) based data preparation and enrichment tools, machine vision/image recognition, voice speech recognition, text analysis and natural language processing, bots , e.g. face book website and virtual digital assistance solutions, social media pattern analysis , sentiment analysis, advanced numerical analysis (e.g. IOT streaming , machine logs), supporting technologies, knowledge base dialog management, Q&A processing etc. different (AI) technology system customer relationship management (CRM) tools.

(AI) (CRM) of activity can include these categories, such as: corporate marketing, marketing operation, field marketing, customer support, digital commerce, customer analytics, customer influenced product or service design, product or service pricing, finance information, presentation, customer billing, inventory , logistics and fulfilment support, partner management etc. different CRM tools.

(AI) technology of CRM has been carrying on plan different stages to achieve CRM personal assistant tool for businesses. The stages are such as, in the beginning stage of (AI) projects in place, implement now, pilot phase next year in the final stage of (AI) customer relationship management tools are foreseeable future. So, this CRM technology has been improved to plan in different stages every year to prepare to achieve full capacity of CRM service quality for businesses to use in the future.

Hence, how to develop an estimate prediction of the economic impact (AI) technologies could have CRM activities, which depends on gathering macroeconomic information on business revenue and the basic marketing of business revenue and the basic markup of business expenses by major functions ( customer support, marketing and sales , production etc.)

An economic impact model that can gather data together and forecast the results how (AI) artificial intelligence technology brings (CRM) customer relationship management benefits to businesses, e.g. surveys investigation includes IT spending by sample countries, GDP and population estimates and forecasts, revenue per employee and ratios of IT spend to GDP. Surveys ( questionnaire questions) of forecast results are influenced by (AI) impact can include: results are projected from surveys and rely on estimates are made by respondents on the expected financial improvements in categories of (AI) –assisted customer relationship management activities. The forecast assumes that these estimates are correct; financial estimates are based on estimates of "first year" improvement from full (AI) implementation; forecasts are from planning to implement any artificial intelligence of customer relationship management (CRM) projects, the improvement forecast is of categories of activity , e.g. corporate marketing , digital commerce, and customer analytics. They are not estimates of ROI for the (AI) software. They rely on conservative estimates to which each of these entities might affect company revenue, expenses or productivity. They also rely on estimates of the penetration of software in customer relationship management activities . Net new jobs created are based on the ratio of new revenue to jobs required to support that revenue . They can assume that 50% of the net new revenue will support increases in labor and the rest will go for capital and other operating expenses that may replace jobs lost to automation.

In the future, some of the ways in micro economic benefits to any organizations. (AI) technology is expected to impact CRM activities include: Spending up sales cycles, improving lead generation and qualification solving customer support problems faster ( raising service quality), helping companies improve brand campaigns and recognition, lowering costs of support calls when increasing resolution rates, lowering the cost of recruiting employees and partners, increasing revenue from optimized product marketing, optimizing price, distribution logistics and preventing loss through fraud detection. So, micro economic benefits view point, it seems that (AI) CRM technology can raise any companies economic benefits for care term.

Artificial intelligence enables machines or the in-build software to behave like human beings which allows these decisions and act. The advent of (AI) is leading , talking, making decisions and act. The advent of (AI) is leading to new technologies advances and transforming the economic and employment opportunities for humans in a positive way. (AI) related technologies can facilitate our live. For example, industrial robotics, robotic medical assistants, smart games, financial forecasting software, big data analysis, algorithms in health and bioinformatics, pilotless cargo places, drone ambulances and general purpose and workplace robots and others. (Disruptors technologies: Advances that will transform life, business and the global economy).

Artificial intelligence also known as computational intelligence is defined as " the human –like intelligence exhibited by machines or software. It is theorized that intelligence of humans can be described and intelligence machines or software can simulate it. These machines software can be reasonable , learn, perceive and process information, like human mind and thus facilitate human life. They can think and act for us. So, artificial intelligence is an interdisciplinary field of study including computer science, neuroscience, psychology, linguistics and philosophy.

However, (AI) research and developments have economically impacted many industries, such as robotics, telecommunications, computer applications , health, finance, heavy manufacturing, transportation, aviation, e-service and e-commerce, military , music and movie, toys and games entertainment etc. industries.

In fact, many ideas, systems and technologies have been developing in the world of (AI) technology. However, which are net called or considered (AI) products, rather which are mentioned with their specific names, such as smart graphics, machine learning, e-commerce etc. ( i.e. this is called (AI) effect).


What is relationship between
(AI) and digital economy
● How can (AI) technology influence digital economy?

Nowadays, (AI) related industrial applications will replace most human power in fields, including call centers, customer services and air cargo transportation. (AI) technologies also help weather forecasting based on repeated rainfall pattern ( data) recognition, through robotics ( i.e. floor cleaning, moving lawns etc.) transporting people and products with unmanned vehicles, sending space unmanned smart shuttles, developing robotic arms, predicting market values in stock exchanges by internet, making homes safer, helping elderly and disabled using robotic servants etc.

Among the (AI) related technologies , there are a few that significance for the impact on society and especially on digital economy . (AI) is particularly influential in machine learning. Such as robotics, transportation, finance, health and bioinformatics, e-commerce , e-games, big online data gathering and internet-of-things. For example, machine e-learning is based in bioinformatics and robots that can learn new skills for better caregiving in healthcare. What is machine e-learning? Machines can e-learn from e-data gathering, coming up generalizations and making decisions to act in certain ways from internet.

There are important applications , such as e-machine perception, electronic online natural language learning processing, online search engines, online bioinformatics, online brain –computer interface, online game playing, online robot locomotion, online advertising, online computations finances, online health monitoring, online DNA classification and decision making, online in chemistry –cheminformatics . So, online machine learning can positively impact productivity and it can enhance information and analytical system from (AI) online channel.

What is robotics? Robotics is one of the most strongly influenced fields in (AI). For example, heavy manufacturing industries, robots and used and man power is replaced for effectiveness, precision, and accuracy, especially in respective or dangerous tasks, including welding, assembling , picking and placing .

So, robots can acquire new skills or adapt the changing dynamic environment. Also, artificial intelligence can be applied in developing transportation. For example, automated vehicles, driver assistance systems , safety systems, collision avoidance systems and public transportation. Moreover, (AI) technology has proven to produce some of the best tools to predict stock market fluctuations from internet data gathering method. It's predictions are based on ever-evolving predictions algorithms and systems learn new models and make connections between historical data and new data to measure stock market trading more accurate from internet data gathering channel.

In health field, especially in health data processing , analysis, decision making support and medical diagnosis. So, online data can show which patients will need what treatment and what alternative drugs could be used more accurate from (AI) online data gathering method. Bioinformatics is an interdisciplinary field combining statistics, (AI) online technology can help in discovering data patterns and modeling through the application of machine learning, artificial neural networks and genetic algorithms. For example, further (AI) technology development of human genome project of online data sequences.

Online shopping can be facilitated by virtual assistants developed through (AI) technology and these assistants can offer the best advice. (AI) online purchase coming after every product image recommendations and personalization bring important revenue to shopping online sites, like Amazon . Smart computer graphics and games, artificial intelligence is useful in smarter computer, graphics, scene modeling , scene rendering processes in order to create, for example, effective human –robot interactions , online machine learning, online strategic games techniques etc. online computer related (AI) software.

So, online big data analysis and big data does have a critical need in the world of online intelligence machines and software in our future. In other words, (AI) offers online technology to enable online big data analysis to provide industrial organizations with valuable information for effective decision making in short time. For example, what IBM's Watson achieved: this machine used 200 million of structured and unstructured content with a special technology of hypothesis generation, massive evidence gathering, analysis and scoring from internet channel.

Finally, (AI) online technology another related internet invention ( internet of things) (IOT) is the network of machines or objects connected through internet. These connected objects can sense their internal and external environment, communicate with each other, can send critical data and finally can make decisions to act or correct their environment from (AI) online technology. For example, factories can monitor and automatically change production processes, hospitals can monitor and regulate the health conditions of their patients , schools can collect data from facilities and cars can send data to car makers from (AI) online technology.

Partner predicts that (IOT) market will create about trillion amount value by 2020 year. Although machines collect big data from their environment, whether which gain an insight or learn from these online data largely depends on the (AI) online machine learning principals and (AI) online technology. In 2013, Mckinsey estimated that disruptive technologies closely related with potential economic impact in 2025 year between $7.1 to $13.1 trillion amount ( automation of knowledge work, advanced robotics, autonomous or near-autonomous vehicles).

What is the relationship between
(AI) and global digital economy development ?


● Could work activities in China be automated
making in the nation with the world's largest automation potential?

Can (AI) technology influence China economy? Could China workers be affected and jobs made up of routine work activities and predictable? Will programmable tasks be particularly impact to China employment market ? When impact on labor market is likely to be gradual at the aggregate level, it can be sudden and dramatic at the level of specific work activities, rending some job obsolete fairly. Overall (AI) technology will raise digital skills when reducing demand for medium incomer inequality for China workers. It seems (AI) technology's effect on productivity could be crucial to China's future economic growth as the population ages are increasing.

In China, some biggest technological companies driving significant investments in research and development. Moreover, China is one of the leading global (AI) technology development county. However, China will need to focus on building its innovation capacity. For example, United States and United Kingdom are currently producing more influential (AI) technological research. However, if China planed to achieve (AI) technology success, it's traditional industries will need to develop technical know-how –to and overcoming implementation costs prepare to develop (AI) . When (AI) technology is introduced into China society, China government needs to raise concerning ethical, legal, technological security etc. business questions. Also, surrounding issues include privacy, discrimination, legal liability and regulation. It aims to encourage overseas investors to choose to invest (AI) technological industry to raise GDP growth and manufacturing industries income growth for long term in China.

If China encouraged overseas (AI) technology investment in its country. It is possible to influence China employment market to be changed. Because (AI) technology will impact to influence China people daily life. Due to (AI) technology is introduced to China society, many rich people will prefer to spend to buy any high (AI) technological products for entertainment or learning or machine man driving etc. daily necessity activities. Then it will raise GDP growth and will raise (AI) manufacturers or related-(AI) technological manufacturers profit. It is beneficial to China because it can become one high knowledgeable and (AI) technological economical society.

But it will bring bad influences to raise unemployment chance for the low skillful labor. In labor economy aspect influence , how (AI) technology can influence China low skillful labor unemployment ratio raising. The raising low skill labor unemployment reason is because China low skillful human labors are argued or are replaced by (AI) technology creating new challenges to introduce to influence China society of simply human manufacturing job nature to be changed to be high (AI) technology manufacturing job nature in any China factories. Moreover, when (AI) technology introduction to China, it will cause other related social challenges in China. The varied (AI) related challenges, including the difficulty of creating safe and reliable hardware for sensing and affecting ( transportation and education), the challenges of gaining public trust, a low resource comities and public safety and security, the challenges of overcoming fears or marginalizing humans in China employment and workplace and the risk of diminishing interpersonal trust because the low skillful labors won't believe any China employers will give chance to employ them , due to (AI) technology will replace their skills and man manufacturing of productivity is much less to compare to (AI) technology manufacturing method.

- How does ( AI) technology influence
the future of employment change?

Are future nature of jobs changed to computerization from (AI) technology? Where are the probability of computing occupations from (AI) technology influence? What is expected impacts of future computing on labor market from (AI) technology influence? John Maynard Keynes's frequently cited prediction of widespread technological unemployment " du to our discovery of means of economic the use of labor outrunning the pace of which we can find new used of labor" ( Keynes, 1933, p.3).

In the future, (AI) technology will impact some nature of occupations to change computing. This chance will also influence some countries' economic change. For example, some factory human labors hand routine manufacturing tasks will be changed to computerization of routine manufacturing tasks by (AI) technological machine men hand manufacturing method. it will cause a structured shift in the labor market, with workers reallocating their labor supply from middle-income manufacturing to low-income service occupations.

Arguably, this is because the manual tasks of service occupations are less computerization, as who require a higher degree of flexibility and physical adaptability. So, (AI) technology will influence the human hand labor skillful occupation nature of task cheaper , such as vehicle manufacturing , ship manufacturing, computer manufacturing, steel manufacturing, television, radio etc. home electronic products of heavy machine industry change. Due to (AI) technology machine man will be proper to be used to manufacturing these electronic products when the (AI) technology innovation can develop to the mature stage. Then, any countries manufacturers will choose to use (AI) technology machine man, instead of human hand production.

Supposing the future prices of computing are fallen, seriously, problem solving skills are becoming relatively productive, explaining the substantial employment growth in manufacturing occupations, involving cognitive tasks where skilled labor has a comparative advantage, as well as the increase education needs for (AI) technology computing of machine man subject study.

Prediction of education needs for (AI) technology student numbers will increase, due to manufacturing industry needs many (AI) technology students in future employment market. Another (AI) technology influence if the future (AI) technological innovation, e.g. machine man manufacturing or machine man service industries will both increase demand, then with more sophistic software technologies will be disrupted labor markets by marketing workers redundant.

For publishing industry, what is striking about the case in paper book publishing industry will be unpopular? Due to the electronic book publishing industry will be popular, e.g. Amazon publish . (AI) technology can influence paper book manufacturing method which is replaced by machine man electronic book manufacturing method as well as it will cause the computerization is no longer confined to routine manufacturing tasks. Due to (AI) machine man manufacturing technology will be proper to be used to manufacture any products in short time efficiently and effectively , e.g. electronic book products. In the future, if it is fact to occur this case, such as ( AI) technological machine man manufacturing method will be adopted ( applied) to manufacture electronic books or any products in possible. (AI) technology will cause many manufacturing workers are unemployed. It is beneficial to employers, who can reduce to spend much wages expenditure to employ manufacturing workers, but it will cause many manufacturing workers loss jobs and reduce income to support whose families lives. It will cause social challenges, e.g. increasing stealing crimes if the manufacturing workers had not other skills to find other jobs to do easily. So, manufacturers need to concern over technological unemployment which will be hardly future phenomenon if who decided to dismiss all manufacturing workers, due to (AI) technology machine men replace to them.

If ( AI) technology can be innovated to produce any kinds of machine man to serve any service or manufacturing industries successfully. Then, it will bring these questions: Can future that workers be influenced to be automation employment and productivity by (AI) technology influence? Does it impact to influence the (AI) technology countries' productivity and growth and natural resources development and labor markets and evolution of global financial markets and economic impact of technology and innovation and urbanization etc. issues? How will automation transform the workplace? What will be the implication for employment? What is likely to be its impact both on productivity in the global economy and on employment?

In fact, automatic of activities can enable businesses to improve performance by reducing errors chance and improving quality and speed, and same cases achieving outcomes that go beyond human capabilities. Some economists indicate (AI) technology would give a needed boost to economic growth and prosperity have of the working age population in many countries. Based on the scenario modeling, they estimate automation could raise productivity growth globally by 0.8 to 1.4 % annually. They also indicated that almost half the activities people are almost $1.6 trillion in wages to do in the global economy have the potential to be automated adapting current demonstrates technology, according to their analysis of more than 2,000 work activities across 800 occupations. When less than 5% of all occupations can be automated entirely using demonstrated technology, about 60% of all occupations have at least 30% of worker made activities, that would be automated. More occupation will change to be automated. They also indicated for business performance benefits of automation are relatively clear, but the issues are more complicated by policy making to attract foreign investors. Beyond technical feasibility, the cost of technology, competition labor will include skills and supply and demand dynamics, performance benefits and beyond labor cost savings and social and regulatory acceptance will affect the automation. Their predictions suggest that half of today work activities could be automated by 2055 year, but this could happen 10 to 20 years earlier or latter depending on the various factors in addition to their wider economic condition.

Some scientists suggest (AI) technology is finally starting to deliver real-life business benefits. Computer power is growing significantly , algorithms are becoming more sophisticated and perhaps most important of all, the world is generating vast quantities of the fuel that powers (AI) technology data billions of gigabytes of it every day. Also, online firms are digital natives, such as Google online search service company is investing on (AI) technology. For new though most of the news if coming from the suppliers of (AI) technologies. And many new users are only in the experimental phase. Few products are on the market or are likely to arrive these soon to drive immediate and widespread adoption. As a result, analysts believe (AI) technology's potential will give true economic benefit in the future. (AI) industry will introduce to suppliers and users to raise economic potential of (AI) technology.

In the future, (AI) technology systems can solve business problems. Some scientists categorized those into five technology systems that are key areas of (AI) technology development: robotics and autonomous vehicles, computer vision language virtual agents and machine learning , which is based on algorithms that learn from data without replying on rules-based programming in order to draw conclusions or direct an action.

Such as computer vision and language includes natural language processing, analytics, speech recognition technology, some are about learning from information, such as about machine learning and others are related to acting on information, such as robotics, autonomous vehicles and virtual agents, which are computer programs that can converse with humans. Machine learning and a subfield called deep learning are artificial intelligence applications.

- Can artificial intelligence impact

global economy growth?

Artificial intelligence (AI) is a term first defined in 1956 year. It is a branch of computer science that aims to create intelligent machines that work and react like humans. In contrast today, 60 years later, (AI) is characterized by a number of applications, including computers playing games against humans and understanding human languages, virtual personal assistants, and robotics which involve computers seeing , hearing and reacting to sensory stimuli.

In the future, technologists predict for (AI) technology ranging from (AI) being used as a tool to aid relatively simple processes for robots with human like mental capabilities, who expect (AI) technology can emulate human performance by learning, coming to mind its own conclusions, understanding complex content, engaging in dialog with people, enhancing human cognitive performance or replacing humans in executing both routine and non-routine tasks. In existing industry, (AI) technology is used , such as targeted advertising and virtual used personal assistant as well as the (AI) technology that my exist in the future, such as robots with human vehicle processing capabilities.

The range of (AI) technology's progress in the future will determine the economic impact future of (AI) technology on the global economy with more limited advances and applications ( i.e. weak (AI) only) corresponding to more limited economic impacts and more substantial progress, i.e. strong (AI) technology is corresponding to more significant economic impact.

(AI) technology learning that automates analytical model, including predicting cause-and-effect relationship from biological data, identifying new drugs, self-driving cars and protecting against fraud etc. functions. Also (AI) learning can improve natural language processing that allows computers to continue to better analyze, understand and generate language to interface with human using the natural human language, virtual personal assistant, helps users by providing scheduling appointment, reminds organizing personal finance and finding providers of various services, machine vision allows (AI) machine man to identify object, scenes and activities in detect pedestrians and bicyclists.

We expect the economic effects of (AI) technology to include both direct GDP growth from sectors that develop or manufacture (AI) technology and indirect GDP growth through increased productivity in existing sectors that employ some from of (AI) technology. If (AI) producing sectors could grow, then it could lead to increase revenues and employment of (AI) technological professionals within these existing firms as well as the potential creation of entirely new economic activities to any countries' societies productivity improvement in existing sectors could be realized through faster and move efficient processes and decision making as well as increased (AI) technological knowledge and access to information available in societies easily.

In the future, if (AI) technology is an increasingly critical component of more products, it will become an integral part of necessary products of many people's lives. The extent of (AI)'s economy effort is also likely to vary from region to region, thought variation may be more dependent on the predominate economic activity of a region and the (AI) ability can influence economic activity, rather then the economic or developmental status of the regions. (AI) technology can move accessibility and can use source development to do international business between one country and another country.

So (AI) technology has the potential to give benefits to different income chooses and to bring significant gains to both developed and developing countries. For agricultural technology, (AI) has the potential to optimize food production around the world by analyzing agricultural regions and identifying what is necessary to improve crop yield. In total, (AI) technology gives greater economic impact to any countries agricultural regions if which implemented (AI) technology to grow crop , fruit etc. food production in the farms.

Investment in (AI) technology is such as capital investment to any countries' public or private enterprises. So, it will have large economic impact to the future . If the (AI) technology is reasonable invested to the different needs aspect by the public or private enterprises in the country. Then, it will have good economic impact to the country in the future. However, when (AI) technology is likely to affect both the productivity and employment components of economic growth in many sectors. Significant public debate has focused on projections of (AI)'s effect on the labor force. However, for instance, some researchers have argued that the rise of (AI) technology and automation will led to significant unemployment as capital is substituted for the low skillful labor. So, they point to the concern that the increasing sophistication of (AI) technology may balance skilled and semi-skilled workers and the reduce the size of the middle class. However, this is not a new argument, due to (AI) technology negatively affecting the labor force and leading to mass unemployment. Because the (AI) technology is the substitution of machinery for human labor. Although, employment in certain industries, has been reduced in the past due to technological advancement. For long term, the labor market has adapted to the introduction of new technology, giving rise to new jobs in new areas. (AI) technology may also be accomplished without a reduction to total employment in the long-term to some Asia countries, such as Hong Kong and Japan. Because Hong Kong and Japan many low skilled labor, e.g. security, cleaner who complaint that employers need them to work long time hours. ( abnormal working hours) e.g. one day 12 to 15 working hour per day. Hence, if (AI) machine means invention technology success. Security or cleaning job can be worked from (AI) machine man in some hours every day in order to reduce the long time working hours cleaners or security workers, e.g. one (AI) machine man works 4 hours for cleaning or security job, one day as well as another cleaner or security labor only needs to work 8 hours one day. So total security or cleaning employers can employ 12 hours machine cleaners or security workers and human cleaners or security workers in one day. For long term benefit, Hong Kong or Japan every security or cleaning worker does not need to work 12 hours minimum working hours one day. They won't feel tried and bore and without private with whose families, so who will accept to do these cleaning or security jobs, even they can raise work efficient and performance when who feel happy and health.

So, (AI) technology of machine man invention can raise low skillful labor efficiency and it can help them to avoid abnormal working hours demand in some busy work life countries, such as Hong Kong and Japan. Before, one Japan female labor feel unhappy to work, due to who often needs to work abnormal working hours for her employer and who has less sleeping and without any private time to enjoy her life with her families every day. So this abnormal working hours factor causes her to do commit suicide behavior, then she is die unlucky. So (AI) technology of machine man invention ought avoid abnormal working hours demand for employer in any countries in the future.

The most important occurrence to any employers, some researchers had attempted to do one experiment to find that private research and development , venture capital and public research and development investment all have strong net effect or economic growth with venture capital funding further having the strongest such effect from (AI) technology. The researchers hypothesize the venture capital investment contributes to economic growth through (AI) technology innovation and by the capacity of an economy to use existing (AI) technology knowledge to increase productivity. They predict the impacts of venture capital, business-research and development and public research and development can raise multi factor productivity from (AI) technology introduction.

Can (AI) technology influence the economic development to developing countries? The developing regions of the world contain most of natural resources. If one day, (AI) technology has invent one kind of machine man which can assist any gas or oil workers to seek any new oil/gas natural resource locations easily. I believe that (AI) technology can help these natural resource exploitation countries will gain economic benefit more easily. So, (AI) driven technology can be used to change to create any new opportunities to address poor management or resources and improve human well being, such as Africa Latin America and India can use (AI) technology machine man to seek any oil/gas natural resource countries exploitation activities to attempt to gain much economic benefits.

● Why will (AI) technology grow economic
development ?

Nowadays, increases in capital and labor are no longer driving the levels of economic growth, such as (AI) technology. The ability of increase in capital investment and in labor of traditional drivers of production, have no longer to be enjoyed in most developed economies ,e.g. developed country, US, UK . However, artificial intelligence has the potential to overcome the physical limitation of capital and labor to avoid missing out on this opportunity. So, policy makers and business leaders must prepare for and work toward a future with artificial intelligence. They must do with the idea that (AI) is another simply method to enhance productivity method . Rather they must see (AI) as the tool that can transform thinking about how growth is created.

Economists have always thought of new technologies are as driving growth their ability to enhancing. It can replace labor and capital factor of production. So, it brings this question: What is the factor of production (AI) technology characteristics. They key factor is to see (AI) technology as a capital-labor .

(AI) can replicate labor activities at much greater scale and speed, and to even perform some tasks began the capabilities of human. For example, by using virtual assistants , 1000 legal documents can be reviewed in a matter of days instead of taking three people six moths to complete. Some (AI) technology may be one kind of factor of production in the future. For another example, people will work in workplace digitalization environment. So, in the future, working environment and information management are automated. Such as Konica camera sale company will use workplace digitalization. So , (AI) technology can provide workplace digitalization in order to raise productivity efficiency. (AI) technology will be one kind of production which is replaced by workplace digitalization and it will grow any organization productivity efficiently. Then, (AI) technology will assist overall social economy growth , due to productivity is raised and products can be produced in short time to prepare to sell in consumption market. So, time will be shortened to increase GDP growth fast for the development of (AI) technology countries.

● How can (AI) technology impact to global
economic and social and psychological
changes?

What will be the development of (AI) technology and predictions concerning the future evolution? The computers and robots will develop conscious, intelligent and minds into humans, enhancing psychological and behavioral abilities and allowing for direct communication with (AI) minds. (AI) technology will be impacted human life by (AI) technology information communicative and environmental influence. A " world brain" and " world mind", this psychological system will be enhanced and enriched the capacities of both individual and collective cognition by (AI) technology of service industries.

(AI) technology with influence these human needs of service industries changes, such as , biological science, finance, entertainment, business, biological science, transportation, communication military etc. The personal computer evolution, the internet and the world wide web which exploded on the scene, linking business, homes, schools, social organizations which were a completely unpredicted phenomenon to influence human life. Kurzweil (1999) predicts that by 2029 year, most human communication will be with machines. According to Person, by 2100 year, there will be human machine convergence.

How can (AI) technology influence environmental protection to make benefits to farming economic growth? (AI) technology can be applied to predict how to solve environmental pollution challenge to avoid to damage any crop or vegetable or rice or fruit etc. food growth. Because environmental experts can gather global environmental pollution data from an environmental database to build a perform a systematic analysis from (AI) technology. The first step is this broad analysis can include understanding, statistical and data gathering techniques to obtain the relevant data, the correlation among the variables involved, and a list of possible models. The next step is to select a set of methods and models that cover all kinds of knowledge and functionalities needed for the decision making process. Once the models are selected, they must be fully implemented by means of machine learning , data mining, statistical or numerical technique. After that, those models must be integrated to build the whole EDSS. The EDSS must be tested to check its performance, accuracy, usefulness and reliability, both from the user's and (AI) technology/computer scientist's point of view. If these is any wrong feature in any development stage, such as model's integration, models' implementation, selection of models, database, problem analysis etc. the developers must come back in the update th required components. When the evaluation phase is all right, the EDSS is ready to be applied to the environment. The great contribution of artificial intelligence to EDSS the integration of several methods complementing the classical statistical models/simulation , statistical analysis, linear models, etc. and numerical models ( control algorithms, optimization techniques etc.) .

This cooperation makes the resulting systems more reliable and powerful in coping with real world environment systems. Date interpretation has been a principal area of research in (AI) technology since the very beginning. The most demanding problem in the environmental assessment context. Knowledge representation permits the definition of the different types of data that the existing methods adapt to the process. There is also a lot of work to clean, repair and transform the huge available quantities of raw data. Apart from this, the availability of meta-information or background knowledge is required to guide the process. Data mining is multi-disciplinary: It covers expert systems, data based technology, statistics, data visualization and unsupervised machine learning. These techniques operate at the level of data and background information, where numerous and often incompatible new commensurate pieces of information from disparate sources have to be brought together ( K, Fedra, 1994).

So, it seems that in the future, (AI) technology with the increasing maturity in particular those related to knowledge and engineering, new dimensions can be assisted to users in environmental decision making are available. For example, many environmental systems are characterized both by incomplete models and by limited data. Hence, in the future, (AI) technology will be applied to predict climate change to reduce crop or fruit etc. food agriculture challenge by climate change bad influence.


● Will (AI) technology influence digital economy change to manufacturing industry ?

To understand how the manufacturing business must adapt to prosper in the technology, we need to understand how (AI) technology will change us to shape our daily habits to satisfy our expectation of products to how we shop and even the immediate of the entire process. For example, taxi services are in the crosshairs as on demand transportation services like, available of the touch of a smart phone button expand. In fact, Yellow lab, US country , san Francisco city's largest taxi company is filing for bankruptcy as the industry starts to change faster than almost anyone expected. However, at this point, its more than an app that is changing, some our taxi passengers renting taxi transportation to catch consumption behavior.

(AI) technology will influence digital economy for taxi passenger's individual customer experience, offering a growing renting taxi to catch of service and feedback opportunities when any one taxi passenger who chooses to use mobile phone app online tool to prepaid to rent any taxi more easily.

Also in the long term, (AI) technology can influence vehicles drive themselves of behavior. Already, companies like Google and GM are working on projects to bring fleets of autonomous vehicles to cities at the path of a button. Moreover, this on-demand service model is beginning to appear across a much broader range of markets. For example , Amazon company is investing in its own fleet of trucks, planes and even drone at the same time as it pushes for same-day delivery of products. As some point, vehicles will be autonomous too. So, it seems that (AI) technique will influence any transportations choose to use digital autonomous driving technology in the future . For Amazon company case, it is not stopping of logistics. It is also aiming to automatically manage the supply of consumer home products with its recently launched Amazon replenishment service, Dash. Dash is a digital service that enables that connected derive to automatically order physical products from Amazon when supplies are running low. So, it seems (AI) technology will be applied to logistic function by digital technology method introduction in the future.


Hence autonomous vehicles will optimize industry supply chains and logistics operations through increased efficiency and flexibility. In fact, fully automated and lean supply chains will keep reduce load sizes and inventory by leveraging smart distribution technologies and smaller autonomous vehicles by machine man assistance. If Amazon continues to grow market share for online sales by reducing effort required by the consumer to place an order, when also contributing the almost immediate delivery of products to the doorstep. So, it will further fuel the trend toward on-demand derive. As Amazon company fuels the on-demand economy, consumers will expect immediacy in more parts of the digital economy. On top of speed, consumers increasing expect more personalization options.

So, (AI) technology will influence digital manufacturing, such as Amazon publishing to monitor every aspect of every process in real -time and communicating to self-optimized deep learning robotics, new methods of high volume and high customization will become possible. Then, as products merge into product platforms and even services, manufacturers have the opportunity to provide components and platforms used by smaller players. So, (AI) technology will influence manufacturing industry to choose automated SMI lines, robots installed, automation engineers.

Another future (AI) technology development can be applied to space science aspect, such as Automation engineering space in manufacturing process to achieve digital manufacturing benefits to any businesses in the future. Such as reducing cost, shortening manufacturing time, raising efficiency, shortening delivery products to client individual time. How can artificial intelligence give the need and advanced fast and evaluation methods benefits for space exploration? When US NASA ( space exploration organization) achieves any space exploration missions, it will answer this question:

When is it useful to have a machine use (AI) technology to achieve a decision? After all, after millions of years of space exploration and rough 10,000 years of civilization, humans are usually quite good at making decisions in complex uncertain environments. Through, Johns Hoplains University's Applied Physical Lab. Research in (AI) technology enabled systems, which has identified three general use cases for (AI) technology to explore space mission:

First, for some tasks (AI) technology is more cost effectiveness than human. Second, (AI) technology is better suited than humans at solving some, but not all problems. Third, (AI) technology allows NASA organization's space exploration mission to develop machines that ate capable of responding faster than when a human is in the decision loop ( D. Scheidt, 2012, A. Castano et. al. 2008).

So, the use of (AI) technology to enable science by observing the pace of rapidly evolving phenomena was demonstrated. It is more effectively coordinating and (AI) technology utilizing to earn economic benefits to use for space exploration mission.

However, (AI) technology also have current risk for space exploration. Today (AI) technology is immature and requires further development to reach its potential. For instance, the (AI) technology algorithms that detected the dust derive could not have identified whether the Martain weather represented a threat to the cover. Also it can not yet use instrument input to determine what, where and how to autonomously make the next space science measurement. An equally important factor limiting (AI)'s deployment is that lacks the methodology and technology to effectively test (AI) technology. So, the challenge will testing (AI) enabled system is how (AI) performance can be measured. It would be NASA organization's difficulty to find (AI) technology to develop to carry on researching any space exploration missions in the future. However, (AI) technology will be a good economic benefit choice for space exploration mission in the future.

● What is artificial intelligence potential benefits and ethical considerations?

The ability of (AI) technology systems to transform vast amounts of complex information into insight has the potential to help solve manufacturing or service challenges for human needs. However, to reap the societal benefits of (AI) systems, humans will need to trust then and make sure that which follow the same ethical principles, moral values, professional codes and social norms that we humans would follow in the same scenario, research and educational efforts as well as carefully designed regulation in order to achieve the most effort of economic benefits goals. For example, international business machines corporation (IBM) is actively engaged both competitors , in global discussions about how to make (AI) ethical and as beneficial as possible for people as social economic benefits. (AI) is usually defined as the " capability of a computer program to perform tasks or reasoning processes " that human usually associate to intelligence in a human being. Often, it has to do with the ability to make a good decision, even when there is uncertainty, too much information to handle. As an example, play chess or complex card games of entertainment activities is believed to need some form of intelligence in a human being, as well as choosing the best medical facilities in a difficult medical case, or creating something new, such as mathematical theorem or even some form of act, or even driving automatic machine man ( self driving vehicle) replacing human driving in the middle of a crowded city.

(AI) needs depends on what we consider being intelligence in the behavior of a human being act a certain point in time. If human belief about human intelligence changes and we don't believe any longer that a certain task requires intelligence, then a computer program performing that task is no longer part of (AI), it becomes just another boring computer program. So, it means that (AI) technology will replace some old computer programs, if human can invent new generation of (AI) software for any functions or activities to satisfy human needs.

As IBM, it argues intelligence. This means that we aim to build systems that enhance and scale human expertise and skills rather than replacing them. We therefore focus on practical applications of (AI) capabilities that assist people in performing well-defined tasks of needs by exploiting and wide range of (AI)-based services. We also use the term " cognitive computing" it is mean a comprehensive net of capabilities based on technology. It comprises the fields of machine learning, reasoning and decision technologies, language, speech and vision recognition and processing technologies, high performance and high efficient functions for any industries or individual consumers needs. For example, robotics, which are usually very good at doing what which are supposed to in any environment, much have public shopping center, factory etc. places which need simply services from the robot ( machine man), such as cleans the floor of our houses to the robot that can work together with humans in production chains, passing through the warehouse, robots can take care of the tasks of an entire warehouse and the companion robots like Nao, Pepper, Aibo and Giraff, who can entertain use, talk to use and help elderly people to stay connected to their friends, relatives and doctors.

Google company is building automatic machine ( self-driving cars ) and has acquired more than 10 robotics companies. Facebook had opened whole new research facility only on (AI) research. Apply computer has developed Siri. Microsoft computer company has built a similar personalized assistant. Google has Deep mind, a UK company whose long term aim is to build general (AI) and has already great potential to win game to the world champion and IBM is investing a huge amount of resources in applying its Watson cognitive computing system to the medical domains to finance and to personalized education. In Europe, IBM is establishing new centers in Munich and Milan focused in the application of cognitive computer capabilities to the internet of things and healthcare respectively.

For example, automatic machine man ( self-driving cars) are all about (AI), which used to be able to see what happens in the street ( signals ,lanes, other cars, pedestrians, traffic lights, which need to able predict what other cars and pedestrians will do, and who need to be able to cope with unforeseen situations. Since, most car accidents are due to human fault, it is estimated that the adoption of self-driving cars will save about half of the lives that are usually last in car accidents.

IBM Watson company has to understand spoken language, make sense of massive amount to text , respond correctly to questions in many categories, as well as assess its own confidence in responding to such questions. In the future, (AI) technology can own question/answering capabilities that would be very useful, for example, in assisting a doctor when trying to some to the correct diagnosis for a patient and to propose the best therapy .

Intelligent machines can also rely on huge amounts of data to be used to learn how to make better decisions. This data comes from all of us over the years Facebook users have uploaded more than 250 billion pictures and every day who upload about 350 million more. Every second, we submit 40,000 google search queries. So, (AI) technology will be connected through the web from appliances to traffic lights from cars to watches. Other tasks that are very easy for humans are physical and manipulation tasks, such as walking , running, picking up an object to make its shape and location, restricted environment. But (AI) machine man technology still not able to have the general physical and manipulation capabilities even of a 6 year old.

So, it brings this question: Why do (AI) scientists need to concern ethics? Because (AI) technology is complex, information into insight has the potential to reveal long held secrets and help solve some of the world's most difficult problems. (AI) systems can potentially be used to help discover insights to treat disease, predict the whether, and manage the global economy. So, ethic issues is important to and (AI) scientists . If any one new (AI) technology research investigation could success, it will be a secret to and the (AI) scientists can not permit to their loyalty to any competitors to damage the fair (AI) technology products trading market. The country ( countries) (AI) technology scientists need to concern ethic issues, who need to keep secrets for their countries economic or/and social benefits. This is moral issues to any countries/country loyalty is whose countries intangible assets. They can not sell (AI) loyalty to any their countries to assist whose economic benefits immorally.

● How can (AI) technology influence to global
health care economy development?

According to (AI) lecturer analysis, when combined key clinical health (AI) application can potentially create $150 billion in annual savings for the US healthcare economy by 2026 year. (AI) technology is re-winning modern conception of healthcare delivery. It enables machines to sense, comprehend, act and learn. So which can perform administrative and clinical healthcare functions (Accenture, 2017).

It will help health care service organizations to reduce health care cost, will improve and raise service quality and access. So, (AI) health market size will be predicted growth. (AI) applications in health care include robot-assisted surgery, virtual nursing assistant, administrative workflow assistant, fraud detection, error reduction connected machines, clinical trial participant identifier, preliminary diagnosis, automated image diagnosis and cybersecurity.

What kind of benefits (AI) technology can contribute to healthcare service? (AI) technology can deliver what many health care organizations need, such as financial and operational of labor costs, digital expectations from patient consumers how to use (AI) technology to solve interoperability challenges in any healthcare organizations. Also (AI) technology can be applied to wellness an d lifestyle management, diagnostics, delivers financially but also way of organizational and workflow improvement. So, (AI) technology will be continue to become most prevalent and adoption to healthcare organizations , which must need to enhance structure to be position to take full advantages of new (AI) technological capabilities. (AI) technology can change the nature of work and employment is rapidly changing to make the best use of both humans and (AI) talent in healthcare industry in the future. For example, (AI) technology offers a way to fill in gaps and the rising labor shortage in healthcare. According to Accenture analysis, the physicians shortage is increasing. However, (AI) technology will manufacture healthcare machine men to replace physicians in future one day(2017). Hence, (AI) technology will be invented to raise health care service staffs work efficiency and performance in any hospitals or clinics in the future.

In conclusion, (AI) technology will raise efficiency for any service or manufacturing industries in the future, although, it is possible that it will also rise low skillful workers unemployment numbers. But, the most important influence to human technological innovation will be risen and it will influence human life will be changed to be better, e.g. self drive cars, health care physician machine men, machine man cleaners etc. intelligent machine men will be manufactured to serve for our daily life. Furthermore, (AI) technological products will influence countries trading, some low technological development countries manufacturing businessmen can choose to buy any (AI) products to raise whose productivity and efficiency and reducing cost to achieve economic cost saving result. Also, GDP of trading growth income will increase to the (AI) products sale countries. Hence, it will be beneficial to economic development to both developed and developing countries both in the future as well as (AI) scientists time and money spending will be valued to continue to invest (AI) technology development for human life and economy benefits for long term.

In conclusion (AI) technology will raise macro economy growth and it can create many (AI) jobs , but it also raise the low level technological worker unemployment change. In the future, (AI) technology can be applied to digital technology to attempt to invent any new undiscovered (AI) and digital technology. So, it needs any scientists to continue to research how digital and (AI) technology can be mixed to satisfy human's future undiscovered needs.

Artificial intelligence and the future of defense

Nowadays, artificial intelligence (AI) is widely knowledge to be one kind of the dramatic technology. However, it is expected to continue, to have a disruptive impact on human's private and public life, so defense and security will be no exception. But how exactly will these be affected ? How will (AI) defense and security is incremental in nature?

To research why artificial intelligence (AI) has possible to be used to cause autonomous weapons by human. We need to understand these three aspects of relationship. They include cybersecurity and artificial intelligence and machine learning and autonomous weapon systems relationship between of them.

Firstly, we need to know what is the mean of artificial intelligence and cyber defense/offense? It means defense of critical networks: real time, pattern finding, anomaly seeking, it must utilize machine (AI) learning algorithms to efficiently, and instantaneously respond to potential network threats as well as it means human on or out of the loop. On the loop : it means anomaly detection: human notified, IT analysis, response. Out of the loop: it means anomaly detection: (AI) decides best method of response: quarantine, honey pot monitoring, hack-back. Thus, it is possible that (AI) can be used , such as autonomous cyber weapon.

What is artificial intelligence and autonomous weapons? Autonomous weapons mean one kind of weapon that can be selected and engaged a target, without intervention by a human operator. Are these machines artificially intelligent? I believe the answer is not, because present weapons systems are not capable of human level reasoning. But, (AI) algorithms are presently employed to process sensor data, monitor system health, take and respond to vocal commands manage data, navigate. This, future autonomous weapons systems will require stronger (AI) to be secure and operationally and cost effective. Moreover, self-aware autonomous cyber systems are crucial.

What is cybersecurity mean? It means the ability to control access to networked systems and the information they contain. It is acted to prevent , detect, recover, react. It is application objects concern people, process, technology and it's application goals are confidentiality, integrity and popular availability. Thus, what is cyber weapon mean? Walware means viruses, Trojans, zero-days, worms ransomware, spyware etc. Does it require a particular objective? E.g. military paramilitary or intelligence. Does it require physical harm? E.g. functional harm or interruption? Mental harm? Is (AI) a technological weapon that it is an object or tool? What about when it is an weapon agent?

In simplicity, (AI) can be one of scientific weapons platform. When one day, it is invented to be applied to control war planes to fly to any countries to attack enemies or it is invented to be seemed to human to replace soldiers to bring guns or any weapons go to other countries to attack. So, it is possible that future any war defense planes, (AI) technological automatic control weapon can be replaced of human soldiers or war plane pilots to control any war defense planes to go to different enemy countries to attack them easily. It is very horror matter to threaten global human's ourselves life in the future , if (AI) automatic control war defense planes or (AI) automatic control machine soldiers were invented successfully.

Hence , when (AI) can be applied to weapons platforms, it structures that launch weapons, i.e. jets, ships, vehicles. (AI) platform and weapon and software architecture components are be done one (AI) technological weapons systems. Thus, human will encounter any (AI) benefits or risks ( threats) causes in the same time as soon as possible. If we can predict when (AI) weapon system will be manufactured or invented successfully. Then, we can reduce (AI) weapon systems risks , if we can threaten any (AI) scientists continue to invent any undiscovered (AI) weapons in any time to avoid the future first time (AI) weapon war occurrence in possible.

The (AI) weapon system risk means autonomy: the ability to problem solve technological war , when (AI) weapon system is manufactured successfully, the power to act, how to damage the (AI) weapon system. The power to chance to stop (AI) weapon system manufacturing processes, ability to create a new goals, how to change the (AI) weapon system inventors' or scientists' minds to avoid to apply (AI) tools to achieve attack goals to change to another positive goal. Due to human can't know a prior what an autonomous (AI) weapon system will do.

Although, human is known what (AI) is , but human is also known when (AI) scientists whose emergent behaviors will do to change to do any negative behaviors from positive behaviors. Whatever (AI) weapon system design we use, there will be cybersecurity, problems arising from computation design/complexity. Due to any one (AI) scientist can manipulate the system to act against itself, or who can utilize traditional " cyber weapons" against the (AI) weapon system, or who can manipulate the system to lie to humans, but also due to complexity, there is no way to know if it is lying or not or bounded rationality : satisficing.

Finally, the most serious (AI) technological invention risks are human is unknown these aspects of (AI) absolutely: They are not simple automatic systems, learning reasoning, communication of " self-aware" systems. Thus, human will face (AI) technological invention risks or threats. We need to find any methods to avoid (AI) weapon system is manufactured successfully to avoid (AI) technological war can occur in future anyone day. (AI) system immoral intention

Why (AI) system can be invented to damage our society ? IS it possible to achieve this (AI) damage system successfully? ON (AI) attribution hand, it can be applied to cars, aircraft, which are subject to regulation designed to protect the public from harm and ensure fairness in economic competition. Thus, (AI) safety issue is important to scientists to consider.

IN general, the approach to regulation of (AI)-enabled products protect public safety issue should be informed by assessment of the aspects of risk that the addition of (AI) way reduce any respects of risk that it may increase. Also, where regulatory responses to the addition of (AI) threaten to increase the cost of compliance, or slow the development or adoption of beneficial innovations, policymakers should consider how those responses could be adjusted to lower costs and barriers to innovation without adversely impacting safety or market fairness.

For example, regulatory challenges that (AI) enabled present are found in the cases of automated vehicles. (AI)s, such as self-driving cars and (AI)-equipped unmanned aircraft systems. IN the long run, self-driving cars will likely save many lives by reducing driver error and increasing personal mobility, it will offer many economic benefits. Thus, public safety must be protected as these technologies are tested and begin to mature. Creating safe spaces and test beds for experimentation , and working with industry and civil society to evolve performance based regulations that will enable more uses as evidence of safe operation accumulates. Thus, it implies that any scientists can also invent (AI) system to control weapon defense planes or (AI) automatic machine human to do any soldier's behaviors to attack to any countries easily, instead of none driver automatic control vehicle invention. Thus, (AI) system can be applied to harm to human or achieve to damage our society aim by ourselves in possible.

The rapid growth of (AI) has dramatically increased the need for people with relevant skills to support and advance the field. AN (AI) –enables would demand a data literate citizenry that is able to read, use, interpret and communicate about data and participate in policy debates about matters affected by (AI). Thus, if (AI) technology is applied to assist human's social development and raising life enjoyment or benefits. It will bring positive impact to influence human's future life. Otherwise, if (AI) technology is unsafe to be applied to threaten human's society. It will bring negative impact to influence human's future life. Thus, (AI) scientists need to consider how to apply (AI) technology.

As (AI) technologies move toward deployment, technical expects, policy analysts and ethicists have raised concerns about unintended, consequences of adoption. Use one (AI) to make consequential decisions about people, often replacing decisions made by human –driven bureaucratic processes, leads to concerns about how to ensure justice, fairness, and accountability, the same concerns of human's safety issue. Thus, )AI) expects have cautioned that there are challenges in trying to understand and predict the behaviors of advanced (AI) systems.

Use of (AI) to control physical-world equipment leads to concerns about safety, especially as systems are exposed to the full complexity of human environment. A major challenge in (AI) safety is building systems that can safety transition from the closed world of the laboratory into the outside open world, when unpredictable things can happen. Adapting to unforeseen situations are difficult necessary for safe operation. Experience in building other types of safety artificial systems and, such as aircraft, power plants, bridges and vehicles has much to teach (AI) practitioners about verification and validation, how to build a safety case for a technology, how to manage risks, and how to communicate with stakeholders about risk. The risk means the harm of human's safety of (AI) damage system control machine invention. Thus, any (AI) scientists need consider moral responsibility when who decide to invent what kind of (AI) system machine to aim to bring human's benefits or attribute to human's welfare intention.

Thus, (AI) products safe invention matter will need any scientists' considerations. Because , if (AI) any products are unsafe or harm human's invention in the manufacturing process, it will bring any human's life danger when the (AI) system damage tools are invented successfully and are provided weapons to humans to use to attack other countries easily. It will cause future global human (AI) technological war occurrence.

I shall recommend the solution is necessary of ethical training for (AI) practitioners and students. Ideally, every student learning (AI) , computer science, or data science would be exposed to curriculum and discussion on related ethics and security topics. However, ethics alone is not sufficient. Ethics can help practitioners understand their responsibilities to all stakeholders, but ethical training should be methods for deciding good intentions into practice by doing the technical work needed to prevent unacceptable or immoral (AI) invention outcomes.

Hence, global human needs to concern (AI) weapon system invention security issue. Nowadays, (AI) has important application is increasing role for both defensive and offensive cyber measures. Currently, designing and operating secure systems requires significant time and attention from experts.

Challenges issues are raised by the potential use of (AI) in weapon systems. The United States has incorporated autonomy in certain weapon systems for decades, allowing for greater precision in the use of weapons and safer, more humane military operations. Nonetheless, direct human control of weapon systems involves some risks and can raise legal and ethical questions concern (AI) manufacturing process intention.

The key to incorporating autonomous and semi-autonomous weapon system into American defense planning is to ensure that U.S. Government entities are always acting in accordance with international humanitarian law, taking appropriate steps to control , to develop standards related to the development and use of such weapon systems. The United States has activity participated in ongoing international discussion on Lethal autonomous weapon systems and anticipates continued robust international discussion of those potential weapons systems. Thus, (AI) scientists have responsibilities to manage the potential to be a major driver of economic growth and social progress only, their (AI) intentions are not the global dominance aims absolutely, if (AI) product industry , civil society, government and the public work together to support (AI) positive development of the technology with thoughtful attention to its potential and to managing its invention threat risks to avoid (AI) products to manufacture to be used weapon tools.

Finally, I recommend that as the technology of (AI) continues to develop, practitioners must ensure that (AI) enables systems are governable, that what their inventions need to be openness to let public to know clearly and understandable; that they can work effectively with people and that their operation will remain consistent with human values and aspirations. Researchers and practitioners have increased their attention to these challenges , and should continue to focus on their future any (AI) inventions.

Hence, (AI) safe system ought to be applied to solve the biggest challenges that society faces, such as mobility for the elderly and those with disabilities, smart buildings may save energy and reduce carbon emissions, precision medicine may extend life and increase quality of life, smarter government may solve citizens more quickly and precisely., better protect those at any immoral invention risk and save money.

Moreover, (AI) enhanced education may help teachers give every child on education that opens doors to a secure and fulfilling life. Thus, these are the future human's potential benefits if the (AI) technology is developed to its benefits and scientists ought avoid to manufacture (AI) tools to cause weapon risks and challenges.

Consequently, the main point is that how experts invent (AI) systems. (AI) systems ought not be advanced weapon systems, it doesn't seem to be thought similar human soldiers mind and behaviors. (AI) system ought be systems that think like humans. (e.g. cognitive architectures and neural networks), systems that act like humans ( e.g. pass the test via natural language process, knowledge representation, automated reasoning, and learning), systems that think rationally , e.g. logic solvers, inference and optimization and systems that act rationally e.g. intelligence software agents and embodies robots that achieve goals via perception, planning reasoning, learning , communicating, decision-making and acting function.

In conclusion, it is horror (AI) scientists will invent (AI) systems to be owned human's (soldier's) mind and attack strategic behavior to attack other countries easily, who must need to consider (AI) system ought be invented to own scientists' creating mind and non manual assistance functions for positive attribution to human's society. I expect that (AI) system can only be invented to create human's welfare in our future.

(AI) soldier weapon ethical, social and economic negative impact

In the future, how human can avoid (AI) technological ethical, social and economic negative impact. Scientists need to concern these questions: how to develop of a good (AI) society, how the role and responsibility of the government, the private sector, and the reserch community( including education), in pursuing such a development, whether how the recommendation to support , such a (AI) system development may be in need of improvement.

However, none appers to deliver a comprehensive explicit vision of the role that (AI) system should play in mature information societies. Thus, (AI) 's potential contribution to social good shoud include an in-depth plan for linking in a comprehensive socio-political design questions of responsibility of the different stakeholders, of cooperation between them and of sharable values to understand of a good (AI) positive impact society, not a bad (AI) negative impact society.

Thus, the notion of mature information societies is introduced to stree the importance of addressing the current ethical challenges that (AI) poses in a comprehensive fashion.

It seems (AI) wil invention will be human's moral societal consideration issue. It concerns our (AI) scientists' moral issue, how who invent (AI) system to apply to which kind aspects. IF (AI) system was one direction on war weapon tools to similar to soldier's personal mind or attacking behavior. Then, it will bring poor social safety and poor economy growth our world, due to (AI) scientists' moral is low level.

Thus, the developed country US (AI) technological leader needs to focuse on the impacts of (AI)-driven customatin on the US job market and economy. It represents three specific policy responses to the perceived impact of (AI) on the US economy. They include these three aspects such as: How to invest in and develop (AI) for its many benefits, how to educate and train Americans for the jobs of the future and how to aid workers in the transition and empower workers to ensure broadly shared growth.

The future of (AI) influenced cyber conflicts need more than just the application of current and past solutions in order to ensure security and stability of societies, and avoid risks of escalation. To achieve this end, efforts to regulate cyber conflicts require an in-depth understanding of this new phenomenon, identify the changes brought about by cyber conflicts and the information revoluation, and defines a set of shared values that will guide the stakeholders operating to avoid the international (AI) war occurrence. This becomes clear when considering for example, cyber deterrence. Deploying conventional (cold war) strategies to deter (AI)-influenced cyber conflicts proves highly problematic and the urgent need to foster and coordinate new solutions able to account for the any kinds of conflicts of the cyber demain and of mature information societies to avoid (AI) technological war occurrence in the future.

We hope that in the on-going international conversations and reviews, the US government with further specify how " (AI) system invention law" fit into their vision of the future of society in this case the future of (AI) technological war and conflicts. Hence, (AI) scientists need to concern ethical issues related to (AI), like fairness, accountability and social justice can be addressed through increasing needs. Such as: how the creation of a new body focused on robotics and related (AI) system development to avoid to intent to apply weapon tools to provide advice on the policy, legl and consumer protection issues arising in these fields should be considered.

How to achieve ethical training of (AI) staff and ethical education of the public is certainly important responsibility for (AI) tools ethical behavior and design to the private sector and the citizens : of unique challenges that (AI) brings to society in terms in fairness, social equity and accountability are addresses. Thus, the development of the (AI) technology and defining good (AI) remains problematic. In particular, the US government's innovation driven approach to defining the potential, positive impact of (AI) shows that more could be done to ensure that the opportunities and advantages brought about by (AI) are shared by all society.

An initial on Robotics, based upon the ethical framework and guiding principles is proposed. It should be complementary to legislaton and comprise ethical codes of conduct for Robotics researchers and designers, codes for research ethics committees as well as licenses ( rights and duties) for designers and users. Thus, (AI) robotics invention of safety issues is very important considertion to any (AI) inventions or researchers. Every country's government ought have legal guiding to control their robotics' manufacturing intention. If their robotics (AI) is applied to seem to be soldiers to attack other countries to threaten their people's safety. Then, those (AI) inventors or researchers need to be punished by law.

In conclusion, I believe (AI) technology will be applied to weapon, when it's technological development is nearly mature to able to learn human's mind to do any behavior. During (AI) technology reachs thie mature stage, I predict the (AI) weapon tool , e.g. (AI) soldiers will have chance to be caused. This (AI) invention mature stage has these characteristics such as:

When (AI) invetion reachs this mature stage, computers and robots will develop conscious, intelligent, personified minds. Further, information technology devices and (AI) systems will be implanted into humans, enhancing, psychological and behavioral abilities and allowing for direct communication with artificial intelligent minds. There will be both artificial intelligence (AI) and intelligence amplification (AI) in the relatively near future stage.

During the (AI) invention reachs this mature stage, these will be an ongoing mulit-faceted integration of information technologies and human life. Humans and information technology will cooperate. Humans will increasingly immerse their lives and minds in (AI) systems of technological intelligence and virtual reality. The distinction between humanity and technology will increasingly close dependence.

During the (AI) invention mature stage reachs that the environment will be infused with information technology, becoming animated, communicative and more intelligent. The destinction between the artificial and the natural will increasing close dependence.

During the (AI) invention mature stage will expand through virtual reality, simulated and virtual reality will increasingly into normal reality, e.g. the (AI) weapons is virtual reality to seem to be soldier weapon.

Finally, during the (AI) invention mature stage is as the global expression of the evolving human-technology integration a " world brain" and " world mind" will emerge on the earth. This psychophysical (AI) weapon system will enhance and enrich the capacities of both individual and collective cogniton. This (AI) weapon system is a potential starting point toward the evolution of a cosmic brain and cosmic mind.

Thus, it is possible that the workship raw data was a unique way in which (AI) could be weaponized to cause war, during the (AI) invention stage reachs the invention mature stage. However, (AI) weapon manufacturing factory will be built possibly. In the future, how will we defins and locate (AI) weapon factories. Especially, as these factories are no longer solely buildings , but a mil of virtual and substantially different facilities, particularly as it shifts from a physical assemly and development model to a distributed and flexible network. Needing minimal raw materials to develop (AI) weapons, the phsysical location of their (AI) factories could be anywhere and their identification from the outside, nearly impossible. Given the expanding uses for intelligent and super-intelligent (AI). How will we tell the different form a location that is manufacturing (AI) for the creation of weapons versus creating (AI) for an innovative new gaming platform?

In conclusion, human needs to consider every (AI) scientist's personal ethical or moral mind and research intention and (AI) system invention of (AI) weapon factories cause. During (AI) invention reachs the mature stage if human expects to avoid (AI) technological war occurrence in future one day. The technological development on autonomous military robots, ideally among relevant social groups and actors including human-rights, activists, researchers developers, engineers, philosophers, policy-makers, military authorities, lawyers, journalists and the publis need to consider when human has effort to invent autonomous military robots successfully in the future one day. Finally, some ambitious countries or dominant global countries must like to apply (AI) autonomous military robots to be machine soldiers more than human soldiers if (AI) technology had reached the mature stage. So, future (AI) autonomous military robots will be the next choice of weapon to follow nuclear weapon. If civilians were used as a human (AI) soldiers, the weapon simply ignored them and targeted anyway. This scenario highlighted the dangers of proliferation and quick replication of autonomous weapons. Unlike nuclear weapon, a piece of code for (AI) artificial intelligent soldier could be obtained on the black market and replicated at little cost and the hardware for this type of weapon doesn't require costly or hard to obtain components and materials. Thus, (AI) artificial intelligent soldiers can be manufactured many at cheaper cost. Otherwise, manufacturing one nuclear bomb weapon will spend too much cost. Hence , it is possible that (AI) artificial intelligent soldier will be future new technological weapon to follow nuclear bomb weapon. Hence, any country government needs to legislate to control any (AI) scientists' inventions whether they are attributed benefits or welfares to human or damage human's safety.

Reference

A. Castano et. al. " Automatic detection of dust devils and clouds at Mars" Machine vision and applications, Oct. 2008, vol. 19, no 5-6, pp. 467-482.

Accenture, " Why artificial intelligence is the future of growth"(2017) <http://www.accenture.com/us-en/ insight-a rtificial-intelligence-future-growth>.

D. Schedidt , Unmanned Air Vehicle Command And Control, Handbook Of Unmanned Air Vehicles, Springer-Verlag, 2014. Facebook (AI) Research Available at https://research.facebook.com/ai, research at google, machine intelligence available at

http://research.google.com/pubs/machineintellige nce.html; micro soft research-machine learning and artificial intelligence available at http://research.microsoft.com/en-us/research- areas/machine-learning-ai.aspx.

International Federation Of Robotics, 2016. IFR press release world robotics report. IFR, org . 29 Sept. Accessed Feb. 01, 2017. http://www.ifr.org/news/ifr-press-release/world-robitics report -2016-8321.

K, Fedra , "GIS and environmental modelling" in environmental modelling with GIS, edited by M.F. Goodchild.B.O. Parks and L.T. Steyaert, Oxford University press, pp. 35-50, 1994.

Keynes, J.M. (1933). Economic possibilities for our grandchildren (1930). Essays in persuasion, pp.358-73.

Mckinsey & Company (2013, May). Disruptive technologies: Advices that will transform life, business and the global economy , USA.

Ministry of economy, trade and industry, Japan, 2015, Japan's robot strategy. Ministry of economy, trade and industry.


Ray Kurzweil , The age of spiritual machines (1999) is cited numerously through this chapter: Kurzweilai.net http://www.kurzweilai.net

Rich, Elaine & Knight, Kevin, Artificial Intelligence Second Edition, 1991, New York; Mc-Graw-Hill.

# Robots development stages brings what advantages and disadvantages to our societies

development first stage
Competitive influences between artificial
intelligence and human job

Although, (AI) technology will be popular to applied to different jobs, but it still needs social acceptance to replace some human jobs. Today, it is increasingly common for people to use robots in various situations at home and in retail stores, hotels and hospitals. Robots are classified into several types based on their functionality ( service and utility robots or those designed to communicate with humans ) and appearance ( humanoid robots or mechanical robots). The types of robot to which every country attaches particular important in the advance of robotics, reflects the sense of values and preferences of its population . Thus, (AI) will be applied to replace human to do these above different kinds of job nature. For example, U.S. has the highest level of robot utilization at home and an retail stores with its people being the most enthusiastic about the future use of robots. Otherwise, Germany shows a strong tendency to consider robots for industrial purposes, and its people feel strong to the presence of robots in their households. Japanese accepts to apply" human aid robot" that can communicate with humans and they have a high level of familiarity with robots.

Hence, it implied those three countries have accept (AI) to replace human to do any these kinds of job duty and it will influence these three countries' workers lose their old occupations and who will unemployed absolutely, due to many (AI) robots replace them to do their job duties in the future. Also, US will have many retail service workers or retail warehouse workers are unemployed. Germany will have many manufacturing industry's workers are unemployed. Japanese will have many communication industry workers are unemployed, such as telephone service, shopping center services etc. different kind of service industry's service staffs . It will cause these kind of workers' competitive abilities are lost in themselves countries' jobs that require such skills include software developers, court judges, nurses, high school teachers, dentists and university lecturers, these occupations are still difficult to be replaced by (AI) robots.

Are robots taking our jobs or making them? In fact, our societies will have unemployment challenges, even (AI) technology has not created before. However, after (AI) robots invention, some of human jobs will be replaced and it can raise many low skillful and low knowledge level worker unemployment number. However, I think that high productivity driven by increasingly powerful IT -enabled machines is the causes of global labor market problems and accelerating technological change will only make those problems worse.

IT technology brings this question: Are robots killing human's jobs or benefiting human's jobs? I suppose that there is a limited amount of labor to be done. The implication is that technology can create unemployment by displacing workers, such as (AI) invention, because the more efficiently worker work ( using machines or (AI) robots), the loss work there is for workers to do. Even, any new jobs will be better done by machines or (AI) robots, and unemployment will still skyrocket. How do we know that humans will always be better at some work, or more importantly, enough work, than machines or (AI) robots, e.g. human drivers drive more safe or careful to compare (AI) robot drivers. But, the challenge is that it is not ensure that (AI) robots drivers must not drive careless to cause the chance of accident occurrences more than human drivers. However, technological change can be beneficial to innovation, automation and increasing productivity for businesses.

Consequently , it may seem machines can hurt wages and job for low skillful, less educated workers. Also, high educated workers are likely as less educated workers to find themselves displaced and devalued, and more education may create as many problems as it solves. Thus, in negative influence, automation effects on particular jobs shift workers to other jobs that are equally or more desirable. Workers may be highly compensated for possessing human capital that is specialized to a labor market. If technological advance is very rapid, such as (AI) invention, causing a large and very rapid drop in demand in a large labor market, the economy may not be able to absorb the sudden surplus of labor in a short period of timer when (AI) robots are popular to replace some workers to do some occupations in global societies.

For example, self-driving vehicles threaten to send truck drivers to the unemployment office. Computer programs can now write journalistic accounts of sporting events and stock price movement. There are even computers that can grade essay revolutionize some part of teaching jobs. Hence, (AI) robots will have possible to replace human brain to do any judgement, argument, and mind job duties. It implies some occupations which need human' mind will be threaten by (AI) robots, e.g. author, accountant, nurse, engineer. Thus, (AI) robots will have possible to replace some professional and high educated workers' jobs in the future.

But, technology can create new nature of jobs in possible. For example, a 60 minutes program indicated technology is putting new categories of jobs in the sites ( sic) of automation, the 60% of the workforce that makes its living gathering and analyzing information. Also, recession: technology kills middle -class jobs that overall technology is eliminating for more jobs than it is creating by (AI) technology. Hence, human's brain work may be assisted by 60% of (AI) gathering and analyzing information for some occupation , e.g. space scientists, ocean scientists, earth scientists etc.

However, I believe the (AI) invention and human job competition may influence global productivity change. Productivity is economic output per unit of input, the unit of on input can be labor hours( labor productivity), but if (AI) robots replace human job, then the unit of input may be (AI) machine hours ( AI) robot productivity or all production factors including labors, machines and energy ( total factor of productivity). Producing more output with less input can take several forms.

The traditional notion of productivity is a form reorganizing production and/or using better or more technology to produce more output per worker hour. But when (AI) robots invention, the form can be reorganizing production and/or using better or more (AI) robots to produce more output per (AI) robot hour. Hence, if the firm apply (AI) robots to produce its products. Then , productivity improvements in the firm may result in less workers employment, due to (AI) robots replace more worker number to achieve more productivity improvement, it has economic benefits ( less factor of production) , but more production in long term.

Thus, (AI) robots can help any firm to achieve productivity improvement in long term, for example, if unproductve farmers move to the city and start working for high-tech. manufacturers. The shift effect can be more dynamic and disruptive as low-productivity industries lose out in the marketplace to high -productivity industries and the compositional mix of the economy changes. Thus, in the long term (AI) robots can also be beneficial to high productivity industries to bring the mix of economy positive changes.

Moreover, automation will also produce some new jobs in firms that sell the new robot or other labor-saving technology. This means that, in general, there will be shift in the economy in the direction of higher-skill and higher wage jobs. Even if the (AI) robot invention country, US becomes a leader in (AI) robots producing productivity-enhancing technology, it will experience a growth in jobs serving foreign (AI) robots product buyers. Hence, (AI) robots can also create (AI) salespeople, (AI) manufacturing workers , (AI) inventors, scientists, (AI) software designer etc. occupations, when if all society does is move workers from insurance firms, restaurants and car factories to robot factories, productivity will have remained the same to create job needs for insurance, restaurant and car manufacturing worker service occupations for (AI) software designer, (AI) service robots manufacturer, (AI) service robot seller etc. related (AI) service robot product occupation created in (AI) robot technology job market. Hence, (AI) invention also create new (AI) technology job chance. (AI) impacts management job market.

In future, organization management will be changed from (AI) introduction. Division of labor will change and collaboration among humans and machines will increase. Companies will have to adapt their training, performance and talent acquisition strategies to account for a new found emphasis on work that hinges on human judgement and skills, including experimentation and colloboration.

How (AI) impacts any organizational administrative management work? (AI) 's greatest impact will be on administrative coordination and control tasks, such as scheduling, resource allocation and reporting, (AI)-driven will place a higher premium on what we call " judgement work", the application of human experience and expertise to critical business decisions and practices when information available is insufficient to suggest a successful course of action. This kind of work will require new skills and mindsets; replacing people with machines is not goal in itself. When, artificial intelligence enables cost-cutting automation of routine work, it also empowers value -adding augmentation of human capabilities.

Thus, administrative and routine tasks, such as scheduling, allocation of resources, and reporting, will within intelligent machines, responsibilities that have long been reserved for humans. For instance, a typical store manager or a lead nurse at a nursing home must constantly juggle shift schedules, accounting for staff members' absense owing to illness, vaction, time or sudden departures. Many of these tasks will be automated by (AI). Imagine (AI) writing management monthly reports, it is not a distant dream. Leading news providers and Wall street banks are now using (AI) report generators to write news and analytical reports by drawing on quantitative data. The associated press, for example, expanded its quarterly earnings reporting from approximately 300 companies to nearly 3,000 with the help of (AI) powered software robots, freeing up journalists to conduct more investigative and interpretive reporting. For another example, Jobalime, a job-placement site, uses intelligent voile analysis algorithms to evaluate job applicants. The algorithm assesses paralinguistic elements of speech, such as tone and inflection, products which emotions a specific voice will elicit, and identifies the type of work at which an applicant will likely excel. In the future , (AI) machines can be applied to assist some kind of office administrative jobs duties. It's attractive to office managers to achieve more accurate judgment to do any administrative matters when who can be assisted from (AI) machines. Thus, managers need to spend time to learn how to apply (AI) machine to assist them to do more accurate judgement, and better informed choices. (AI) robots can be applied to improve the speed quality and cost of available products and services, instead of applying on productivity improvement and administrative improvement aspects. Thus, they may also displace large numbers of workers. This, possibility challenges the traditional benefits model of trying health care and retirement savings to jobs.

In an economy that employs dramatically fewer workers to deliver benefits to displaced workers. For example, the worldwide number of industrial robots has increased rapidly over the past few years. The fall prices of robots, which can operate all day without interruption, make them cost- competitive with human workers. In special consideration, in the service sector, computer algorithums can execute stock trades in a fraction of a second, much faster than any human. As those technologies become cheaper, more capable, and more widespread, they will find even more applicants in an economy.

Consequently, (AI) technology brings unemployed number increasing many businesses continued automating their operations rather than hiring additional workers. A trend among technology companies that receive massive valuations with relatively few workers. For example, in 2014 year Google was valued at $370 billion with only 55,000 employees, a tenth the size of AT & T's workforce in the 1960 year. Hence, if automation technologies like robots and artificial intelligence make jobs less secure in the future, there needs to be a way to deliver benefits outside of employment " flexi security" or flexible security is one idea for providing healthcare, education and housing assistance whether or not someone is formally employed.

In conclusion, (AI) and robots technology will raise unemployment to some occupations when (AI) replaces same industries' workers job duties in our societies in the future, but it also create new jobs to raise employment in any related (AI) robots and automated machine products in (AI) manufacturing. (AI) design, (AI) sale self-related industry, when (AI) replaces same industries' workers' job duties.

(AI) journalism, media publishing, digital
communication technology trend

How to apply (AI) technology in digital communication journalism media, publishing industry? Some scientists indicate future (AI) and digital technology may consist such as: voice driven assistants, emerge. For example, Amazon e book publish applying digital technology and (AI) auto printing technology to sell e books to let readers to listen any e book content by (AI) voice driven speaker when they turn on computer to read e book contents; capable phones start to unlock the possibilities of 3D image of mobile story telling. New smart wearables include ear buds that handle instant translation and glasses that talk and hear. China and India will become a key focus for digital growth with innovations around payment online identity, and artificial intelligence. Thus, future (AI) technology can be applied to 3D image mobile story telling, online payment method to dealt online transaction publishing industry.

Thus, future (AI) technology can be applied to online e book publishing industry to make sound books to let readers feel more attractive . Such as Amazon publish has published sound e books to attract readers to choose to read any its books from online. Also, (AI) technology can also be applied to communication industry. For example, some online pure-play news, opinion and entertainment websites. It is a digital communication media, e.g. online journalism blog (AI) technology can be applied to visual storytellers to let online book readers to enjoy to listen to watch and send any online electronic book contents more attractive. Thus, future (AI) technology will be popular to assist any electronic book publishers to publish visual and sound talking storybook to let readers who can watch motive image and listen and read words from e books more attractive.

Thus, (AI) technology can be applied to internet ecommerce publishing or media industry to help any electronic book publishers to publish sound, image motion electronic book to attract global readers to read, even (AI) technology can be applied to digital entertainment industry, e.g. electronic 3D image virtual video games, computer games. It can be also applied to education industry, e.g. the first true digital native generation and are the native speakers of the digital language of computers to let student to learn different languages or translate words to compare to classroom learning more easily. It can be also applied to communication industry, e.g. (AI) mobile phone. Hence, it seems (AI) technology can be applied to publishing, communication, education , entertainment etc. different industries in the future. (AI) technology will be one kind of tool to satisfy human's daily life needs in the future and these industries has one characteristics is that they need to apply internet to operate to operate to do online business.

Thus, it has three trends of (AI) technology and internet technology need to be linked to achieve one kind of attractive technological business to satisfy client's needs. These three trends as below: All consumer trends involve the internet. It will be many consumer's online habits, shopping, working, socializing, watching TV, studying, travelling, listening. Thus, (AI) music, eating and exercising are just a few examples. This is happening because human usually use mobile broadband or Wi-Fi, rather than cables. Thus, (AI) technology will be applied to mobile to satisfy client's need absolutely.

The mobile phone can be more popular to be used more than computer or laptop tools. The reasons are because women dive the smartphone market by defining mass-market use. But as the speed of technology adoption increases mass market use becomes much quicker then before. Successful new technological products and services , such a (AI) mobile phone products now reach the mass market in popular use. It means that the time period when early adopters influence others is shorter than before. Also, since new products and services increasingly use the internet mass markets are not only faster , but are also more important than ever to consumer themselves. Most internet services become more valuable to individuals when many use them. Thus, it causes why (AI) mobile phone will be popular to be used.

Since, new products and services increasingly use the internet, in the future several trends focus on (AI) smart phone users. Consumers' familiarity with using smartphone apps. Essentially, the technologies will bring other related (AI) and internet service needs, e.g. sound and image emotion e book needs, (AI) mobile communication needs, e-virtual games or e-3D image virtual games etc. entertainment activities needs with such a large part of the world's population now online, it is clear that there is strength in numbers.

Thus, (AI) imagines , if future any (AI) and internet related services or products new technology is easy to use and inexpensive, when the latest products reach the mass market almost as quickly as they reach the early adopters and industry experts. I believe that any (AI) and internet related products or services must be popular to accept to consume for entertainment or useful aim. For example, with major players including Apply, Facebook and Google had invested (AI) technology to develop their businesses. (AI) technology has the potential to disrupt everything in the coming years, from the lives of connected consumers to every industry (AI) will be an alternative route for brands to reach consumers with convincing and relevant messages. Digital technology will assist of the future, then it can improve technology to bring this effect, such as sophisticated software machine learning and speech recognition effective. Hence, Google, Facebook , Yahoo web site service companies can apply (AI) technology to help other companies to advertise their businesses, such as travel, retail, and education etc. industries more attractive. (AI) technology can be applied to internet company to be aware and familiar enough to drive among mainstream consumers, it can create online experience to travel, retail , education and other entertainment needs to online consumers to seek their entertainment needs more easily. Hence, in the future (AI) technology and internet related entertainment service needs will be raised in this (AI) and online consumption market.

● (AI) healthcare service industry development

In the future, (AI) medical internet technology tool can be applied to assist individual's health at the center of their focus, e.g. smartwatch compatible mobile app. patients can let personalized reminders for taking their medication snap pictures of their prescriptions to expedite refills, and scan their insurance card. So that, store clerks are prepared with up-to-date patients' information . (AI) owned health operated technological clinics can help patients to receive treatment for minor illnesses, flu shots, cholesterol screenings and more than a dozen other medical services, all of which can be patients who can't make it to a physical location. (AI) healthcare services organizations can provide various telemedicine services. So, patients can receive care via phone or video chat.

For example, one London-based intelligent Brewing company has developed an (AI) system to continuously collect and incorporate customer feedback, which the system itself uses to brew ne various of the company's beers. Thus, the beer clients can give feedback to talk to the algorithm (AI) machine, whenever or anywhere who're drinking the beer. It is such any healthcare services organizations can apply (AI) machine to collect patient's feedback to talk to the algorithum (AI) machine whenever or anywhere who're eating any medicines. So doctors can know every patient's health conditions any time. If the patients feel uncomfortable, the doctor can know from (AI) machine notification to decide whether the patient needs to eat another new medicine or keep to eat same medicine is better. Hence, (AI) medial internet technological body check report machine will be proper to be needed to serve any hospitals' patients in the future.

However , it brings this question. How can (AI) medical internet technological body check report machine apply to hospital more efficient? The essential new medicine co-workers for the health service digital age health service leaders need apply (AI) medical report machines and artificial intelligence to the newest recruits to the workforce bringing new skills to help health service staffs do new jobs and reinventing what's possible, building the health service workforce for today's digital health service demands for patients. Thus, technology-driven health service model innovation from the health service organization outside in and providing digital health service ecosystems for patients to use the (AI) health service equipment will be popular to be accepted to be used.

● I Robot and internet things future machine
men invention

Nowadays, there are some company, which apply internet and (AI) I Robot technology to do any similar human job nature. For fishing industry example, one company, known for creating the Roomba, I Robot is now working with marine conservationists to launch an ocean-patrolling intelligent robot to hunt and manage invasive species, protecting native populations. And evolved industries like precision agriculture are ramping of our increasing population. Area of practice that once seemed impossible to digitize are fundamentally changing because of the impacts of (AI), internet of things capabilities and big data analytics, which have many potentially positive impactions for society.

For textile industry example, automation is nothing new, it has shaped the workplace to replace human jobs to boost productivity in the textile industry. Textile machines have had a generally positive impact over gears, creating value and allowing textile workers to take up more rewarding age will likely continue to create opportunities and lead to new textile industries, companies and textile occupations. It may also compensate for a demographically driven slowdown in the growth of the textile workforce. The future impact of textile (AI) and automatic and internet link is somewhat uncertain. It seems textile industry will be trend to accept (AI) textile workers and internet of thing to replace traditional manual textile workers to produce any shirts, cloths etc. wearing products in factories popularly, during the (AI) textile machine and internet thing technology can be invented to reach the mature stage in the future.

For factory worker transportation job example, they have also expanded their influence, migrating from the factory floor to the service sector and taking the place of humans in a range of activities from financial transactions to transport route optimization. Further (AI) machines and robots are increasingly programmed to learn, meaning they improve with time and undertake cognitive activities. Hence, (AI) machines and internet technology enable automation of work activities to raise factory workers' efficient and performances, also factories can reduce manual worker numbers, due to (AI) machine workers' assistance.

Future, (AI) robotics technologies and internet technique have these different kinds of characteristics: For soft robotics example, it is non-rigid robots construct with soft and deformable materials that can manipulate items of varying size, shape and weight with a single device. For swarm robotics, it coordinated multi-robot systems often involving large numbers of mostly physical robots. For touch/factile robotic example, it robotic body pails (often biologically inspired hands) with capability to sense, touch , dexterity robots example, serpentine robots with many internal degrees of freedom to threat through tightly packed spaces for humanoid robots example, robots physical is similar to human being often bi-pedal that investigate variety capable of performing human tasks , including movement across terrains, object recognition, speech sensing etc. For autonomous cars and trucks example, it is capable of operating with a human pilot, e.g. the unarmed general atomics Predator XPUAV with roughly half the wingspan of a Boeing 737 can fly autonomously for up to 35 hours from take-off to landing, for unmanned aerial vehicles example, flying vehicles capable of operating without a human pilot, the unarmed general atomics predator -XPUAV , with roughly half the wingspan of a Boing 737, and fly autonomously for up to 35 hours from take off to landing, for (AI) chat bots example, (AI) systems designed to simulate conversation with human users, particularly those integrated into massaging apps.

In Dec. 2015. the general service administration of the US Govt. described how it used a chat bot named Mrs. Landingham (a character from the television show the west wing) to help onboard new employees. Finally, for robotic process automation example, class of software robots that replicates the actions of a human being interacting with the user interfaces of both software systems. Enables the automation of many back-office work flows without requiring expensive IT integration . Hence, future (AI) robot machine men will have different functions to be applied to different industries to use in possible.

Statistics Denmark shows that (AI) automation potential robots will influence few jobs are completely automatable , but close to half consists of 40% automatable tasks: It showed example occupations include share of automated, such as brewing machine operators are more than 80%, logging equipment operators are more than 50%, roofers , stock tasks clerks, travel agent are more than 50%, farmers , nursing assistants are more than 30%, physicians, teachers , managers are more than 10%.

For example, humans perform a wide variety of tasks from planting corn to examine spreadsheets, meeting clients and lifting crates in a store. Each of these actions requires a combination of innate or acquired capabilities, internet technique assistance, ranging from social perceptiveness to fine motor skills and natural language understanding. To understand and map automation feasibility by existing technology. Mc Kinsey has developed a framework of 18 technical capabilities that can substitute tasks performed by humans. The capabilities are grouped in five categories: sensory, cognitive, language, social and emotional and physical. So, it seems (AI) robot machine men and internet technique will have possible combination to invent to own human's emotion , language, learning, task skill abilities.

Mckinsey global institute analysis also showed current technologies have achieved different levels of human performance across 18 capabilities include: sensory perception, autonomously infer and integrate complex input using sensors, cognitive capabilities reorganizing known patterns/categories supervised learnings, generating novel, logical reasoning/problem solving, optimization and planning, creative, information retrieval, coordination with multiple agents, output articulation/presentation, national language processing, social and emotional capabilities-natural language understanding, social and emotion sense, reasoning output, physical capabilities-fine motor skills, navigation mobility. Hence, it seems (AI) robots and internet technological will combine to invent to own human' some skills to replace human to do some kind of tasks in possible.

In conclusion, future (AI) robot and internet will be needed to link to cooperate together to raise human's work efficiency in popular.

How artificial intelligence replaces human job possibility

What is the risk of automation for jobs to replace human job? In recent years, there has been a revival of concerns that automation and digitalization night after all result in jobless future. As I argue, this might lead to an overestimation of job (AI) automate , as occupations labelled as high-risk occupations often still contain a substantial share of tasks that are hard to automate.

For example, when the share of (AI) automatable jobs is 6% in Korea, the corresponding share is 12% in Australia. Differences between countries may reflect general differences in workplace organization, differences in previous investments into (AI) automation technologies as well as differences in the education of workers across countries. I also discover that (AI) automation and digitalization are unlikely to destroy large numbers of jobs. But, however, low qualified labors are likely to raise costs as the (AI) automate of their jobs is higher compared to highly qualified workers.

In fact, (AI) technology will influence some new technology to replace some human's job, such as driverless car, the largely autonomous smart factory , service robots or 3D printing. These technologies are driven by advances in computing power, robotics and artificial intelligence and ultimately redefine what type of human capabilities machines are able to do.

Hence, question brings whether (AI) invention will influence general human jobs to be replaced by (AI) autonomous jobs? Whether will the potential foe automation with actual employment loss? In particular, the technical possibility to use (AI) machines rather tasks need not mean that the substitution of humans by machines actually takes place.

Whether (AI) technology replaces human's some job, it is beneficial to our society or not. Instead, machines are increasingly capable of performing non-routine cognitive tasks, such as driving or legal writing . In particular, advances in the field of machine learning (ML), e.g. computational statistics and visions, data mining, artificial intelligences allow for automating cognitive task, when the use of (ML) in mobile robotics (MR) also allows for automating certain manual tasks. So, it seems, (AI) technology can replace some labor job, e.g. warehouse transportation, even mind's job, e.g. legal writing, driving in possible.

For example, if (AI) automatic non-manual driving can reduce hurt or death risk, it is beneficial to our society, or (AI) automatic robots can more any heavy things ( products) in warehouse safely. Then, it can reduce the warehouse labor's bodies hour risk, it is beneficial to the workers. Even, if (AI) robot can write any legal documents, no any word errors in short time. It is beneficial to the law companies , but it also bring unemployment chance, due to these jobs can be replaced by (AI) robots to do in the future. Hence, it will cause some occupation to be disappeared, due to (AI) robots can do our these kinds of jobs in the future.

Frey & Osborne (2013) reported these kinds of occupations will be replaced by (AI) robots in possible. They include computer, engineering, financial, management, legal , art and medium, community service, education, healthcare practitioners and technical service, sales and related, office and administrative support, farming, fishing and forestry, construction and extraction, installation, maintenance, and repair , production, transportation and material moving. It seems our future some professional occupations will have possible to the replaced by (AI) robots to replace, instead of labor jobs. Hence, (AI) robots technology will have much trend to replace high knowledge or low knowledge skillful labors in the future.

In conclusion, it implies that only using information on task-usage at the individual level leads to significantly lower estimates of jobs " at risk", some workers in occupations with according to high automate nevertheless often perform tasks with are hard to automate. Why can (AI) replace human to do some kinds of jobs? (AI) artificial intelligence refers to the ability of a computer or a computer enable robotic system to process information and produce outcomes in a manner similar to the thought process of humans in learning, decision making and solving problem. By extension, the goal of (AI) systems is to develop systems to capable of tasking complex problems in ways similar to human's logic and reasons who feel in our future. Hence, it means future (AI) robots has effort to replace human to do any jobs in possible.

(AI) development second stage
(AI) directions for future non-manual
control road vehicles market

Future road vehicle products and technologies must meet social, economic and environmental protection and driving safety goals , and satisfying market requirements for mobility, accident reducing, performance, cost desirability. Thus, (AI) auto-non manual control vehicles need to be followed this direction to invent. To satisfy future driver's safety of needs, enhanced vehicle speed desired functional performance of road transportation system, required and desired technological response, including research needs. It is long term up to 20 years vision, for (AI) auto non-manual research. Thus, (AI) auto non manual vehicle manufacturers need often to revise their (AI) vehicles functions to raise to improve their system performance and driving industry driver's needs, e.g. private drivers need or public transportation driver's need or business client's need. Hence, future (AI) transportation will need have individual driving consumer and business driving consumer both targets.

Thus, future (AI) automation manual control vehicles need to deliver high impact technology solutions to meet social , economic and environmental and safe goals. Engine needs to be improved efficiency , performance, drivability, reliability, durability and speed-to-market together with reduced emissions and cost; hybrid, electric and alternatively fuel (AI) non manual control vehicle technology development, leading to new fuel and power systems, such as hydrogen, fuel cells and batteries, which satisfy future social, economic and environmental and safe goals. Software, sensors, electronics and telematics technology development are needed to be lead to improve vehicle performance, control and adaptability, intelligent , mobility and security, structure and materials technology development, leading to improved safety, performance and leading to flexibility with reduced cost and environmental pollution to achieve the (AI) non manual drivers to feel (AI) vehicle performance, auto control and adaptability is better to compare traditional manual driving vehicles.

In fact, in traditional manual driving market, Japan and USA had had over 80% of world car production by six major global groups. In the future, it is possible only USA can dominate (AI) non manual auto driving vehicle manufacturing market if Japan had no effort to manufacture any (AI) auto non manual control vehicles. So, it means that it is only Japan is USA potential (AI) auto non manual control vehicle manufacturing competitors. Also, it means that it is only USA has effort to export (AI) auto non manual control vehicles to global (AI) auto non manual control vehicle market.

Thus, in long term, (AI) non manual control vehicle product market development, USA (AI) vehicle manufacturers will have these requirement to win new technological competition to traditional manual control vehicle. The requirements include: low cost fuel, low carbon, fuel cell and telematics technologies, the technological roadmap function, such as detailed consideration clear provision of other important areas to the drivers. When the (AI) non manual auto drivers are sitting in the non manual control auto vehicle. Although, who does not need to drive, but who need to know how to go to anywhere by electric road map show clearly. So, the driver won't lose direction and he/she can know the (AI) non manual control vehicles is driving to anywhere in any time, even when who is sleeping.

In the future, the (AI) non manual control vehicles need to be invented to satisfy any business , transportation clients' needs, instead of individual clients needs, e.g. cans, trucks, buses, emergency and utility vehicles, trains, trams etc. Hence, technological road mapping is one important tool to help any business, transportation ( AI) non manual control vehicle clients. Technological roadmap is a technique that is used in industry to support strategic planning for (AI) non manual driving vehicles in the future. Electronic road maps generally take the form of multi-layered time based charts, linking technology developments to future (AI) non manual control vehicle market requirements.

Technology road mapping is a flexible technique and the roadmap architecture and process for developing the roadmap most generally be customized to meeting the particular aims. Why technology roadmap will be popular to

(AI_ non manual control vehicles. It's advantages include: It is a technology solutions and options that can enable the performance targets to be achieved engine hybrid, electric and alternatively fueled vehicles, software, sensors electronics and telematics, structures and materials design and manufacturing process. It is road transport system performance measures and targets tool, in response to the trends and get (AI) non manual control vehicle drivers to get society, economy, environment protection, low cost driving benefits, also it can help any transportation clients to know how to go to anywhere clearly. Hence, technological road map will be one good tool to assist (AI) non manual control auto vehicle to develop future road driving market.

Reference(source)

Frey & Osborne (2013), The future of employment: How susceptible are jobs to computerization? University of Oxford.

Mckinsey Global Institute Analysis

Statistics Denmark, Global automation impact model, Makinsey analysis

(AI) -driven automation industry development

(AI) -driven automation industry will create wealth and expand economy growth to any countries, but it will be accompanied by changed in the skills that workers need to learn. One of main ways that technology increases productivity is by decreasing the number of labor hours needed to create a unit of output. It implies (AI) technology will influence low educated and low skillful labor number to be decreased ( reduction employment number).

In contrast, technological change tended to work in a different direction throughout the nowadays. The advance of computer and the internet raised the relative productivity of higher skilled workers. So, routine-intensive occupations that focused on predictable tasks disappearance, such as switch board, operators, filming checkers, travel agents and assembling line workers etc. were particularly replaced by new technologies.

However, today, it may be challenging to predict exactly which jobs will be most immediately affected by (AI) driven-automation. The reason is because (AI) is not a single technology, but rather a collection of technologies that are felt unevenly through the economy to influence job changing both negatively and positively. In positively view point, (AI) driven-automation will make many workers more productive and increase demand for certain skills. Consequently, new jobs are likely to be directly create in areas , such as the development and supervision of (AI) as well as indirectly created in a range of areas throughout the economy as higher incomes lead to expanded demand. Otherwise, in negatively view point, many traditional human needed ( demand) skillful jobs will be threatened by automation are highly concentrated among lower-paid, lower-skilled and less -educated workers. It means automation will cause pressure on demand for this group, pressure and employment, if (AI) can replace the low skilled and less educated workers' jobs. Thus, (AI) will have negative influence to impact on the labor market.

(AI) capabilities will enable automation of some tasks that have long required human labor. Why can (AI) replace some simple human jobs? For example, advances in robotics are expanding machines' abilities to interact with and sharp the physical world. Combined , (AI) and robotics will give rise to smarter machines that can perform more sophisticated functions than ever before and brings more advantages that humans have exercised. This will permit automation of many tasks now performed by human workers and could change the shape of the labor market and human activity.

5.1 How (AI) influences labor market

Today, it may be challenging to predict exactly which jobs will be most immediately affected by (AI)-driven automation. Because (AI) is not a single technology, but rather a collection of technologies that are applied to specific tasks.

Some specific predictions are possible based on the current (AI) technology. For example, driving jobs and house cleaning jobs, bank counter service jobs, telephone enquiry service operators. Restaurant cooking jobs, simple accounting record service jobs etc. that require relatively less education to perform. Advancements in computer vision and related technologies have made the feasibility of fully appear more likely, potentially displacing some workers in driving-dominant professions. Seemingly similar robot, for which the operational tasks is less specific of navigating to a specific destination when following a set of given rules and preserving safety.

In the future, the effects of (AI) on the labor market in the decade ahead will continue the trend toward skill-biased change that computerization and communication innovations have driven in recent decades. Thus, some human driving occupation will be disappeared or replaced by (AI) automation driven. For example, bus drivers, light truck or delivery services drivers, heavy and tractor-trailer truck drivers, school drivers, tax drivers, travel bus drivers.

However, (AI) technology could enable some workers to focus time on other job responsibilities, boosting their productivity, and actually raised wage growth among those still holding the reshaped jobs. For example, salespeople, who currently spend a considerable amount of time driving could find themselves able to do other work when a car drives them from place to place, or inspectors and appraisers could fill out paperwork, when their car drives itself. This (AI) -driven technology should make these workers more productive, with (AI) -driven technology serving as a complement, not a substitute. New jobs will also likely be created, both in existing occupations cheaper transportation costs with lower prices and increase demand for products and all the related occupations, such as service and fulfillment, and in new occupations not currently foreseeable.

What kind of jobs will be created by (AI) technology? Predicting future job growth is extremely difficult, due to it depends on technologies or substitute for existing today as well as they may complement or substitute for existing human skills and jobs. However, (AI) will also lead to substantial indirect job creation to the degree it raises productivity and wages, it may also lead to higher consumption that would support additional jobs from high-end draft production to restaurant and retail. The future(AI) " augmented intelligence", the technology's role is as assisting and expanding the productivity of individuals rather than replacing human work. Thus, based on the biased-technical change framework, demand for labor will likely increase the most in the areas where humans complement (AI) automation technologies. For example, (AI) technology , such as IBM's Watson may improve early detection of some cancers or other illnesses, but a human healthcare professional is needed to work with patients to understand and translate patients' symptoms, inform patients of treatment options, and guide patients through treatment plans. Shipping companies may also partner workers who pick up and deliver products over the last feet with (AI) enabled autonomous vehicles that move workers efficiently from site to site. In such cases, (AI) augments what a human is able to do and allows individuals to either be move effective in their specially task or to operate on a larger scale. Thus, it seems (AI) technology will also create new jobs, raise productivities and workers' efficiencies.

5.2 Redefining management in
the workforce of artificial intelligence

In the future, due to artificial intelligence influences to some kind of human jobs nature. So, the kind of human jobs of management methods will also need to change to adapt the artificial intelligence technology input to their organizations. It will cause challenges for every executive and manager if who won't have effort to manage their teams how to apply artificial intelligence technology to work efficiently and easily. For example, division of labor will change among humans and machines will increase. Thus, companies will have to adapt their training performance and talent strategies how to emphasize on work that how to make human judgment and skills and experimentation. Thus, (IA)'s greatest impact will be on administrative coordination and control tasks, such as scheduling , resource allocation.

In fact, mangers will encounter this challenges: How to apply human experience and expertise to judge critical business decisions and practices when the information available is insufficient to suggest a successful course of action? Due to this kind of work will require new skills and mindsets. I shall indicate these change management methods to adapt (AI) technology. Such as: administration and routine tasks, scheduling , allocation of resources and reporting will fall within the intelligence machines, responsibilities that have long been reserved for humans. For example, a typical store manager or a lead nurse at a nursing home most constantly arrange shift schedules, accounting for staff members' absences owing to illness, vacation time or sudden departures.

Thus, the managers need to learn how to arrange new division of labor within the organizations after (AI) technology had been implemented to the organization. Artificial intelligence is currently influencing into once considered exclusive to humans: assessing and acting on human emotions and personality traits. The influences to managers need to change their strategies to adapt (AI) technology implements include such as below:

Firstly, managers need to spend the bulk of their time on coordination and control tasks from intelligent system implements. Their time spending on these major three aspects from impact of intelligent system: coordinate and control, solve problems and collaborate and people and community , strategy and innovation three aspects. Thus (AI) will influence managers need to change their judgment method to teach whose teams how to adapt the (AI) system operations in any organizations.

Secondly, (AI) will influence top, middle and low level management needs to change to adapt the (AI) technology operations to any owned (AI) technology organizations in the future. Intelligent machines must be trained in context. Just like humans , on-the-job training is a requirement for such machines because they typically arrive with only very general capabilities. To get the most from (AI), managers at all levels must participate in the instructional experience and in the learning process and provides managers' familiarity with such systems on these aspects, e.g. How the system works and generate advice, how the system has a proven track record , how the system provides convincing explanations , how the system can make simple rule- based decisions.

Thirdly, managers need to learn how to make judgment more accurate (AI) systems assistance. Although (AI) will invariably take on more routine work and even augment human decision-making, it won't judgment work, the application of human experience and expertise to critical business decisions when the information available is insufficient to suggest a successful course of action or reliable enough to suggest an obvious course of action. For a sense of the nature of judgment work, consider big data marketing and sales analytics. Such analytics often provide insights that can inform promotional campaigns, including predicting which promotions will generate desired sales brand further into the future, marketing executives need use judgment, combining analytics with their own and others' insight and experience.

The application of experience and expertise to critical business decisions and practice represents the real value of human judgment. But, when artificial intelligent machines are implemented to any organizations to assist the low, middle and top level management to make any business judgment. These forms of judgment work that managers can gather data interpretation, idea development more absolute from (AI) machine assistance. Thus, why these level management executives need to learn how to apply (AI) machines to help them to make any business judgment more accurate.


5.3 How (AI) influences organizational change

Consequently creative and social intelligence will be in even greater demand as (AI) makes in management and the workforce. This development will represent a long term trend in labor markets , one characterized by intensifying demand and reward for social skills with a growing desire for creative capabilities, managers will seek to fashion of ideas and hypotheses from inside and outside of the enterprise to shape solutions to their most pressing business problems. Thus, (AI) will influence overall organizational team members who have chance to participate any decision to make more accurate business judgment.

Many managers mistakenly view judgment work as only an individual discipline, failing to appreciate that it can also involve decide interpersonal and organizational practices. In more complex settings, judgment is typically a collective outcome of individuals' and teams' diverse perspectives, insights and experiences. And often , the resulting choices are better informed than decisions that an individual would have arrived at on his or her own.

Thus, when any organizations apply (AI) technology to assist managers to gather data and ideas to make any judgment. In these cases, organizations can create the conditions for effective collective judgment by establishing structures , such as " shadow advisory boards" that prompt managers and employees to source and synthesize multiple perspectives. Thus, a traditional organization (firm) might freshen its thinking is t put together a shadow advisory board, comprised of young, digital people who can apply (AI) machine assistance to make judgment work more accurate whether related to people development, problem-solving or strategizing and innovating for considerable degrees of creative and social intelligence.

Thus, on the one hand, (AI) technology machine augmentation and automation can give these advantages to human (organization managers) , e.g. developing people and community, solving problems and collaborating, coordinating and controlling work, shaping strategy and leading innovation. Besides, on the other hand, the next generation managers need have these individual attitude to treat intelligent machines to be as colleagues.

When, judgment is a human skill, intelligent machines can accelerate human learning that supports it, assisting in data -driven simulations, scenarios and search and discovery activities. Focuses on judgment work, some decisions require insight beyond what data can tell them. This is the sweet sport for human judgment, the application of experience and expertise to critical business decisions and practices. Thus, managers will also need to find ways to learn how to use digital (AI) technologies to tap into the knowledge and judgment of partners, customer external stakeholders and role models in other industries after the (AI) machine had been implemented to the organization.

5.4 Future works change:

Automation, employment

and productivity

Human future " micro to macro" industry trends will be affected business strategy and public policy by (AI) technology. In the future (AI) technology will influence those six themes: productivity and growth, natural resources, labor markets, the evolution of global financial markets, the economic impact of technology and innovation and urbanization. However, (AI) technology will bring economic benefits of tackling gender inequality, a new global competition, Chinese innovation and digital globalization.

Nowadays, advances in robotics artificial intelligence, and machine learning are in a new age of automation, as machines match or outperform human performance in a development to any countries. For example, automation of activities can enable businesses to improve performance by reducing errors and improving quality and speed, and in some cases achieving outcomes that go beyond human capabilities. For example, some research indicated automation could raise productivity growth globally by 0.8 to 1.4 % annually; more than 2,000 work activities across 800 occupations. When less than 5% of all occupations can be automated using demonstrated technologies about 60% of all occupations have at least 30% of constituent activities that could be automated. Many occupations will change that will be automated away: Activities most susceptible to automation involve physical activities, in highly structured and predictable environments, as well as the collection and processing of data. They are most prevalent in manufacturing , accommodation and food service and retail trade and include some middle-skill jobs. For example, such as natural language processing is a key factor. Beyond technical feasibility, the cost of technology competition with labor including skills and supply and demand dynamics, performance benefits including and beyond labor cost savings, and social and regulatory acceptance will be affected by (AI) automation technology. Thus, (AI) automation will impact to influence global employment in those aspects as below:

Firstly, assuming that people are displaced by automation will find other employment. The anticipated shift in the activities in the labor force is of a similar order as the long-term shift away from agriculture and decreases in manufacturing share of employment. Both of manufacturing and agriculture industries which would be accompanied by the creation of new types of work not foreseen at the time.

Secondly, for business, the performance benefits of automation are relatively clear. Thus, the businessmen have opportunities for their micro economies to benefits from the productivity growth potential and macro economies to benefit to encourage continued progress and innovation , investment and market incentives. At the same time, employers must innovate policies to help workers and institutions adapt to the impact on employment.

This will likely include rethinking education and training, income support and safety nets , as well as support for those dislocated, when employees need to leave themselves homes to move to other cities to learn new (AI) automation works. Thus, individuals in the workplace will need to engage move comprehensively with machines as part of their everyday activities, and acquire new skills that will be in demand in the new automation age. Consequently , the scale of shifts in the labor force over many decades that automation technologies can be a similar order to the long -term technology -enables shifts in the developed countries' workforces away from agriculture in the 21 th century. Those shifts did not result in long-term mass unemployment because they were accompanied by the creation of new types of work not foreseen at the time. However, human will still be needed in the workforce when the total productivity gains are caused by (AI) technology.

5.5 What occupations will be influenced by (AI) technology.

In the future, scientists predict that these occupations will be influenced by (AI) technology mostly. They include : retail salespeople, food and beverage service workers, language or translation teachers, health practitioners. Since these work activities have a more relevant occupations are made up of a range of activities with different potential for (AI) automation . For example, a retail salesperson will spend more time interacting with customers, stocking shelves , or ringing up sales. Each of these activities is distinct and requires different capabilities to perform successfully.

Thus, these job activities have similar simple control characteristics. Simple activities include greet customers, answer questions about products and services, clean and maintain work areas, demonstrate product feature process sales and transactions. All these activities can have similar simple activities in order to (AI) machines can be learn how to do these activities from (AI) technology . For example, the capability perception includes sensory perception, cognitive capabilities, such as retrieving automation, recognizing known patterns( supervised learning), logical reasoning problem solving.

Thus, (AI) machine is such human, which has feeling and emotion, such as social and emotional sensing, judgement reasoning methods, natural language understanding and physical capabilities, such as mobility , navigation, gross motor skill, fine motor skills. It seems that the future, (AI) human invents machines which will have these human characteristics to do human similar behavioral job duties more easily and efficiently. It implies these above human occupations will be replaced by (AI) human invention machines in the future. Due to (AI) creation, it is possible to cause unemployment number of these above workers will increase because (AI) machines can do their similar job behavioral activities.

Consequently, employers won't need to employ many of these skillful labor. Otherwise, they can buy less number (AI) machines to attempt to do whose job activities more easily and efficiently. So, it seems (AI) machines will have more high work performance to replace these occupation workers' work performance. Finally, these occupation worker unemployment number will only increase when the (AI) machines had been invented to achieve to do their work behavioral activities absolutely success in the future.

5.6 Whether (A) technology machine labor will replace human worker more or assist human worker more

There is no single agreed definition of a robot how outcome of a task that is completed without human intervention. When some definitions require the task to be completed by a physical machine moves and respond to its environment, other definitions use the term robot in connection with tasks completed by software , without physical embodiment.

However, to answer the question : Whether (AI) technology machine labor will replace human worker more or assist human worker more. I shall indicate some examples to let readers to judge whether (AI) technology can create new jobs or reduce old jobs.

Firstly, I shall explain what (AI) function is. (AI) is a service robot that performs useful tasks for humans or equipment excluding industrial automation application . Thus, the classification of a robot into industrial robot or service robot is done according to its intended application. It is also a personal service robot or a service robot for personal used for a non commercial task, usually by lay persons . Examples are domestic servant robot, and pet exercising robot. It is also a professional service robot or a service robot for professional used for a commercial task, usually operated by a properly trained operator. Examples, are cleaning robot for public places, delivery robot in offices or hospitals, fire-fighting robot, rehabilitation robot and surgery robot in hospitals. Thus, these functions will be future (AI) application to our daily life necessaries or business necessaries.

However, some authors agree (AI) will bring negative outcomes of automation, due to raise competiveness, reduce human job nature. Otherwise, other authors argue (AI) will bring positive outcomes of automation, due to raise productivities, job creation, assist humans work.

On the positive outcome hand, robots can increase productivity . This is particularly important for small-to medium sized businesses both are in developed and developing countries economies. It also enables large companies to increase their competitiveness through faster product development and delivery. Increased use of robot is also enabling companies in high cost countries to re shore, or bring back to their domestic base parts of the supply chain that will have previously outsourced to sources of cheaper labor. Currently , the greater threat to employment is not a automation, but an inability to remain competitive. Automation has led overall to an increase in labor demand and positive impact on wages. The reason is that the middle-income/middle-skilled jobs have reduced as a proportion of overall contribution to employment and earnings leading to fears of increasing income inequality, the skills range within the middle income bracket is large. Thus, robots are driving an increase in demand for workers at the higher -skilled and with a positive impact on wages. This issue is how to enable middle-income earners in the lower-income range to unskilled or retain. Finally, the (AI) positive impact supporter who argue the future will be robots and humans can work together.

However, on the negative outcome hand, robots can substitute labor activities, but don't replace jobs. They believe that less than 10% of jobs are fully automatable. Increasingly , robots are used to complement and augment labor activities, the net impact on jobs and the quality of work is positive. Automation can provide the opportunity for humans to focus on higher-skilled, higher-quality and higher-paid tasks. Robots can improve productivity when they are applied to tasks that which perform more efficiently and to a higher and more consistent level of quality than humans. For example, increased productivity is enabling some firms, such as Whirlpool, Caterpillar and Ford Motors company in the US restructure their supply chains, bringing back parts of the manufacturing process to the country of origin. Thus, productivity gains due to robotics and automation are important not just at the company level, but also for build industry and nation competitiveness.

I suppose that productivity can be raised. What are the impacts of robots on employment? Firstly, the main focus of development has been on personal entertainment, which does not drive worker productivity ( manufacturing production). When the internet ( information and communication technology (ICT)) innovation. This is borne and by findings that manufacturing productivity, which has been driven by innovations in automation rather than consumer technologies, has government strongly than productivity in the services sectors of the economy in most nature economies. It seems (AI) automation will create many jobs in internet communication entertainment game industry. For example, many young people like to use internet to play any electronic games from computer or mobile at home or outside home conveniently. Thus, (AI) automation will increase demand to be invented to any new entertainment game from internet channel. It will need to employ many (AI) entertainment game inventors to create many automation entertainment games. Thus, (AI) automation in internet entertainment game industry will need human (AI) entertainment game inventors to invent the knowledge-based capital of (AI) automation entertainment games. The (AI) entertainment game inventors will need own research and development skills, form specific skills, organizational know-how skills, databased knowledge, design and various forms of intellectual property to do these (AI) automation entertainment game invention occupations in the future.

International Federation Of Robotics(2016) indicated that China will be as a major robotics manufacturer and user of robots, benefiting from jobs created by robot manufacturing and productivity gains from robot use. Chins had sold of robots to any one single market every year since 2017 year. The Chinese government has included a focus on robotics in its 10 year strategy. In order to achieve its target of a robot density of 150 units per 10, 000 workers by 2020 year. Thus, Chinese companies will have to install around 650,000 new industrial robots between 2016 to 2020 year, 2.5 times more than installed globally in 2015 year.

Hence, China (AI) manufacturing industry will need to employ many workers . It implies (AI) manufacturing industry will create many new occupations in China. Also, ministry of economy, trade and industry (2015) also showed that Japan currently has the largest stock of industrial robots in operations, primarily in the automation industry. Driven by a rapidly aging population and low productivity rates, the Japanese government has sights on a 20-fold increase in the use of robots in the non-manufacturing sector and a three-fold growth rate of labor productivity in the service sector both by 2020 year. Thus, it also implies Japan will need many robots to be provide to service industry. Due to robots will provide to serve any businessmen's clients. Thus, it is possible that the service workers won't be dismissed as well as it is depended on the serving job nature to decide whether Japan's service workers can still serve to their employer when the service (AI) robots are applied to whose employers. Consequently, it seems that (AI) can create employment, Ministry of economy, trade and industry (2015) showed that such as China will develop the major (AI) automation manufacturing industry. The (AI) employers will need to employ many workers to manufacture any these different kinds of (AI) robots to satisfy China or overseas individual or business buyers needs. But, (AI) can also cause unemployment to the low skillful service workers. Such as if Japan some service businesses choose to buy any (AI) service robots to replace their service staffs to serve their clients. It is possible that the service staffs will be dismissed, due to (AI) robots can do such as their same service job duties to achieve better service performance.

Thus, today, it is increasingly common for people to use robots in various situations at home and in retail stores, hotels and hospitals these service industries. Robots are classified into server types based on their functionality ( service and utility robots or those designed to communicate with humans) and appearance ( humanoid robots or mechanical robots). The type of robot, to which each country allocated particular importance in the advance of robotics, reflects the sense of values and preferences of its population. Thus, if the country has high population needs to use robots, then they will influence either more new jobs creation or more old job loss in the country's (AI) manufacturing or (AI) service industries both. For example, Japan respondents often associate the term " robot " with humanoid robots that can communicate with human and they have a high level of familiarity with robot. The US has the highest level of robot utilization at home and in retail stores with its people being the most enthusiastic about the future use of robots. Germany shows a strong tendency to consider robots for industrial purposes and its people feel strong effort to the presence of robots in their households.

In conclusion, to judge whether how (AI) will influence the country's employment to be better or worse. It will depend on the country home buyers (users) or business buyers (users) how to use (AI) for their daily needs. If the country , such as US retail stores need to use (AI) , it will have possible to reduce some or many retail service workers. Even, if the country , such as Japan has many home users need to use (AI) , it will not influence the employment market. Otherwise, it will raise (AI) salespeople numbers. Even, if the country, such as Germany and China will have many (AI) manufacturers, then it will create many (AI) manufacturing occupations for these (AI) manufactory workers.

Consequently, (AI) robots manufacturing and service needs will have positive or negative impact to any country's employment. It will depend on the (AI) service provision and service workers' job nature as well as the manufacturing workers of (AI) knowledge level to decide their employment chance in their country's employment market.

What does artificial intelligence(AI) mean
● What (AI) function is?
Some scientists explain that artificial intelligence means which is an expert system, computer software that embodies a portion of the specialized knowledge of a human portion in a specific, narrow domain, owns decision making ability of human expert. The (AI)technology is based on the premise that what makes a person an expert is years of experience that enables who recognizes certain patterns in a problem as being similar to pattern. For example, in the future artificial intelligence system can be applied to control air traffic, design to computer configuration, medical diagnosis, instruction/training, speech/interpretation, monitoring to ( nuclear plant), planning to mission, factory scheduling, prediction weather, repairing telephone, automatic driving etc. different industries.

Artificial intelligence characteristics include: creative, adaptive , common sense, fact processing, quick replication, broad focus permanent and consistent skill. Otherwise, traditional computer expert system disadvantage includes perishable, unpredictable, slow reproduction, expensive, slow reproduction, slow processing lacks inspiration, needs instruction, narrow focus only machine knowledge. So, artificial intelligence is a branch of computer science devoted to creating computer to influence software and hardware to attempt to create human intelligence or human intelligent behavior. It is learning from experience, responds flexibility in situation that are, new or not anticipated.

Thus, (AI) can be learnt programmed knowledge to solve problems, using reasoning in solving problem, understanding and inferring facts and rules, recognizing the relative importance of different elements in a situation.

In summary, artificial intelligence is concerned with two basic ideas mainly: The first idea, it involves studying the thought processes of humans to understand what intelligence is; the second idea, it deals with representing thought processes using companies to create artificially intelligent entities for testing the theories of intelligence.

- Can (AI) impact human job nature?

Human need concern this question: Will artificial intelligence (AI) reduce some human jobs in order to instead of replacing machines to do? Due to artificial intelligence is the ability of machines to do thing, that people would require intelligence. For example, artificial intelligence machine man driving( self-driver), it (AI) machine man driving research is an attempt to discover and describe aspects of human intelligence that can be simulated by driving machine functions. Alternatively, (AI) mathematical research may be another viewed as an attempt to develop a mathematical theory function to describe the abilities and actions of things ( natural or man-made) exhibiting intelligent behavior and server as a design of intelligent calculation machine function.

Why do humans need artificial intelligence machines to instead of traditional human service job? For example, can artificial intelligence machine man (self-driving) driver drive to replace human driver? I shall compare the differences between humans and computers : The characteristics of humans are good at recognizing various things, either seen before or not, recognizing the relationship patterns between things. Human thinking is common sense reasoning, combining all types of sensory input, acting appropriately in novel situations, learning new things and changing behavior patterns, making decisions , even when given incomplete information, working with noisy, incomplete information gathering behaviors . However, characteristics of computers are good at: The tasks humans do naturally are extremely difficult for a computer program as intelligent, which must be able to do the same kind of tack as humans do naturally.

Hence, (AI) is an combination of many different success and technologies: Linguistics - computational and socio, philosophy-logic, philosophy of mind and of language, electronical engineering -image and speech processing, pattern recognition, robotics, machine learning, neural networks, optimization scheduling, management information system and decision making. So, it is possible that (AI) can impact human job nature to instead of human working behavior in the future.

- How can human society job nature

to be changed to artificial intelligent society?

From the first intelligent perspective reason view point, artificial intelligence is making machines " intelligent" acting as humans expect people to act. Artificial intelligence has ability to distinguish computer responses from human responses, it owns knowledge to solve expert problem. From another research perspective reason view point, artificial intelligence is the study of how to make computers do things which, at the moment, people do better ( Rich & Knight, 1991, p.3).

(AI) researchers are native in a variety of domains, e.g. formal tasks ( mathematics, games), tasks ( perception, robotics, natural language, common sense reasoning), expert tasks ( financial analysis, medical diagnostics, engineering, scientific analysis and other areas).

From the second business perspective reason view point, (AI) is a set of many powerful tools, and methodologies for using those tools to solve business problems. From a programming perspective reason view point, (AI) includes the study of symbolic programming problem solving and search .

From the third human technological perspective reason view point, today's computer can do many well-defined tasks, for example, arithmetic operations, are much faster and more accurate than human beings. However, the computers' interaction with their environment is not very sophisticated yet. How can human test whether a computer has reached the general intelligence level of a human being? Can a computer convince a human interrogator that it is a human? But before thinking of such advanced kinds of machines, human will start developing our own extremely simple " intelligent" machines.

So, it is possible that human society job nature will to be changed to artificial intelligent society when (AI) technology is developed to the mature stage in the future.

- Why does human need artificial intelligence machines?

One of major division in (AI) is between humans who think (AI) is the only serious way of finding out how we (

human) work and human who want companies to do very smart things, independently of how we ( human) work. This is the important distinction between cognitive scientists vs engineers. One of another major division in (AI) is between symbolic (AI), which represents information through symbols and their relationships. Specific Algorithms are used to process these symbols to solve problems or deduce new knowledge and connectionist. So ( AI) , which represents information in network. Biological processes underlying learning, task performance and problem solving are imitated from human mind behaviors.

Thus, it is possible that artificial intelligence machines can do the better judgicious behavior to compare human.

● How does artificial intelligence influence future working changing in automation employment and productivity aspects?

In the automation changing influence aspect, as companies increasingly use robots on production lines or algorithms to optimize their logistics manage inventory, any carry out other core business functions. Technological advances are creating a new automation age in which ever-smarter and more flexible machines will be deployed on an ever larger scale in the marketplace. However, researching artificial intelligence with how influences human working nature. We need to answer these questions: How will automation transform the workplace? What will the implications for employment? And what is likely to be its impact both on productivity in the global economy and on employment?

Advances in robotics, artificial intelligence, and machine learning are growing in a new age of automation as machines match or outperform human performance in a range of work activities, including ones requiring cognitive capabilities. What factors are determined the changing in workplace adoption by artificial intelligence innovation? What advantages are automation? Automation of activities can be enabled businesses to improve performance by reducing errors and improving quality and speed, and achieving outcomes that go beyond human capabilities.

Some scientists indicated based on their scenario modeling. They estimated automation could raise producing growth globally by 0.8 to 1.4 percent annually. Almost, the activities people are paid almost \$16 trillion in wages to do in global economy have the potential to be automated by adopting currently demonstrated technology. According to their analysis of more than 2,000 work activities across 800 occupations. When less than 5% of all occupations have of least 30% of activities that could be automated. They also indicated that technical economic and social factors will determine automation. Continued technical progress, for example, in areas such as natural language processing is a key factor beyond technical feasibility , the cost of technology, competition with labor including skills, and supply and demand dynamics, performance benefits including and beyond labor cost savings and social and regulatory acceptance will affect ( alter) the scope of automation.

Other some scientists also indicate U.S. country for example, the anticipate shift in the activities in labor force of a similar order of magnitude as the long term sight away from agriculture and decreases in manufacturing. Share of employment in the United States both which were achieved. So, those factors can influence why artificial intelligence technology needs. So, it is possible that future agriculture and manufacturing both industries will apply (AI) technology manufacturer-kind of job nature to raise productivity instead of farmers, fruit picking workers, farming transportation labours as well as factory manufacturing workers and supervisors etc. human-kind of job nature.

● Is artificial intelligence possible to replace labor ?

Not just intelligence, but also debating, if machines are capable of having a conscious minds. Artificial intelligence has those characteristics as below:

On functionalism aspect, artificial intelligence inputs mental states, sensory inputs, ( beliefs, desires being in pain feeling) and behavioral outputs. Since mental states are identified by a functional role, which are thoughts to be manifested in various systems. Even, perhaps computers which are physical devices with electronic substrate that inform computations on inputs to give outputs similar to brains which are artificial intelligence composed of part any intrinsic relationship to each other. Thus, artificial intelligence activities is not the whole itself, but into parts or on external influence on the parts.

On dualism aspect, artificial intelligence is a set of views about the relationship between mind are matter. On materialism aspect, it builds the only thing that exists is matter, including consciousness.

On biological naturalism aspect, it is similar a human brain than feels pains makes mental situation. So, artificial intelligence is similar biologist which might to be excited to human labor work.

Hence, it seems artificial intelligence can change ( alter) or replace human labor work of nature in possible in the future.

● Can (AI) technology replace human labour nature of work?

On technological innovation reason view point, the history development of artificial intelligence studying the intelligence is one of most ancient scientific discipline. The history development of artificial intelligence what aims to achieve human use to sense, learn remember and think, logic probability, decision making and calculation develop from mathematics, instead of replacement human labor functions.

Artificial intelligence history development aim is the scientific analysis of skills in connection and practice with the appearance of computers from 1950 year beginning. The artificial intelligence (AI) can deal with the ultimate challenges. How can ( either biological or electronic) mind sense, understand and manipulate a world that is much simple and more complex than itself? And what if would human like to construct something with such capabilities?

The general-purpose software of the early period of (AI) were only able to solve simple tasks effectively and failed when which should be used in a wider range or an more difficult tasks. One of the sources of difficulty was that early software had very few or mix knowledge about the problems which handled, and activities successes by simply syntactic manipulation. Moreover, the other difficulty was that many problems that were tried to solve by the (AI) were untreatable.

The early (AI) software whether trying step sequences based on the basic facts about the problem that should be solved, experimented with different combinations till which found a solution. From the end the 1960 year, developing the so-called expert systems were emphasized. These systems had ( sue-based) knowledge base about the field which handled. Till to the beginning of the 1970 year, ( Prolog) the logical programming language was born, which was built in the computation realization of a version of the resolution calculus. ( Prolog) is a remarkably prevalent tool in developing expert systems ( on medical, judiciary and other scopes), but natural language parsers were implemented in this language. Then, in 1981 s, the Japanese announced the fifth generation computer system project, a 10 years plan to build an intelligent computer system that use the ( Prolog) language as a machine code. Nowadays, (AI) can be applied any industries, such as car manufacturing industry can use (AI) technological machine-men manufacture car, instead of replacing human labors in factory. Even, in the future, using (AI) machine-men drivers can drive any private cars or public transportation tools, instead of replacing human drivers, e.g. bus, train, tram, ferry etc. Also in the future, machine-men can replace housewives to serve families to do housekeeping clean job , e.g. cleaning toilets, bathrooms, kitchens, even cooking functions at home. So (AI) machine-man can reduce housewives works at home. Moreover, (AI) machine man can take care old people , when who are living at homes or elder care centers.

So, it seems artificial intelligence (AI) will be possible developed to manufacture a new generation machine-man to assist ( serve) families to do any simply cleaning or cooking jobs at homes. Moreover, the overall demand of ( AI) general social needs will also rise, such as security, driving transportation tools, restaurant cleaning, elder centers care service etc. So, it seems that individual or families or social needs of (AI) will be increase in the future. Thus, it will influence macro economy growth (GDP) if there are large house family consumer group and hotel or bus or taxis or ferry etc. different business consumer group demand any artificial intelligence machine numbers increasing. Then, the artificial intelligence products and material manufacturers must need to buy many artificaial intelligence materials to produce any kinds of artificial intelligence machines to prepare to satisfy consumer individual needs. Consequently, macro economy will grow to the owned artificial intelligence development countries, e.g. US, China, UK.

● Why can artificial intelligence satisfy human needs?

First, On machine-man satisfactory demand aspect view point, it makes computers that think, it is the automation of activities. We associate with human thinking: like decision making, learning. It is the act of creating machine that perform function that require intelligence when performed by people. It is the study of mental faculties through the use of computational models. It is the study of computations that make it possible to perceive, reason and act. It is a branch of computer science that is concerned with the automation of intelligent behavior. It is anything in computing service that human don't yet know how to do property.

Second, on thought aspect artificial intelligence means systems thank think like humans, systems that think rationally.

Third, on behavioral aspect, artificial intelligence systems that act like human and that systems act rationally. However, the basic objective of (AI) is to represent human's thought processes in computation . These machines are supposed to exhibit behavior that. It is performed by a human being, would be considered intelligent. However, some authors feel (AI) has disadvantages, such as it is not creative, it is excited in the use of sensory devices, it can't make use of a very wide context of experiences and it does not use common sense.

For speech recognition and understanding function needs example, (AI) can be applied in speech recognition and understanding function, which (AI) speech or voice recognition is a data input method. For example, the computer recognizes and understands one ( or a few) word commands. Speech understanding on the other hand is the computer's ability to understanding a spoken language. That is , the computer understands the meaning of sentences, an paragraphs through (AI).

So, (AI) can be attempted to learn human language how to speak. It is similar to translate human language skill, instead of actual human speaking skill. Also, (AI) can assist handicap learning or language student how to listen different languages by machine-man sounds from computers more accurately.

So, it seems that it (AI) can replace human language teachers speaking function and can change teaching language nature of job in language speaking and listening education industry.

● Is artificial intelligence one good choice for human future technological benefit?

Nowadays, new technology development is popular. However, artificial intelligence is one kind of new technology choice among different technologies innovation. So it brings this question: Is artificial intelligence technology value to invest? To answer this question. I shall indicate some other new technology developments to compare (AI) technology development to judge which has urgent needs to achieve human expectation nowadays.

For example, why is green peace interested in new technologies? New technologies features prominently in our ongoing campaigns against genetic modified crops and number power. However, which are also an integral part of our solutions to environmental challenges, including renewable energy technologies, such as solar, wind and wave ( water) power energy as well as waste treatment technologies, such as mechanical, biological treatment.

It seems humans need concern how to apply (AI) technology to solve environment pollution challenges in our future. So, environment protective, agriculture, natural energy technology will be popular demand to attempt to apply (AI) technology to solve their challenges or apply (AI) to assist to develop their industry.

What is relationship between
(AI) and economy growth?

● How can artificial intelligence technology influence economy?

Advances in artificial intelligence (AI) technology and related fields have opened up new markets and new opportunities progress in critical areas, such as health, education, energy, economic development, social welfare and the environment pollution.

(AI) automation will continue to create wealth and expand the global economy development in the future. However, when many will benefits that growth won't be costless and will be accompanied by changes in the skills, that workers need to increase productivity in the economy and structural changes in the economy. So, in the skills that workers need to succeed in the economy and structural changes.

I shall indicate why aggressive policy action will be needed to help Americans who are disadvantaged by these changes , due to (AI) technology is caused. For automation industry change example, artificial intelligence (AI) capabilities will enable automation of some tasks that have long required human labor. These artificial intelligence technology introduction can increase new opportunities for individuals. The economy and society, but (AI) has also the potential to disrupt be current livelihoods of many Americans. However, (AI) leads to unemployment and increase in inequality over the long run depends not only on the (AI) technology itself, but also on the institutions and policies that are changed.

Thus, it is possible that (AI) technology will raise some countries unemployment number if the employer apply (AI) technology workers to work instead of human labor in their factories, but it can also raise productivities for these employers.

● Can (AI) influence global economy growth?

Technological progress is main driver of growth of GDP per capita, allowing output to increase faster than labor and capital . However, technology can increase productivity, but also decrease the number of labor hours needed to create a unit of output. So (AI) causes unequal to labor wage decreases, even reduces the number of labor to manufacture, e.g. artificial intelligence technology of automation car manufacturing industry; clothing manufacturing industry; plane manufacturing etc. high technology of artificial intelligence manufacturing method. But (AI) should be potential environment benefit, although it raises unemployment ratio. Moreover, it can rise production , due to many skilled craft were replaced by the combination of machines and lower-skilled labor. The result of (AI) technology introduction , it causes output per hour risen when inequality declined, driving up average living standards, but the labor of some high-skill workers was no longer as valuable in the market. Otherwise, if (AI) technology is continue developed to be success. Some routine intensive occupations will be loss, which focused on predictable, e.g. easily programmable tasks, such as switchboard operators, filing clerks, travel agents, and assembly line workers would be particularly replaced by new (AI) technology. However, at the same time, (AI) technology development will bring these benefits: improvement in education ( training (AI) technology scientists) , due to (AI) manufacturing technology needs are raising to businesses and institutional changes, such as the reduction in unionization and raising in the minimum wage to the (AI) manufacturing technology skilled labor in factories.

Because (AI) technology is not a single technology, but rather a collection of technologies that are applied to specific tasks, the effects of (AI) will be felt unevenly though the economy. It will bring some tasks will be most easily automated than others , and some jobs will be affected more than others, both negatively and positively. Finally, new jobs are likely to be directly created in areas , such as the development and supervision of (AI) as well as indirectly created in a range areas though out the economy as higher incomes lead to expanded demand.

However, if (AI) technology could dominate global labor markets. If labor productivity increases, do not influence into wage increases, then the large economic gains brought about by (AI) technology could be increased wealth inequality, due to employers can reduce production cost, but workers ( labors) wages will not be increased, even will be decreased. Hence, it seems the (AI) technology will bring disadvantages to labor market to cause unemployment or reduce wages in possible, although it can reduce employer individual salary (wage) expenditure and it can raise productivity.

● How can artificial intelligence impact global economy growth?

Artificial intelligence (AI) technology is a branch of computer science that aims to create intelligent machines that work and react like humans. So, (AI) is a technology that appears to impact ( influence) human preference by learning, understanding complex contents, enhancing humans in executing both routine and non-routine tasks. In the future, (AI) technology that can be virtual personal assistant, as well as it may exist, such as robots with human-like processing capabilities.

How can (AI) technology impact global economy growth over the next 10 years? During this time period, (AI) technology is predicted to have wide-ranging applications including: Machine learning that automates analytical model building by using algorithms that allow machines to operate without human assistance.

In global education aspect, potential applications include predicting cause-and-effect relationships from biological data, identifying new drugs, self-driving cars, and protecting against fraud, improved natural language processing that allows computers to continue to better analysis, understand and generate language to interface with humans using natural human languages. For example, transcribing notes dictated by physicians, automatically drafting articles and translating text and speech. So (AI) technology can be applied to education aspect to improve humans' knowledge level.

In visual art aspect, (AI) machine vision that allows computers to identify objects, scenes and activities in images. Current applications of (AI) machine vision include providing objective descriptions for the blind seeing( visual) needs.

We except the economic effects of (AI) technology to include both direct GDP growth from sectors that develop or manufacture. (AI) technology and indirect GDP growth through increased productivity in existing sectors that employ some form of (AI). If (AI) technology is an increasingly critical component of more products, it will become an integral part of many people's lives. Thus, (AI)'s ability to influence economic activity, rather than the economic or development status of the region. (AI) has the potential to impact income classes and to bring significant gains to both developed and developing countries. For example, (AI) has the potential to optimize good production around the world by analyzing agricultural regions and identifying what is necessary to improve crop yields.

In estimating the future economic effects by (AI) technology innovation, it is important to note that it is challenging to accurately predict which applications of (AI) will ultimately be commercially successful. In micro level economic influence, we need to apply methodologies to estimate the economic effects of investment in firms developing (AI) technology since investment levels in a technology are a telling sign of the future potential of that (AI) technology.

- How can (AI) influence GDP of high income countries in the next ten years?

How (AI)'s development may affect the global economy over the next ten years. In fact, (AI) technology has the potential to affect business across the global in a wide range of industries in ways only a number of technologies have done in the parts. For example, (AI) technology's expected to be a useful tool for enhancing human capabilities and in some instances replacing functions, such as driving a car, adoption of broadband internet, mobile telephone, industrial robotic automation have served to enhance human capabilities.

However, significant public debate has focused on projections of (AI) technology's effect on the labor force. However, large companies prefer to invest in (AI) technological industry. For example, face book's (AI) research lab., google machine intelligence lab. and micro soft machine learning and artificial intelligence research division are all making advances in (AI) technology and investing in the industry's top talent. Additionally, between 2010 year and 2015 year, nearly $5 billion in venture capital funding invested in firms across the global developing and employing (AI) technology ( Facebook (AI) Research).

- How can artificial intelligence impact on workplace?

Modern information technologies and the labor economy growth of machines is powered by artificial intelligence have already strongly influenced the world of work in the 21 ST century. Computers, algorithms and software simplify every tasks and it is impossible to image how most of our life could be managed without them. How can be the information economy characterized by exponential growth replaces the most production industry based on economy of scales? What will the future world of work look like and how long will it take to get? Will the future world of work be a world where humans spend less time earning their livelihood? Alternatively, are mass unemployment, mass poverty and social distortions also possible scenario for the future, where robots, artificial intelligence systems play an increasingly central role? These questions concern how artificial intelligence further development . Can influence labor economy growth on workplace ? When the labor market has widespread impact on intelligence property, information technology, product liability, competition and labor and employment laws.

How (AI) technology impacts on labor workplace.

The future influence any organizations how labor economies use of (AI) can be analyzed, such as deep machine learning is based on a set of model high level data. Unlike human workers, the machines are connected the whole time in workplace. If one machine makes a mistake, all autonomous systems will keep this in mind and will avoid the same mistake the next time.

Over the long run intelligent machines will win against every human expert. Production robots have been replacing employees because of the (AI) technology. They work more precisely than humans and cost loss. Creative solutions like 3D printers and the self learning ability of these production robots will replace human workers, the automatic data recording and data processing, traditional back office activities are no longer in demand. Autonomous software will collect necessary information and will send it to the employee who needs it. Additionally, dematerialization leads to the phenomenon that traditional physical products are becoming software. For example, CD or DVDs are being replaced by streaming services. The replacement of traditional event ticket, e-travel ticket service products or hard cash will be the next step, due to the possibility of payment by smartphone. So, (AI) technology will impact human's daily life consumption behaviors in the future. For another example, transportation tools, such as boats and ferries and private vehicles will use sensors and navigating without human input. Taxi and truck drivers will become obsolete, the stock store applies to stock managers and postal carriers of the delivery is distributed by (AI) machine delivery method.

What is the relationship between (AI) and (CRM)?

● Can (AI) technology impact on customer relationship management (CRM) ?

Nowadays , (AI) is a technology almost as old as the computer industry itself, it is similar with the advent of personal assistants function to businesses and personal promotion channel, such as ( Amazon's Alexa, Apple's Siri, Google's Assistant) image recognition ( face book), personalized recommendations ( Netflix , Amazon). Those innovations have been driven by a increase in processing power, lower cost hardware, and the exploding creation and availability of data. It seems, (AI) technology can impact global customer service management method.

How to forecast economic impact modeling to (AI) will affect global economy? Can human forecast business revenue growth and job creation ( or destruction) based on (AI) applied to customer relationship management (CRM) activities? In addition to the economic impact on (AI) or (CRM) which can include an estimate of the economic impact attributable to sales forces customer base. What can economic benefits be brought to (CRM) from (AI) technology?

Artificial intelligence(AI) comprises a set of technologies that use natural language processing, machine learning, knowledge graphs, and other tools to answer questions, discover insights and provide recommendations. Computer systems can use (AI) hypothesize and formulate possible answers based on available evidence can be trained through the ingestion of vast amounts of content, and automatically adapt and learn from (AI) self mistakes and failures.

So, any business organizations (customer service departments) can provide efficient and effective customer relationship management of excellent customer service quality if which applied (AI) technology system. The different type of (AI) systems include: (AI) system platforms, machine learning (AI) based data preparation and enrichment tools, machine vision/image recognition, voice speech recognition, text analysis and natural language processing, bots , e.g. face book website and virtual digital assistance solutions, social media pattern analysis , sentiment analysis, advanced numerical analysis (e.g. IOT streaming , machine logs), supporting technologies, knowledge base dialog management, Q&A processing etc. different (AI) technology system customer relationship management (CRM) tools.

(AI) (CRM) of activity can include these categories, such as: corporate marketing, marketing operation, field marketing, customer support, digital commerce, customer analytics, customer influenced product or service design, product or service pricing, finance information, presentation, customer billing, inventory , logistics and fulfilment support, partner management etc. different CRM tools.

(AI) technology of CRM has been carrying on plan different stages to achieve CRM personal assistant tool for businesses. The stages are such as, in the beginning stage of (AI) projects in place, implement now, pilot phase next year in the final stage of (AI) customer relationship management tools are foreseeable future. So, this CRM technology has been improved to plan in different stages every year to prepare to achieve full capacity of CRM service quality for businesses to use in the future.

Hence, how to develop an estimate prediction of the economic impact (AI) technologies could have CRM activities, which depends on gathering macroeconomic information on business revenue and the basic marketing of business revenue and the basic markup of business expenses by major functions ( customer support, marketing and sales , production etc.)

An economic impact model that can gather data together and forecast the results how (AI) artificial intelligence technology brings (CRM) customer relationship management benefits to businesses, e.g. surveys investigation includes IT spending by sample countries, GDP and population estimates and forecasts, revenue per employee and ratios of IT spend to GDP. Surveys ( questionnaire questions) of forecast results are influenced by (AI) impact can include: results are projected from surveys and rely on estimates are made by respondents on the expected financial improvements in categories of (AI) –assisted customer relationship management activities. The forecast assumes that these estimates are correct; financial estimates are based on estimates of "first year" improvement from full (AI) implementation; forecasts are from planning to implement any artificial intelligence of customer relationship management (CRM) projects, the improvement forecast is of categories of activity , e.g. corporate marketing , digital commerce, and customer analytics. They are not estimates of ROI for the (AI) software. They rely on conservative estimates to which each of these entities might affect company revenue, expenses or productivity. They also rely on estimates of the penetration of software in customer relationship management activities . Net new jobs created are based on the ratio of new revenue to jobs required to support that revenue . They can assume that 50% of the net new revenue will support increases in labor and the rest will go for capital and other operating expenses that may replace jobs lost to automation.

In the future, some of the ways in micro economic benefits to any organizations. (AI) technology is expected to impact CRM activities include: Spending up sales cycles, improving lead generation and qualification solving customer support problems faster ( raising service quality), helping companies improve brand campaigns and recognition, lowering costs of support calls when increasing resolution rates, lowering the cost of recruiting employees and partners, increasing revenue from optimized product marketing, optimizing price, distribution logistics and preventing loss through fraud detection. So, micro economic benefits view point, it seems that (AI) CRM technology can raise any companies economic benefits for care term.

Artificial intelligence enables machines or the in-build software to behave like human beings which allows these decisions and act. The advent of (AI) is leading , talking, making decisions and act. The advent of (AI) is leading to new technologies advances and transforming the economic and employment opportunities for humans in a positive way. (AI) related technologies can facilitate our live. For example, industrial robotics, robotic medical assistants, smart games, financial forecasting software, big data analysis, algorithms in health and bioinformatics, pilotless cargo places, drone ambulances and general purpose and workplace robots and others. (Disruptors technologies: Advances that will transform life, business and the global economy).

Artificial intelligence also known as computational intelligence is defined as " the human –like intelligence exhibited by machines or software. It is theorized that intelligence of humans can be described and intelligence machines or software can simulate it. These machines software can be reasonable , learn, perceive and process information, like human mind and thus facilitate human life. They can think and act for us. So, artificial intelligence is an interdisciplinary field of study including computer science, neuroscience, psychology, linguistics and philosophy.

However, (AI) research and developments have economically impacted many industries, such as robotics, telecommunications, computer applications , health, finance, heavy manufacturing, transportation, aviation, e-service and e-commerce, military , music and movie, toys and games entertainment etc. industries.

In fact, many ideas, systems and technologies have been developing in the world of (AI) technology. However, which are net called or considered (AI) products, rather which are mentioned with their specific names, such as smart graphics, machine learning, e-commerce etc. ( i.e. this is called (AI) effect).


● How can (AI) technology influence digital economy?

Nowadays, (AI) related industrial applications will replace most human power in fields, including call centers, customer services and air cargo transportation. (AI) technologies also help weather forecasting based on repcatcd rainfall pattern ( data) recognition, through robotics ( i.e. floor cleaning, moving lawns etc.) transporting people and products with unmanned vehicles, sending space unmanned smart shuttles, developing robotic arms, predicting market values in stock exchanges by internet, making homes safer, helping elderly and disabled using robotic servants etc.

Among the (AI) related technologies , there are a few that significance for the impact on society and especially on digital economy . (AI) is particularly influential in machine learning. Such as robotics, transportation, finance, health and bioinformatics, e-commerce , e-games, big online data gathering and internet-of-things. For example, machine e-learning is based in bioinformatics and robots that can learn new skills for better caregiving in healthcare. What is machine e-learning? Machines can e-learn from e-data gathering, coming up generalizations and making decisions to act in certain ways from internet.

There are important applications , such as e-machine perception, electronic online natural language learning processing, online search engines, online bioinformatics, online brain –computer interface, online game playing, online robot locomotion, online advertising, online computations finances, online health monitoring, online DNA classification and decision making, online in chemistry –cheminformatics . So, online machine learning can positively impact productivity and it can enhance information and analytical system from (AI) online channel.

What is robotics? Robotics is one of the most strongly influenced fields in (AI). For example, heavy manufacturing industries, robots and used and man power is replaced for effectiveness, precision, and accuracy, especially in respective or dangerous tasks, including welding, assembling , picking and placing .

So, robots can acquire new skills or adapt the changing dynamic environment. Also, artificial intelligence can be applied in developing transportation. For example, automated vehicles, driver assistance systems , safety systems, collision avoidance systems and public transportation. Moreover, (AI) technology has proven to produce some of the best tools to predict stock market fluctuations from internet data gathering method. It's predictions are based on ever-evolving predictions algorithms and systems learn new models and make connections between historical data and new data to measure stock market trading more accurate from internet data gathering channel.

In health field, especially in health data processing , analysis, decision making support and medical diagnosis. So, online data can show which patients will need what treatment and what alternative drugs could be used more accurate from (AI) online data gathering method. Bioinformatics is an interdisciplinary field combining statistics, (AI) online technology can help in discovering data patterns and modeling through the application of machine learning, artificial neural networks and genetic algorithms. For example, further (AI) technology development of human genome project of online data sequences.

Online shopping can be facilitated by virtual assistants developed through (AI) technology and these assistants can offer the best advice. (AI) online purchase coming after every product image recommendations and personalization bring important revenue to shopping online sites, like Amazon . Smart computer graphics and games, artificial intelligence is useful in smarter computer, graphics, scene modeling , scene rendering processes in order to create, for example, effective human –robot interactions , online machine learning, online strategic games techniques etc. online computer related (AI) software.

So, online big data analysis and big data does have a critical need in the world of online intelligence machines and software in our future. In other words, (AI) offers online technology to enable online big data analysis to provide industrial organizations with valuable information for effective decision making in short time. For example, what IBM's Watson achieved: this machine used 200 million of structured and unstructured content with a special technology of hypothesis generation, massive evidence gathering, analysis and scoring from internet channel.

Finally, (AI) online technology another related internet invention ( internet of things) (IOT) is the network of machines or objects connected through internet. These connected objects can sense their internal and external environment, communicate with each other, can send critical data and finally can make decisions to act or correct their environment from (AI) online technology. For example, factories can monitor and automatically change production processes, hospitals can monitor and regulate the health conditions of their patients , schools can collect data from facilities and cars can send data to car makers from (AI) online technology.

Partner predicts that (IOT) market will create about trillion amount value by 2020 year. Although machines collect big data from their environment, whether which gain an insight or learn from these online data largely depends on the (AI) online machine learning principals and (AI) online technology. In 2013, Mckinsey estimated that disruptive technologies closely related with potential economic impact in 2025 year between $7.1 to $13.1 trillion amount ( automation of knowledge work, advanced robotics, autonomous or near-autonomous vehicles).

What is the relationship between (AI) and global digital economy development ?


● Could work activities in China be automated making in the nation with the world's largest automation potential?

Can (AI) technology influence China economy? Could China workers be affected and jobs made up of routine work activities and predictable? Will programmable tasks be particularly impact to China employment market ? When impact on labor market is likely to be gradual at the aggregate level, it can be sudden and dramatic at the level of specific work activities, rending some job obsolete fairly. Overall (AI) technology will raise digital skills when reducing demand for medium incomer inequality for China workers. It seems (AI) technology's effect on productivity could be crucial to China's future economic growth as the population ages are increasing.

In China, some biggest technological companies driving significant investments in research and development. Moreover, China is one of the leading global (AI) technology development county. However, China will need to focus on building its innovation capacity. For example, United States and United Kingdom are currently producing more influential (AI) technological research. However, if China planed to achieve (AI) technology success, it's traditional industries will need to develop technical know-how –to and overcoming implementation costs prepare to develop (AI) . When (AI) technology is introduced into China society, China government needs to raise concerning ethical, legal, technological security etc. business questions. Also, surrounding issues include privacy, discrimination, legal liability and regulation. It aims to encourage overseas investors to choose to invest (AI) technological industry to raise GDP growth and manufacturing industries income growth for long term in China.

If China encouraged overseas (AI) technology investment in its country. It is possible to influence China employment market to be changed. Because (AI) technology will impact to influence China people daily life. Due to (AI) technology is introduced to China society, many rich people will prefer to spend to buy any high (AI) technological products for entertainment or learning or machine man driving etc. daily necessity activities. Then it will raise GDP growth and will raise (AI) manufacturers or related-(AI) technological manufacturers profit. It is beneficial to China because it can become one high knowledgeable and (AI) technological economical society.

But it will bring bad influences to raise unemployment chance for the low skillful labor. In labor economy aspect influence , how (AI) technology can influence China low skillful labor unemployment ratio raising. The raising low skill labor unemployment reason is because China low skillful human labors are argued or are replaced by (AI) technology creating new challenges to introduce to influence China society of simply human manufacturing job nature to be changed to be high (AI) technology manufacturing job nature in any China factories. Moreover, when (AI) technology introduction to China, it will cause other related social challenges in China. The varied (AI) related challenges, including the difficulty of creating safe and reliable hardware for sensing and affecting ( transportation and education), the challenges of gaining public trust, a low resource comities and public safety and security, the challenges of overcoming fears or marginalizing humans in China employment and workplace and the risk of diminishing interpersonal trust because the low skillful labors won't believe any China employers will give chance to employ them , due to (AI) technology will replace their skills and man manufacturing of productivity is much less to compare to (AI) technology manufacturing method.

- How does ( AI) technology influence

the future of employment change?

Are future nature of jobs changed to computerization from (AI) technology? Where are the probability of computing occupations from (AI) technology influence? What is expected impacts of future computing on labor market from (AI) technology influence? John Maynard Keynes's frequently cited prediction of widespread technological unemployment " du to our discovery of means of economic the use of labor outrunning the pace of which we can find new used of labor" ( Keynes, 1933, p.3).

In the future, (AI) technology will impact some nature of occupations to change computing. This chance will also influence some countries' economic change. For example, some factory human labors hand routine manufacturing tasks will be changed to computerization of routine manufacturing tasks by (AI) technological machine men hand manufacturing method. it will cause a structured shift in the labor market, with workers reallocating their labor supply from middle-income manufacturing to low-income service occupations.

Arguably, this is because the manual tasks of service occupations are less computerization, as who require a higher degree of flexibility and physical adaptability. So, (AI) technology will influence the human hand labor skillful occupation nature of task cheaper , such as vehicle manufacturing , ship manufacturing, computer manufacturing, steel manufacturing, television, radio etc. home electronic products of heavy machine industry change. Due to (AI) technology machine man will be proper to be used to manufacturing these electronic products when the (AI) technology innovation can develop to the mature stage. Then, any countries manufacturers will choose to use (AI) technology machine man, instead of human hand production.

Supposing the future prices of computing are fallen, seriously, problem solving skills are becoming relatively productive, explaining the substantial employment growth in manufacturing occupations, involving cognitive tasks where skilled labor has a comparative advantage, as well as the increase education needs for (AI) technology computing of machine man subject study.

Prediction of education needs for (AI) technology student numbers will increase, due to manufacturing industry needs many (AI) technology students in future employment market. Another (AI) technology influence if the future (AI) technological innovation, e.g. machine man manufacturing or machine man service industries will both increase demand, then with more sophistic software technologies will be disrupted labor markets by marketing workers redundant.

For publishing industry, what is striking about the case in paper book publishing industry will be unpopular? Due to the electronic book publishing industry will be popular, e.g. Amazon publish . (AI) technology can influence paper book manufacturing method which is replaced by machine man electronic book manufacturing method as well as it will cause the computerization is no longer confined to routine manufacturing tasks. Due to (AI) machine man manufacturing technology will be proper to be used to manufacture any products in short time efficiently and effectively , e.g. electronic book products. In the future, if it is fact to occur this case, such as ( AI) technological machine man manufacturing method will be adopted ( applied) to manufacture electronic books or any products in possible. (AI) technology will cause many manufacturing workers are unemployed. It is beneficial to employers, who can reduce to spend much wages expenditure to employ manufacturing workers, but it will cause many manufacturing workers loss jobs and reduce income to support whose families lives. It will cause social challenges, e.g. increasing stealing crimes if the manufacturing workers had not other skills to find other jobs to do easily. So, manufacturers need to concern over technological unemployment which will be hardly future phenomenon if who decided to dismiss all manufacturing workers, due to (AI) technology machine men replace to them.

If ( AI) technology can be innovated to produce any kinds of machine man to serve any service or manufacturing industries successfully. Then, it will bring these questions: Can future that workers be influenced to be automation employment and productivity by (AI) technology influence? Does it impact to influence the (AI) technology countries' productivity and growth and natural resources development and labor markets and evolution of global financial markets and economic impact of technology and innovation and urbanization etc. issues? How will automation transform the workplace? What will be the implication for employment? What is likely to be its impact both on productivity in the global economy and on employment?

In fact, automatic of activities can enable businesses to improve performance by reducing errors chance and improving quality and speed, and same cases achieving outcomes that go beyond human capabilities. Some economists indicate (AI) technology would give a needed boost to economic growth and prosperity have of the working age population in many countries. Based on the scenario modeling, they estimate automation could raise productivity growth globally by 0.8 to 1.4 % annually. They also indicated that almost half the activities people are almost $1.6 trillion in wages to do in the global economy have the potential to be automated adapting current demonstrates technology, according to their analysis of more than 2,000 work activities across 800 occupations. When less than 5% of all occupations can be automated entirely using demonstrated technology, about 60% of all occupations have at least 30% of worker made activities, that would be automated. More occupation will change to be automated. They also indicated for business performance benefits of automation are relatively clear, but the issues are more complicated by policy making to attract foreign investors. Beyond technical feasibility, the cost of technology, competition labor will include skills and supply and demand dynamics, performance benefits and beyond labor cost savings and social and regulatory acceptance will affect the automation. Their predictions suggest that half of today work activities could be automated by 2055 year, but this could happen 10 to 20 years earlier or latter depending on the various factors in addition to their wider economic condition.

Some scientists suggest (AI) technology is finally starting to deliver real-life business benefits. Computer power is growing significantly , algorithms are becoming more sophisticated and perhaps most important of all, the world is generating vast quantities of the fuel that powers (AI) technology data billions of gigabytes of it every day. Also, online firms are digital natives, such as Google online search service company is investing on (AI) technology. For new though most of the news if coming from the suppliers of (AI) technologies. And many new users are only in the experimental phase. Few products are on the market or are likely to arrive these soon to drive immediate and widespread adoption. As a result, analysts believe (AI) technology's potential will give true economic benefit in the future. (AI) industry will introduce to suppliers and users to raise economic potential of (AI) technology.

In the future, (AI) technology systems can solve business problems. Some scientists categorized those into five technology systems that are key areas of (AI) technology development: robotics and autonomous vehicles, computer vision language virtual agents and machine learning , which is based on algorithms that learn from data without replying on rules-based programming in order to draw conclusions or direct an action.

Such as computer vision and language includes natural language processing, analytics, speech recognition technology, some are about learning from information, such as about machine learning and others are related to acting on information, such as robotics, autonomous vehicles and virtual agents, which are computer programs that can converse with humans. Machine learning and a subfield called deep learning are artificial intelligence applications.

- Can artificial intelligence impact

global economy growth?

Artificial intelligence (AI) is a term first defined in 1956 year. It is a branch of computer science that aims to create intelligent machines that work and react like humans. In contrast today, 60 years later, (AI) is characterized by a number of applications, including computers playing games against humans and understanding human languages, virtual personal assistants, and robotics which involve computers seeing , hearing and reacting to sensory stimuli. In the future, technologists predict for (AI) technology ranging from (AI) being used as a tool to aid relatively simple processes for robots with human like mental capabilities, who expect (AI) technology can emulate human performance by learning, coming to mind its own conclusions, understanding complex content, engaging in dialog with people, enhancing human cognitive performance or replacing humans in executing both routine and non-routine tasks. In existing industry, (AI) technology is used , such as targeted advertising and virtual used personal assistant as well as the (AI) technology that my exist in the future, such as robots with human vehicle processing capabilities.

The range of (AI) technology's progress in the future will determine the economic impact future of (AI) technology on the global economy with more limited advances and applications ( i.e. weak (AI) only) corresponding to more limited economic impacts and more substantial progress, i.e. strong (AI) technology is corresponding to more significant economic impact.

(AI) technology learning that automates analytical model, including predicting cause-and-effect relationship from biological data, identifying new drugs, self-driving cars and protecting against fraud etc. functions. Also (AI) learning can improve natural language processing that allows computers to continue to better analyze, understand and generate language to interface with human using the natural human language, virtual personal assistant, helps users by providing scheduling appointment, reminds organizing personal finance and finding providers of various services, machine vision allows (AI) machine man to identify object, scenes and activities in detect pedestrians and bicyclists. We expect the economic effects of (AI) technology to include both direct GDP growth from sectors that develop or manufacture (AI) technology and indirect GDP growth through increased productivity in existing sectors that employ some from of (AI) technology. If (AI) producing sectors could grow, then it could lead to increase revenues and employment of (AI) technological professionals within these existing firms as well as the potential creation of entirely new economic activities to any countries' societies productivity improvement in existing sectors could be realized through faster and move efficient processes and decision making as well as increased (AI) technological knowledge and access to information available in societies easily.

In the future, if (AI) technology is an increasingly critical component of more products, it will become an integral part of necessary products of many people's lives. The extent of (AI)'s economy effort is also likely to vary from region to region, thought variation may be more dependent on the predominate economic activity of a region and the (AI) ability can influence economic activity, rather then the economic or developmental status of the regions. (AI) technology can move accessibility and can use source development to do international business between one country and another country.

So (AI) technology has the potential to give benefits to different income chooses and to bring significant gains to both developed and developing countries. For agricultural technology, (AI) has the potential to optimize food production around the world by analyzing agricultural regions and identifying what is necessary to improve crop yield. In total, (AI) technology gives greater economic impact to any countries agricultural regions if which implemented (AI) technology to grow crop , fruit etc. food production in the farms.

Investment in (AI) technology is such as capital investment to any countries' public or private enterprises. So, it will have large economic impact to the future . If the (AI) technology is reasonable invested to the different needs aspect by the public or private enterprises in the country. Then, it will have good economic impact to the country in the future. However, when (AI) technology is likely to affect both the productivity and employment components of economic growth in many sectors. Significant public debate has focused on projections of (AI)'s effect on the labor force. However, for instance, some researchers have argued that the rise of (AI) technology and automation will led to significant unemployment as capital is substituted for the low skillful labor. So, they point to the concern that the increasing sophistication of (AI) technology may balance skilled and semi-skilled workers and the reduce the size of the middle class. However, this is not a new argument, due to (AI) technology negatively affecting the labor force and leading to mass unemployment. Because the (AI) technology is the substitution of machinery for human labor. Although, employment in certain industries, has been reduced in the past due to technological advancement. For long term, the labor market has adapted to the introduction of new technology, giving rise to new jobs in new areas. (AI) technology may also be accomplished without a reduction to total employment in the long-term to some Asia countries, such as Hong Kong and Japan. Because Hong Kong and Japan many low skilled labor, e.g. security, cleaner who complaint that employers need them to work long time hours. ( abnormal working hours) e.g. one day 12 to 15 working hour per day. Hence, if (AI) machine means invention technology success. Security or cleaning job can be worked from (AI) machine man in some hours every day in order to reduce the long time working hours cleaners or security workers, e.g. one (AI) machine man works 4 hours for cleaning or security job, one day as well as another cleaner or security labor only needs to work 8 hours one day. So total security or cleaning employers can employ 12 hours machine cleaners or security workers and human cleaners or security workers in one day. For long term benefit, Hong Kong or Japan every security or cleaning worker does not need to work 12 hours minimum working hours one day. They won't feel tried and bore and without private with whose families, so who will accept to do these cleaning or security jobs, even they can raise work efficient and performance when who feel happy and health.

So, (AI) technology of machine man invention can raise low skillful labor efficiency and it can help them to avoid abnormal working hours demand in some busy work life countries, such as Hong Kong and Japan. Before, one Japan female labor feel unhappy to work, due to who often needs to work abnormal working hours for her employer and who has less sleeping and without any private time to enjoy her life with her families every day. So this abnormal working hours factor causes her to do commit suicide behavior, then she is die unlucky. So (AI) technology of machine man invention ought avoid abnormal working hours demand for employer in any countries in the future.

The most important occurrence to any employers, some researchers had attempted to do one experiment to find that private research and development , venture capital and public research and development investment all have strong net effect or economic growth with venture capital funding further having the strongest such effect from (AI) technology. The researchers hypothesize the venture capital investment contributes to economic growth through (AI) technology innovation and by the capacity of an economy to use existing (AI) technology knowledge to increase productivity. They predict the impacts of venture capital, business-research and development and public research and development can raise multi factor productivity from (AI) technology introduction.

Can (AI) technology influence the economic development to developing countries? The developing regions of the world contain most of natural resources. If one day, (AI) technology has invent one kind of machine man which can assist any gas or oil workers to seek any new oil/gas natural resource locations easily. I believe that (AI) technology can help these natural resource exploitation countries will gain economic benefit more easily. So, (AI) driven technology can be used to change to create any new opportunities to address poor management or resources and improve human well being, such as Africa Latin America and India can use (AI) technology machine man to seek any oil/gas natural resource countries exploitation activities to attempt to gain much economic benefits.

● Why will (AI) technology grow economic
development ?

Nowadays, increases in capital and labor are no longer driving the levels of economic growth, such as (AI) technology. The ability of increase in capital investment and in labor of traditional drivers of production, have no longer to be enjoyed in most developed economies ,e.g. developed country, US, UK . However, artificial intelligence has the potential to overcome the physical limitation of capital and labor to avoid missing out on this opportunity. So, policy makers and business leaders must prepare for and work toward a future with artificial intelligence. They must do with the idea that (AI) is another simply method to enhance productivity method . Rather they must see (AI) as the tool that can transform thinking about how growth is created.

Economists have always thought of new technologies are as driving growth their ability to enhancing. It can replace labor and capital factor of production. So, it brings this question: What is the factor of production (AI) technology characteristics. They key factor is to see (AI) technology as a capital-labor .

(AI) can replicate labor activities at much greater scale and speed, and to even perform some tasks began the capabilities of human. For example, by using virtual assistants , 1000 legal documents can be reviewed in a matter of days instead of taking three people six moths to complete. Some (AI) technology may be one kind of factor of production in the future. For another example, people will work in workplace digitalization environment. So, in the future, working environment and information management are automated. Such as Konica camera sale company will use workplace digitalization. So , (AI) technology can provide workplace digitalization in order to raise productivity efficiency. (AI) technology will be one kind of production which is replaced by workplace digitalization and it will grow any organization productivity efficiently. Then, (AI) technology will assist overall social economy growth , due to productivity is raised and products can be produced in short time to prepare to sell in consumption market. So, time will be shortened to increase GDP growth fast for the development of (AI) technology countries.

● How can (AI) technology impact to global
economic and social and psychological
changes?

What will be the development of (AI) technology and predictions concerning the future evolution? The computers and robots will develop conscious, intelligent and minds into humans, enhancing psychological and behavioral abilities and allowing for direct communication with (AI) minds. (AI) technology will be impacted human life by (AI) technology information communicative and environmental influence. A " world brain" and " world mind", this psychological system will be enhanced and enriched the capacities of both individual and collective cognition by (AI) technology of service industries.

(AI) technology with influence these human needs of service industries changes, such as , biological science, finance, entertainment, business, biological science, transportation, communication military etc. The personal computer evolution, the internet and the world wide web which exploded on the scene, linking business, homes, schools, social organizations which were a completely unpredicted phenomenon to influence human life. Kurzweil (1999) predicts that by 2029 year, most human communication will be with machines. According to Person, by 2100 year, there will be human machine convergence.

How can (AI) technology influence environmental protection to make benefits to farming economic growth? (AI) technology can be applied to predict how to solve environmental pollution challenge to avoid to damage any crop or vegetable or rice or fruit etc. food growth. Because environmental experts can gather global environmental pollution data from an environmental database to build a perform a systematic analysis from (AI) technology. The first step is this broad analysis can include understanding, statistical and data gathering techniques to obtain the relevant data, the correlation among the variables involved, and a list of possible models. The next step is to select a set of methods and models that cover all kinds of knowledge and functionalities needed for the decision making process. Once the models are selected, they must be fully implemented by means of machine learning , data mining, statistical or numerical technique. After that, those models must be integrated to build the whole EDSS. The EDSS must be tested to check its performance, accuracy, usefulness and reliability, both from the user's and (AI) technology/computer scientist's point of view. If these is any wrong feature in any development stage, such as model's integration, models' implementation, selection of models, database, problem analysis etc. the developers must come back in the update th required components. When the evaluation phase is all right, the EDSS is ready to be applied to the environment. The great contribution of artificial intelligence to EDSS the integration of several methods complementing the classical statistical models/simulation , statistical analysis, linear models, etc. and numerical models ( control algorithms, optimization techniques etc.) .

This cooperation makes the resulting systems more reliable and powerful in coping with real world environment systems. Date interpretation has been a principal area of research in (AI) technology since the very beginning. The most demanding problem in the environmental assessment context. Knowledge representation permits the definition of the different types of data that the existing methods adapt to the process. There is also a lot of work to clean, repair and transform the huge available quantities of raw data. Apart from this, the availability of meta-information or background knowledge is required to guide the process. Data mining is multi-disciplinary: It covers expert systems, data based technology, statistics, data visualization and unsupervised machine learning. These techniques operate at the level of data and background information, where numerous and often incompatible new commensurate pieces of information from disparate sources have to be brought together ( K, Fedra, 1994).

So, it seems that in the future, (AI) technology with the increasing maturity in particular those related to knowledge and engineering, new dimensions can be assisted to users in environmental decision making are available. For example, many environmental systems are characterized both by incomplete models and by limited data. Hence, in the future, (AI) technology will be applied to predict climate change to reduce crop or fruit etc. food agriculture challenge by climate change bad influence.

● Will (AI) technology influence digital economy change to manufacturing industry ?

To understand how the manufacturing business must adapt to prosper in the technology, we need to understand how (AI) technology will change us to shape our daily habits to satisfy our expectation of products to how we shop and even the immediate of the entire process. For example, taxi services are in the crosshairs as on demand transportation services like, available of the touch of a smart phone button expand. In fact, Yellow lab, US country , san Francisco city's largest taxi company is filing for bankruptcy as the industry starts to change faster than almost anyone expected. However, at this point, its more than an app that is changing, some our taxi passengers renting taxi transportation to catch consumption behavior.

(AI) technology will influence digital economy for taxi passenger's individual customer experience, offering a growing renting taxi to catch of service and feedback opportunities when any one taxi passenger who chooses to use mobile phone app online tool to prepaid to rent any taxi more easily.

Also in the long term, (AI) technology can influence vehicles drive themselves of behavior. Already, companies like Google and GM are working on projects to bring fleets of autonomous vehicles to cities at the path of a button. Moreover, this on-demand service model is beginning to appear across a much broader range of markets. For example , Amazon company is investing in its own fleet of trucks, planes and even drone at the same time as it pushes for same-day delivery of products. As some point, vehicles will be autonomous too. So, it seems that (AI) technique will influence any transportations choose to use digital autonomous driving technology in the future . For Amazon company case, it is not stopping of logistics. It is also aiming to automatically manage the supply of consumer home products with its recently launched Amazon replenishment service, Dash. Dash is a digital service that enables that connected derive to automatically order physical products from Amazon when supplies are running low. So, it seems (AI) technology will be applied to logistic function by digital technology method introduction in the future.

Hence autonomous vehicles will optimize industry supply chains and logistics operations through increased efficiency and flexibility. In fact, fully automated and lean supply chains will keep reduce load sizes and inventory by leveraging smart distribution technologies and smaller autonomous vehicles by machine man assistance. If Amazon continues to grow market share for online sales by reducing effort required by the consumer to place an order, when also contributing the almost immediate delivery of products to the doorstep. So, it will further fuel the trend toward on-demand derive. As Amazon company fuels the on-demand economy, consumers will expect immediacy in more parts of the digital economy. On top of speed, consumers increasing expect more personalization options.

So, (AI) technology will influence digital manufacturing, such as Amazon publishing to monitor every aspect of every process in real -time and communicating to self-optimized deep learning robotics, new methods of high volume and high customization will become possible. Then, as products merge into product platforms and even services, manufacturers have the opportunity to provide components and platforms used by smaller players. So, (AI) technology will influence manufacturing industry to choose automated SMI lines, robots installed, automation engineers.

Another future (AI) technology development can be applied to space science aspect, such as Automation engineering space in manufacturing process to achieve digital manufacturing benefits to any businesses in the future. Such as reducing cost, shortening manufacturing time, raising efficiency, shortening delivery products to client individual time. How can artificial intelligence give the need and advanced fast and evaluation methods benefits for space exploration? When US NASA ( space exploration organization) achieves any space exploration missions, it will answer this question:

When is it useful to have a machine use (AI) technology to achieve a decision? After all, after millions of years of space exploration and rough 10,000 years of civilization, humans are usually quite good at making decisions in complex uncertain environments. Through, Johns Hoplains University's Applied Physical Lab. Research in (AI) technology enabled systems, which has identified three general use cases for (AI) technology to explore space mission:

First, for some tasks (AI) technology is more cost effectiveness than human. Second, (AI) technology is better suited than humans at solving some, but not all problems. Third, (AI) technology allows NASA organization's space exploration mission to develop machines that ate capable of responding faster than when a human is in the decision loop ( D. Scheidt, 2012, A. Castano et. al. 2008).

So, the use of (AI) technology to enable science by observing the pace of rapidly evolving phenomena was demonstrated. It is more effectively coordinating and (AI) technology utilizing to earn economic benefits to use for space exploration mission.

However, (AI) technology also have current risk for space exploration. Today (AI) technology is immature and requires further development to reach its potential. For instance, the (AI) technology algorithms that detected the dust derive could not have identified whether the Martain weather represented a threat to the cover. Also it can not yet use instrument input to determine what, where and how to autonomously make the next space science measurement. An equally important factor limiting (AI)'s deployment is that lacks the methodology and technology to effectively test (AI) technology. So, the challenge will testing (AI) enabled system is how (AI) performance can be measured. It would be NASA organization's difficulty to find (AI) technology to develop to carry on researching any space exploration missions in the future. However, (AI) technology will be a good economic benefit choice for space exploration mission in the future.

chooses to use mobile phone app online tool to prepaid to rent any taxi more easily.

Also in the long term, (AI) technology can influence vehicles drive themselves of behavior. Already, companies like Google and GM are working on projects to bring fleets of autonomous vehicles to cities at the path of a button. Moreover, this on-demand service model is beginning to appear across a much broader range of markets. For example , Amazon company is investing in its own fleet of trucks, planes and even drone at the same time as it pushes for same-day delivery of products. As some point, vehicles will be autonomous too. So, it seems that (AI) technique will influence any transportations choose to use digital autonomous driving technology in the future . For Amazon company case, it is not stopping of logistics. It is also aiming to automatically manage the supply of consumer home products with its recently launched Amazon replenishment service, Dash. Dash is a digital service that enables that connected derive to automatically order physical products from Amazon when supplies are running low. So, it seems (AI) technology will be applied to logistic function by digital technology method introduction in the future.

Hence autonomous vehicles will optimize industry supply chains and logistics operations through increased efficiency and flexibility. In fact, fully automated and lean supply chains will keep reduce load sizes and inventory by leveraging smart distribution technologies and smaller autonomous vehicles by machine man assistance. If Amazon continues to grow market share for online sales by reducing effort required by the consumer to place an order, when also contributing the almost immediate delivery of products to the doorstep. So, it will further fuel the trend toward on-demand derive. As Amazon company fuels the on-demand economy, consumers will expect immediacy in more parts of the digital economy. On top of speed, consumers increasing expect more personalization options.

So, (AI) technology will influence digital manufacturing, such as Amazon publishing to monitor every aspect of every process in real -time and communicating to self-optimized deep learning robotics, new methods of high volume and high customization will become possible. Then, as products merge into product platforms and even services, manufacturers have the opportunity to provide components and platforms used by smaller players. So, (AI) technology will influence manufacturing industry to choose automated SMI lines, robots installed, automation engineers.

Another future (AI) technology development can be applied to space science aspect, such as Automation engineering space in manufacturing process to achieve digital manufacturing benefits to any businesses in the future. Such as reducing cost, shortening manufacturing time, raising efficiency, shortening delivery products to client individual time. How can artificial intelligence give the need and advanced fast and evaluation methods benefits for space exploration? When US NASA ( space exploration organization) achieves any space exploration missions, it will answer this question:

When is it useful to have a machine use (AI) technology to achieve a decision? After all, after millions of years of space exploration and rough 10,000 years of civilization, humans are usually quite good at making decisions in complex uncertain environments. Through, Johns Hoplains University's Applied Physical Lab. Research in (AI) technology enabled systems, which has identified three general use cases for (AI) technology to explore space mission:

First, for some tasks (AI) technology is more cost effectiveness than human. Second, (AI) technology is better suited than humans at solving some, but not all problems. Third, (AI) technology allows NASA organization's space exploration mission to develop machines that ate capable of responding faster than when a human is in the decision loop ( D. Scheidt, 2012, A. Castano et. al. 2008).

So, the use of (AI) technology to enable science by observing the pace of rapidly evolving phenomena was demonstrated. It is more effectively coordinating and (AI) technology utilizing to earn economic benefits to use for space exploration mission.

However, (AI) technology also have current risk for space exploration. Today (AI) technology is immature and requires further development to reach its potential. For instance, the (AI) technology algorithms that detected the dust derive could not have identified whether the Martain weather represented a threat to the cover. Also it can not yet use instrument input to determine what, where and how to autonomously make the next space science measurement. An equally important factor limiting (AI)'s deployment is that lacks the methodology and technology to effectively test (AI) technology. So, the challenge will testing (AI) enabled system is how (AI) performance can be measured. It would be NASA organization's difficulty to find (AI) technology to develop to carry on researching any space exploration missions in the future. However, (AI) technology will be a good economic benefit choice for space exploration mission in the future.


● What is artificial intelligence potential
benefits and ethical considerations?

The ability of (AI) technology systems to transform vast amounts of complex information into insight has the potential to help solve manufacturing or service challenges for human needs. However, to reap the societal benefits of (AI) systems, humans will need to trust then and make sure that which follow the same ethical principles, moral values, professional codes and social norms that we humans would follow in the same scenario, research and educational efforts as well as carefully designed regulation in order to achieve the most effort of economic benefits goals. For example, international business machines corporation (IBM) is actively engaged both competitors , in global discussions about how to make (AI) ethical and as beneficial as possible for people as social economic benefits. (AI) is usually defined as the " capability of a computer program to perform tasks or reasoning processes " that human usually associate to intelligence in a human being. Often, it has to do with the ability to make a good decision, even when there is uncertainty, too much information to handle. As an example, play chess or complex card games of entertainment activities is believed to need some form of intelligence in a human being, as well as choosing the best medical facilities in a difficult medical case, or creating something new, such as mathematical theorem or even some form of act, or even driving automatic machine man ( self driving vehicle) replacing human driving in the middle of a crowded city.

(AI) needs depends on what we consider being intelligence in the behavior of a human being act a certain point in time. If human belief about human intelligence changes and we don't believe any longer that a certain task requires intelligence, then a computer program performing that task is no longer part of (AI), it becomes just another boring computer program. So, it means that (AI) technology will replace some old computer programs, if human can invent new generation of (AI) software for any functions or activities to satisfy human needs.

As IBM, it argues intelligence. This means that we aim to build systems that enhance and scale human expertise and skills rather than replacing them. We therefore focus on practical applications of (AI) capabilities that assist people in performing well-defined tasks of needs by exploiting and wide range of (AI)-based services. We also use the term " cognitive computing" it is mean a comprehensive net of capabilities based on technology. It comprises the fields of machine learning, reasoning and decision technologies, language, speech and vision recognition and processing technologies, high performance and high efficient functions for any industries or individual consumers needs. For example, robotics, which are usually very good at doing what which are supposed to in any environment, much have public shopping center, factory etc. places which need simply services from the robot ( machine man), such as cleans the floor of our houses to the robot that can work together with humans in production chains, passing through the warehouse, robots can take care of the tasks of an entire warehouse and the companion robots like Nao, Pepper, Aibo and Giraff, who can entertain use, talk to use and help elderly people to stay connected to their friends, relatives and doctors.

Google company is building automatic machine ( self-driving cars ) and has acquired more than 10 robotics companies. Facebook had opened whole new research facility only on (AI) research. Apply computer has developed Siri. Microsoft computer company has built a similar personalized assistant. Google has Deep mind, a UK company whose long term aim is to build general (AI) and has already great potential to win game to the world champion and IBM is investing a huge amount of resources in applying its Watson cognitive computing system to the medical domains to finance and to personalized education. In Europe, IBM is establishing new centers in Munich and Milan focused in the application of cognitive computer capabilities to the internet of things and healthcare respectively.

For example, automatic machine man ( self-driving cars) are all about (AI), which used to be able to see what happens in the street ( signals ,lanes, other cars, pedestrians, traffic lights, which need to able predict what other cars and pedestrians will do, and who need to be able to cope with unforeseen situations. Since, most car accidents are due to human fault, it is estimated that the adoption of self-driving cars will save about half of the lives that are usually last in car accidents.

IBM Watson company has to understand spoken language, make sense of massive amount to text , respond correctly to questions in many categories, as well as assess its own confidence in responding to such questions. In the future, (AI) technology can own question/answering capabilities that would be very useful, for example, in assisting a doctor when trying to some to the correct diagnosis for a patient and to propose the best therapy .

Intelligent machines can also rely on huge amounts of data to be used to learn how to make better decisions. This data comes from all of us over the years Facebook users have uploaded more than 250 billion pictures and every day who upload about 350 million more. Every second, we submit 40,000 google search queries. So, (AI) technology will be connected through the web from appliances to traffic lights from cars to watches. Other tasks that are very easy for humans are physical and manipulation tasks, such as walking , running, picking up an object to make its shape and location, restricted environment. But (AI) machine man technology still not able to have the general physical and manipulation capabilities even of a 6 year old.

So, it brings this question: Why do (AI) scientists need to concern ethics? Because (AI) technology is complex, information into insight has the potential to reveal long held secrets and help solve some of the world's most difficult problems. (AI) systems can potentially be used to help discover insights to treat disease, predict the whether, and manage the global economy. So, ethic issues is important to and (AI) scientists . If any one new (AI) technology research investigation could success, it will be a secret to and the (AI) scientists can not permit to their loyalty to any competitors to damage the fair (AI) technology products trading market. The country ( countries) (AI) technology scientists need to concern ethic issues, who need to keep secrets for their countries economic or/and social benefits. This is moral issues to any countries/country loyalty is whose countries intangible assets. They can not sell (AI) loyalty to any their countries to assist whose economic benefits immorally.

● How can (AI) technology influence to global
health care economy development?
According to (AI) lecturer analysis, when combined key clinical health (AI) application can potentially create $150 billion in annual savings for the US healthcare economy by 2026 year. (AI) technology is re-winning modern conception of healthcare delivery. It enables machines to sense, comprehend, act and learn. So which can perform administrative and clinical healthcare functions (Accenture, 2017).

It will help health care service organizations to reduce health care cost, will improve and raise service quality and access. So, (AI) health market size will be predicted growth. (AI) applications in health care include robot-assisted surgery, virtual nursing assistant, administrative workflow assistant, fraud detection, error reduction connected machines, clinical trial participant identifier, preliminary diagnosis, automated image diagnosis and cybersecurity.

What kind of benefits (AI) technology can contribute to healthcare service? (AI) technology can deliver what many health care organizations need, such as financial and operational of labor costs, digital expectations from patient consumers how to use (AI) technology to solve interoperability challenges in any healthcare organizations. Also (AI) technology can be applied to wellness an d lifestyle management, diagnostics, delivers financially but also way of organizational and workflow improvement. So, (AI) technology will be continue to become most prevalent and adoption to healthcare organizations , which must need to enhance structure to be position to take full advantages of new (AI) technological capabilities. (AI) technology can change the nature of work and employment is rapidly changing to make the best use of both humans and (AI) talent in healthcare industry in the future. For example, (AI) technology offers a way to fill in gaps and the rising labor shortage in healthcare. According to Accenture analysis, the physicians shortage is increasing. However, (AI) technology will manufacture healthcare machine men to replace physicians in future one day(2017). Hence, (AI) technology will be invented to raise health care service staffs work efficiency and performance in any hospitals or clinics in the future.

In conclusion, (AI) technology will raise efficiency for any service or manufacturing industries in the future, although, it is possible that it will also rise low skillful workers unemployment numbers. But, the most important influence to human technological innovation will be risen and it will influence human life will be changed to be better, e.g. self drive cars, health care physician machine men, machine man cleaners etc. intelligent machine men will be manufactured to serve for our daily life. Furthermore, (AI) technological products will influence countries trading, some low technological development countries manufacturing businessmen can choose to buy any (AI) products to raise whose productivity and efficiency and reducing cost to achieve economic cost saving result. Also, GDP of trading growth income will increase to the (AI) products sale countries. Hence, it will be beneficial to economic development to both developed and developing countries both in the future as well as (AI) scientists time and money spending will be valued to continue to invest (AI) technology development for human life and economy benefits for long term.

In conclusion (AI) technology will raise macro economy growth and it can create many (AI) jobs , but it also raise the low level technological worker unemployment change. In the future, (AI) technology can be applied to digital technology to attempt to invent any new undiscovered (AI) and digital technology. So, it needs any scientists to continue to research how digital and (AI) technology can be mixed to satisfy human's future undiscovered needs.

AI thid stage development

Artificial intelligence education development

or the future of defense choice

Nowadays, artificial intelligence (AI) is widely knowledge to be one kind of the dramatic technology. However, it is expected to continue, to have a disruptive impact on human's private and public life, so defense and security will be no exception. But how exactly will these be affected ? How will (AI) defense and security is incremental in nature?

To research why artificial intelligence (AI) has possible to be used to cause autonomous weapons by human. We need to understand these three aspects of relationship. They include cybersecurity and artificial intelligence and machine learning and autonomous weapon systems relationship between of them.

Firstly, we need to know what is the mean of artificial intelligence and cyber defense/offense? It means defense of critical networks: real time, pattern finding, anomaly seeking, it must utilize machine (AI) learning algorithms to efficiently, and instantaneously respond to potential network threats as well as it means human on or out of the loop. On the loop : it means anomaly detection: human notified, IT analysis, response. Out of the loop: it means anomaly detection: (AI) decides best method of response: quarantine, honey pot monitoring, hack-back. Thus, it is possible that (AI) can be used , such as autonomous cyber weapon.

What is artificial intelligence and autonomous weapons? Autonomous weapons mean one kind of weapon that can be selected and engaged a target, without intervention by a human operator. Are these machines artificially intelligent? I believe the answer is not, because present weapons systems are not capable of human level reasoning. But, (AI) algorithms are presently employed to process sensor data, monitor system health, take and respond to vocal commands manage data, navigate. This, future autonomous weapons systems will require stronger (AI) to be secure and operationally and cost effective. Moreover, self-aware autonomous cyber systems are crucial.

What is cybersecurity mean? It means the ability to control access to networked systems and the information they contain. It is acted to prevent , detect, recover, react. It is application objects concern people, process, technology and it's application goals are confidentiality, integrity and popular availability. Thus, what is cyber weapon mean? Walware means viruses, Trojans, zero-days, worms ransomware, spyware etc. Does it require a particular objective? E.g. military paramilitary or intelligence. Does it require physical harm? E.g. functional harm or interruption? Mental harm? Is (AI) a technological weapon that it is an object or tool? What about when it is an weapon agent?

In simplicity, (AI) can be one of scientific weapons platform. When one day, it is invented to be applied to control war planes to fly to any countries to attack enemies or it is invented to be seemed to human to replace soldiers to bring guns or any weapons go to other countries to attack. So, it is possible that future any war defense planes, (AI) technological automatic control weapon can be replaced of human soldiers or war plane pilots to control any war defense planes to go to different enemy countries to attack them easily. It is very horror matter to threaten global human's ourselves life in the future , if (AI) automatic control war defense planes or (AI) automatic control machine soldiers were invented successfully.

Hence , when (AI) can be applied to weapons platforms, it structures that launch weapons, i.e. jets, ships, vehicles. (AI) platform and weapon and software architecture components are be done one (AI) technological weapons systems. Thus, human will encounter any (AI) benefits or risks ( threats) causes in the same time as soon as possible. If we can predict when (AI) weapon system will be manufactured or invented successfully. Then, we can reduce (AI) weapon systems risks , if we can threaten any (AI) scientists continue to invent any undiscovered (AI) weapons in any time to avoid the future first time (AI) weapon war occurrence in possible.

The (AI) weapon system risk means autonomy: the ability to problem solve technological war , when (AI) weapon system is manufactured successfully, the power to act, how to damage the (AI) weapon system. The power to chance to stop (AI) weapon system manufacturing processes, ability to create a new goals, how to change the (AI) weapon system inventors' or scientists' minds to avoid to apply (AI) tools to achieve attack goals to change to another positive goal. Due to human can't know a prior what an autonomous (AI) weapon system will do.

Although, human is known what (AI) is , but human is also known when (AI) scientists whose emergent behaviors will do to change to do any negative behaviors from positive behaviors. Whatever (AI) weapon system design we use, there will be cybersecurity, problems arising from computation design/complexity. Due to any one (AI) scientist can manipulate the system to act against itself, or who can utilize traditional " cyber weapons" against the (AI) weapon system, or who can manipulate the system to lie to humans, but also due to complexity, there is no way to know if it is lying or not or bounded rationality : satisficing.

Finally, the most serious (AI) technological invention risks are human is unknown these aspects of (AI) absolutely: They are not simple automatic systems, learning reasoning, communication of " self-aware" systems. Thus, human will face (AI) technological invention risks or threats. We need to find any methods to avoid (AI) weapon system is manufactured successfully to avoid (AI) technological war can occur in future anyone day. Consequently, why (AI) scientists do not choose to invent (AI) machine men own human reading, writing, speaking, judgement, analyizing abilities to develop human's future education industry, but they choose to invent (AI) machine men own human weapon to attack enemy ability.

Hence, (AI) scientists have moral responsibilities to avoid to invent (AI) machine men to own human's attack abilities. I shall indicate future what aspects of (AI) machine men can be applied to different education industry aspects as below:

8.1 Online technology and online book
technology influences artificial intelligence
mind development

Nowadays, online technological invention bring online book technological development. Also, artificial intelligent technological machine men had been invented to link internet to do any jobs, e.g. children can find any data from artificial intelligent machine men when the artificial intelligent machine man had been installed internet and computer function, then children can find any online books to read from the artificial intelligent machine man. Such as Japan artificial intellgent machine men had installed computer and internet function, the Japan family children can find any online books to read from the artificial intelligent machine man at Japan any families' homes conveniently. Hence, it implies that future one day, artificial intelligent machine has possible to be invented to own human's reading and/or writing abilities.

For example,online book publishing is one kind of popular internet technology. For example, Amazon publish is as a business model with many potential advantages, relative to a physical operation. It held out the potential of lower book inventing and distribution costs and reduced overhead. Consumers could find the books, they were looking for more easily and a variety book topic choices could be offered for sale. It can accept and fulfill orders from almost any domestic location with equal ease. And most purchasers made on its site would be exempt from sales tax. One Amazon strategy hand, it would have to make its returns and redress processes transparent and reliable, and offer other ways for clients to learn, as much about the book possible before buying. Future online book market development trend, such as Amazon, Barnes & Noble etc. online book shops.

Hence, online book store technology can be applied to artificial intelligent technology. Such as artificial intelligent machine men can apply computer technology to learn the abilities of reading and/or writing any books either on paper or on computer. Hence, it is possible that artificial intelligent machine men will have similar human's writing and/or reading books ability when they own human's mind ability. However, it bring this questions: Can artificial intelligent machine men own human's mind abilities? If they own human's mind abilities, is it mean that they can write and/or read any books? Can artificial intelligent machine men own human's mind abilities to create to write any books? Can artificial intelligent machine men own human's mind abilities to read and make any judgements or decisions more accurate than human's judgements or decisions? To answer these questions? I shall indicate that online book reading and writing technology can be applied to artificial intelligent machine men reading and writing technology. Because they are similiar computer mind technological development. So, I believe that future artificial intelligence machine men can be invented to own similar human's reading and writing's mind abilities in future one day.

I believe artificial intelligence and online technological reading abilities are very similiar. Nowadays, computer can be invented to attempt to read and write any books by human. Why can not artificial intelligent machine men replace computer to read and write any books? Artificial intelligent machine men can replace human to attempt to write or/and read books, due to artificial intelligent machine men had invented to own human mind to do some jobs and their mind had been invented to be similiar to human behavioral abilities to do these behaviors, e.g. cooking, driving, playing games, singing songs, speaking, listening, frighting etc. different human's abilities. So, it seems that artificial intelligent will be possible to be invented to own human's mind abilities to do any writing or reading behaviors or functions.

8.2 Prediction of artificial intelligence
    reading and writing abilities

development

What is future trend of artificial intelligence reading and writing abilities development? To answer this question, we need to know what benefits of artificial intelligent machine men can attribute to human's needs when they can own any human's mind to read or/and write any books.

I shall indicate e-books reading and writing example, if artificial intelligent machine men can be invented to own human's mind to write and/or read e-books on computer. Then, it brings this question: Can artificial intelligent machine men assist human to learn to do judgement to solve any challenges?
    I believe that when artificial intelligent machine men can be invented to own human mind to write or/and read any books, then they will own human's mind ability to make judgement to solve any challenges more accurately, even their decisions can be more accurate to compare to human's decisions. So, artificial intelligent machine mens' writing and reading ability is the main factor to cause their mind to do any judgement in order to make any decisions more accurately. Consequently, in future one day, artificial intelligent machine mens' writing and reading ability will be invented to similar human's reading and writing abilities as well as their minds can also be invented to similar human's minds as well as their judgement abilities can be invented to similar to human's judgement abilities to make any decisions more accurate.

8.3 The influences when AI is invented to own
human's mind and judgement abilities
    Finally, I shall discuss what are the influences when AI is invented to own human's mind and judgement abilities in our future job market. The achievement of artificial intelligent (AI) machine men achievement requirement of owning human's mind and judgement abilities which requires extensive manual labor, and by augmenting the calling process with machine learning, the process where speed and accuracy are needed to close to human's mind and judgement abilities. Expert human race callers now have better information at artificial intelligent machine men at their fingertips faster.

Hence, if the above those requirements are achieved to satisfy artificial intelligent machine men ind and judgement abilities demand to close or exceed humans' mind and judgement abilities. Then, I believe that future human's some simple jobs must be replaced by (AI) machine men. Even, human's some professional jobs, e.g. lawyer, accountant, administator, typing etc. professional skillful jobs, which will be either replaced or will be assisted by (AI) machine men. For example, (AI) machine men learn how to type english or other language words to do typing job ; they can learn how to apply accounting knowledge to record any firm's income and expenditure record of accounting job; they can also learn how to assist architects to design any architectural building drawing plans to do architect jobs; they can learn how to analyze any court evidences to judge any criminal or civil cases and assist lawyers to give legal advices to achieve more reasonable judgement for any legal cases; they can also learn how to assist firm's managers or administrators to manage any organization teams efficiently.

Consequently, when (AI) machine men can be invented to achieve to exceed human's mind and judgement abilities level. Then, I believe that they can do instead of human' simple jobs, which can do even human's more difficult and more judgement requirement of professional skillful jobs. So, (AI) machine men must need to achieve to do any jobs, they are same, even exceed to human professionals' abilities. Then, it will cause a lot of human's jobs to be disappeared or some human's jobs will be replaced by owning judgement and mind abilities of (AI) machine men to do.

Hence, future many human's jobs will be replaced by technological labors. Employers choose to buy (AI) machine men to replace human labors. The reasons include (AI) machine men have none unhappy, angry emotin to influence their low efficiencies and low productivities. Their judgement and mind abilities can exceed human's abilities or do any jobs to compare better performance to human's abilities. Consequently, different occupation labors need to prepare to learn how to co-operate with (AI) machine men to let future employers feel (AI) machine men will be human's assistant to assist human to do jobs efficiently when human and (AI) machine men work together. It aims to avoid future employers feel (AI) machine men's judgement and mind abilities can exceed any low knowledgeable and skilful occupation labors, even high knowledge and skilful occupation labors. It means that (AI) machine men are only labors' assistant if (AI) machine mens' judgement and mind abilities are below under to human labors' judgement and mind abilities.

Consequently, to avoid (AI) machine men can replace human to do any simple or complex jobs to cause any future any occupation labors' competitiors. I recommend that it is right time labors ought prepare to learn different skills. So, every individual labor does not only concentrate on one kind of skill. Because supposing one kind of the occupation labor's job duties are replaced by (AI) machine men. If the employee had owned more than one kind of occupation skill. Then, I believe that who can avoid the unemployment threat more easier than the employee only owned one kind of occupation skill, when (AI) machine men had invented to own human's mind and judgement abilities in future one day. The most important, I believe that (AI) invention will be applied to be teach how to learn human's skills and mind ability. Such as education industy, teacher won't be replaced by (AI), otherwise, (AI) will be teacher's assistant to help them to do education data gather or teaching jobs. So, teachers won't be replaced by (AI), otherwise, teachers will depend on (AI) data gather or teaching job to give them opinions how to solve student's teaching challenges as well as teachers can concentrate on researching education jobs for schools' benefits if (AI) technology can be invented to on human's mind and judgement and reading and writing abilities in the future.

8.4

Artificial intelligence and the future of defense

or teaching choice

Nowadays, artificial intelligence (AI) is widely knowledge to be one kind of the dramatic technology. However, it is expected to continue, to have a disruptive impact on human's private and public life, so defense and security will be no exception. But how exactly will these be affected ? How will (AI) defense and security is incremental in nature? If (AI) technological machine men are applied to teach students in education aspect, is it better to my next generation learning develpment more than they are applied to war attack aspect.

To research why artificial intelligence (AI) has possible to be used to cause autonomous weapons by human. We need to understand these three aspects of relationship. They include cybersecurity and artificial intelligence and machine learning and autonomous weapon systems relationship between of them. Basic on (AI) machine can be invented to learn any new knowledge, so if (AI) machine men are taught how to attack enemy, which will be such as human soldier function. But, if (AI) machine men are taught how to learn university knowledge to teach students. Then, they will be such as human lecturer function. So, when (AI) is invented to own human mind and judgement and learning abilities, then they will be either human's enemy or human's assistant, such as university lecturer's assistant.

Firstly, we need to know what is the mean of artificial intelligence and cyber defense/offense? It means defense of critical networks: real time, pattern finding, anomaly seeking, it must utilize machine (AI) learning algorithms to efficiently, and instantaneously respond to potential network threats as well as it means human on or out of the loop.

On the loop : it means anomaly detection: human notified, IT analysis, response. Out of the loop: it means anomaly detection: (AI) decides best method of response: quarantine, honey pot monitoring, hack-back. Thus, it is possible that (AI) can be used , such as autonomous cyber weapon. If (AI) is applied to make the decision best method of response to learning aspect, such as univeristy different subject knowledge. Then, it will be one good technological educational tool to teach university students.

In simplicity, (AI) can be one of scientific weapons platform or one of university teaching tool. When one day, it is invented to be applied to control war planes to fly to any countries to attack enemies or it is invented to be seemed to human to replace soldiers to bring guns or any weapons go to other countries to attack. So, it is possible that future any war defense planes, (AI) technological automatic control weapon can be replaced of human soldiers or war plane pilots to control any war defense planes to go to different enemy countries to attack them easily. It is very horror matter to threaten global human's ourselves life in the future , if (AI) automatic control war defense planes or (AI) automatic control machine soldiers were invented successfully. Otherwise, when one day, (AI) machine lecturer is invented to be applied to learn university different subjects knowledge to replace lecturers to copy lecturer's every prepared lecturer course to speak to let students to listen when they are sitting in university halls as well as the (AI) machine lecturer can make analysis and judgement response to answer every student's enquire immediately after it had speaking all courses to students to listen in lecturer hall every time. Then, it can let human lecturer does any education job duty, e.g. research education work. So, (AI) machine lecturer will be future human lecturer's assistant in future one day.

Finally, the most serious (AI) technological invention risks are human is unknown these aspects of (AI) absolutely: They are not simple automatic systems, learning reasoning, communication of " self-aware" systems. Thus, human will face (AI) technological invention risks or threats if human invent (AI) machine man to learn how to attack enemy. Otherwise, if human invent (AI) machine man to learn how to teach univerity student. I believe that my future university students can raise learn ability and writing ability and reading ability from (AI) machine lecturer teaching more than human lecturer teaching.


8.5

Online technology and online book
technology influences artificial intelligence
mind development

Nowadays, online technological invention bring online book technological development. Also, artificial intelligent technological machine men had been invented to link internet to do any jobs, e.g. children can find any data from artificial intelligent machine men when the artificial intelligent machine man had been installed internet and computer function, then children can find any online books to read from the artificial intelligent machine man. Such as Japan artificial intellgent machine men had installed computer and internet function, the Japan family children can find any online books to read from the artificial intelligent machine man at Japan any families' homes conveniently. Hence, it implies that future one day, artificial intelligent machine has possible to be invented to own human's reading and/or writing abilities.

For example,online book publishing is one kind of popular internet technology. For example, Amazon publish is as a business model with many potential advantages, relative to a physical operation. It held out the potential of lower book inventing and distribution costs and reduced overhead. Consumers could find the books, they were looking for more easily and a variety book topic choices could be offered for sale. It can accept and fulfill orders from almost any domestic location with equal ease. And most purchasers made on its site would be exempt from sales tax. One Amazon strategy hand, it would have to make its returns and redress processes transparent and reliable, and offer other ways for clients to learn, as much about the book possible before buying. Future online book market development trend, such as Amazon, Barnes & Noble etc. online book shops.

Hence, online book store technology can be applied to artificial intelligent technology. Such as artificial intelligent machine men can apply computer technology to learn the abilities of reading and/or writing any books either on paper or on computer. Hence, it is possible that artificial intelligent machine men will have similar human's writing and/or reading books ability when they own human's mind ability. However, it bring this questions: Can artificial intelligent machine men own human's mind abilities? If they own human's mind abilities, is it mean that they can write and/or read any books? Can artificial intelligent machine men own human's mind abilities to create to write any books? Can artificial intelligent machine men own human's mind abilities to read and make any judgements or decisions more accurate than human's judgements or decisions? To answer these questions? I shall indicate that online book reading and writing technology can be applied to artificial intelligent machine men reading and writing technology. Because they are similiar computer mind technological development. So, I believe that future artificial intelligence machine men can be invented to own similar human's reading and writing's mind abilities in future one day.

I believe artificial intelligence and online technological reading abilities are very similiar. Nowadays, computer can be invented to attempt to read and write any books by human. Why can not artificial intelligent machine men replace computer to read and write any books? Artificial intelligent machine men can replace human to attempt to write or/and read books, due to artificial intelligent machine men had invented to own human mind to do some jobs and their mind had been invented to be similiar to human behavioral abilities to do these behaviors, e.g. cooking, driving, playing games, singing songs, speaking, listening, frighting etc. different human's abilities. So, it seems that artificial intelligent will be possible to be invented to own human's mind abilities to do any writing or reading behaviors or functions.

8.6 Prediction of artificial intelligence
reading and writing abilities

development

What is future trend of artificial intelligence reading and writing abilities development? To answer this question, we need to know what benefits of artificial intelligent machine men can attribute to human's needs when they can own any human's mind to read or/and write any books.

I shall indicate e-books reading and writing example, if artificial intelligent machine men can be invented to own human's mind to write and/or read e-books on computer. Then, it brings this question: Can artificial intelligent machine men assist human to learn to do judgement to solve any challenges?

I believe that when artificial intelligent machine men can be invented to own human mind to write or/and read any books, then they will own human's mind ability to make judgement to solve any challenges more accurately, even their decisions can be more accurate to compare to human's decisions. So, artificial intelligent machine mens' writing and reading ability is the main factor to cause their mind to do any judgement in order to make any decisions more accurately. Consequently, in future one day, artificial intelligent machine mens' writing and reading ability will be invented to similar human's reading and writing abilities as well as their minds can also be invented to similar human's minds as well as their judgement abilities can be invented to similar to human's judgement abilities to make any decisions more accurate.

8.7 The influences when AI is invented to own
human's mind and judgement abilities

Finally, I shall discuss what are the influences when AI is invented to own human's mind and judgement abilities in our future job market. The achievement of artificial intelligent (AI) machine men achievement requirement of owning human's mind and judgement abilities which requires extensive manual labor, and by augmenting the calling process with machine learning, the process where speed and accuracy are needed to close to human's mind and judgement abilities. Expert human race callers now have better information at artificial intelligent machine men at their fingertips faster.

Hence, if the above those requirements are achieved to satisfy artificial intelligent machine men ind and judgement abilities demand to close or exceed humans' mind and judgement abilities. Then, I believe that future human's some simple jobs must be replaced by (AI) machine men. Even, human's some professonal jobs, e.g. lawyer, accountant, administator, typing etc. professional skillful jobs, which will be either replaced or will be assisted by (AI) machine men. For example, (AI) machine men learn how to type english or other language words to do typing job ; they can learn how to apply accounting knowledge to record any firm's income and expenditure record of accounting job; they can also learn how to assist architects to design any architectural building drawing plans to do architect jobs; they can learn how to analyze any court evidences to judge any criminal or civil cases and assist lawyers to give legal advices to achieve more reasonable judgement for any legal cases; they can also learn how to assist firm's managers or administrators to manage any organization teams efficiently.

Consequently, when (AI) machine men can be invented to achieve to exceed human's mind and judgement abilities level. Then, I believe that they can do instead of human' simple jobs, which can do even human's more difficult and more judgement requirement of professional skillful jobs. So, (AI) machine men must need to achieve to do any jobs, they are same, even exceed to human professionals' abilities. Then, it will cause a lot of human's jobs to be disappeared or some human's jobs will be replaced by owning judgement and mind abilities of (AI) machine men to do.

Hence, future many human's jobs will be replaced by technological labors. Employers choose to buy (AI) machine men to replace human labors. The reasons include (AI) machine men have none unhappy, angry emotin to influence their low efficiencies and low productivities. Their judgement and mind abilities can exceed human's abilities or do any jobs to compare better performance to human's abilities. Consequently, different occupation labors need to prepare to learn how to co-operate with (AI) machine men to let future employers feel (AI) machine men will be human's assistant to assist human to do jobs efficiently when human and (AI) machine men work together. It aims to avoid future employers feel (AI) machine men's judgement and mind abilities can exceed any low knowledgeable and skilful occupation labors, even high knowledge and skilful occupation labors. It means that (AI) machine men are only labors' assistant if (AI) machine mens' judgement and mind abilities are below under to human labors' judgement and mind abilities.

Consequently, to avoid (AI) machine men can replace human to do any simple or complex jobs to cause any future any occupation labors' competitiors. I recommend that it is right time labors ought prepare to learn different skills. So, every individual labor does not only concentrate on one kind of skill. Because supposing one kind of the occupation labor's job duties are replaced by (AI) machine men. If the employee had owned more than one kind of occupation skill. Then, I believe that who can avoid the unemployment threat more easier than the employee only owned one kind of occupation skill, when (AI) machine men had invented to own human's mind and judgement abilities in future one day.

8.8

Why does AI machine lecturer can raise
education quality?

When (AI) machine men can own human reading and writing and judgement and analytical abilities, then they can replace university lecturers to teach students to raise students' learning abilities absolutely. Then, it bring this question: Why does I machine lecturer can raise education quality? Why do universities prefer to apply (AI) machine lecturer to teach teachers more than human lectuer in university lecturer hall learning environment? Will (AI) university lecturers replace human lecturers to teach students to learn at university lecturing halls popularly? Can (AI) university lecturers replace university human lecturers to teach students more easily and it can let students feel more easily to learn when they are listening what (AI) university lecturers are teaching to them every time university lecture.

In university today, nearly all students need to attend university lecturing hall to listen their lecturer's teaching in every time course. However, many students do not feel interesting to attend university halls to listen human lecturer's teaching. The reasons include, they are busy, so no time to attend lecturer's hall to listen lecturer's teaching; or they feel bore to listen their lecturer's teaching; they feel difficulty to learn; they have confidence to exam and do their assignments, so they feel that they do not need to go to lecturing halls to listen their human lecturer's teaching. However, if one day, (AI) machine lecturers are invented to teach university students to learn and solve their learning difficulties. Can it raise student individual learning interest, due to (AI) machine lecturers' education quality is better than human lecturers' education quality?

What will influence to university students if (AI) machine lecturer can invented to replace human lecture? The influences will include such as below:

First reason: the only way is going to be useful to university lecturers are if all (AI) machine lecturers are well-informed and fully supported to assist human lectuers to teach whose students to let them to listen whose teaching absolutely. So, human lecturers can concentrate on doing any education research and data gathering jobs to prepare for (AI) machine lecturers to help them to explain human lecturers' every time prepared course contents more efficiently. So, (AI) machine lectuers can help human lecturers to share whose teaching time in lecturing halls. Human lecturers' can spend whose hall lecturing time to do whose educational research or other educational gathering jobs absolutely.

The second reason, the human lecturer (Human capital) has ability and efficiency of concentrating on education data gatehering research jobs to prepare to write whose books. When (AI) machine lecturer replace the human lecturer to spend time to attend lecturing hall to teach students. Fo long term, the human lecturer can raise education productivity growth and education quality raising, due to who only concentrate on searching or gathering data to prepare to write whose books to raise their education level.

In macro and micro economic view, the well (AI) machine lecturer educated labor ( human capital) is often replaced to human lecturer as one of the critical factors to influence rapid education productivities and educational quality growth to the Asia developing countries' any regions or cities. Because any of these Asia developing countries, such as China, Korea, Philippines etc. countries which need have well educated and knowledgeable lecturer labors to raise any universities' educational productivities and educational qualities growth. So (AI) assistant lecturer factors which ought have close relationship to cause the good or bad future student learning effectiveness and education or learning qualities raising in these any one of Asia developing countries.

The third reason, for the big population of student growth number example, China's student growth rate is larger than school growth rate. If China expect every students have enough chance to study in schools, but university human lectuer numbers are not enough to supply to universities to teach their students. I believe that (AI) machine lecturer is only one kind of teaching method to solve these big population countries' university lectuer number shortage challenge.

In conclusion, in long term, (AI) machine lecturers can solve university human lecturer shortage challenge as well as they can attract many students to attend lecturing halls and human lecturers can raise education quality when they can concentrate on searching or gathering data to prepare their education career, when (AI) machine lectuers replace them to spend time to attend univesity halls to teach students in every university lecturing time.

8.9

Future AI machine education market

I believe that when AI (artificial intelligent machine men) which can invented to own to similar to human mind, learning, language, analytical, judgement abilites. Then, which can be applied to any education market service industy. (AI) potential education market service industy includes such as below:

● (AI) university lecture assistant

Future (AI) machine men can assist univerity lectuers to attend university halls to attempt to teach university students for different subjects, e.g. english, math, economic, math, engineering, art, architect etc. different subjects. It depends on the human lectuter who prepares to spend time to teach the (AI) machine lecturer to remember whose teaching subject. For example, the economic lecturer spend one year time to teach the (AI) machine lecturer to learn all economic knowledge. Then, the (AI) machine lecturer can use its machine brain to remember all the human lecturer's economic concepts and prepared teaching economic contents within the one year. Hence, after one year the (AI) machine lecturer can remember all the human lecturer's economic concepts and economic theories and economic contents to prepare to attend university lecturing halls to teach all first year undergraduated first year economic students confidently. It means that the human lecturer's job duties will change to teach (AI) machine lecturer to learn whose economic knowledge to prepare to let the (AI) machine lecturer to replace whom to teach whose university students; so the human lectuer can spend more time to do other research job for whose university education development. Hence, the (AI) machine lecturer can share the human lecturer teaching job as well as the human lectuer can concentrate on spending time to do whose research jobs for whose university education development. This is one both win strategy to university and the lecturer if (AI) machine lecturer is invented to assist future university lecturer's teaching jobs

● (AI) secondary and primary teacher assistant

In the future (AI) technolgical development, instead of (AI) machine men can be applied to university education aspect. Future (AI) machine men can also be applied to secondary and primary teaching aspect. For example, primary and secondary schools do not need attend classroom to teach students. (AI) machine teachers can replace them to attempt to do teaching job. They only need to spend one year time to prepare to teach (AI) machine teacher to learn how to apply their teaching skill concern their subjects who need to teach to their students, e.g. english language writing and reading and spelling skill, sing song skill, drawing picture skill, calculation skill etc. different studying skill. Then, the (AI) primary or secondary machine teacher can apply the primary or sendary human teacher skills to attempt yo teach whose students. Hence, the primary and secondary human teacher whose duties will change to learn how to teach whose teaching skills to let the (AI) primary or seondary machine lecturer to remember how to apply human skills to teach whose students for different subjects, such as, english writing and reading and spelling language skills, singing songs language skills, math calculation skills etc. Hence, future primary or secondary school teachers who responsibilities will change to learn how to teach (AI) machine teacher teaching skills to prepare to replace them to teach their students in classroom.

● (AI) scientific research assistant

Future (AI) machine men can be applied to science research aspect, instead of school education job. For example, (AI) machine men can be any scientist's assistant, e.g. space scientist, earth or ocean scientist, human or animal behavioral psychological scientist, climate scientist, chemical scientist, drug scientist etc. How can (AI) machine men can be any kind of scientist to assist scientists to do research jobs ? I shall indiate such as below:

For space science example, the (AI) machine space scientist can assist human space scientist to gather space data to assist space scientist to research any undiscovered material to cause our earth, even space. Hence, the space scientist only need to teach the (AI) machine scientist to learn how to help them to apply space technological tools to gather data and then enter all data to computer to record, even the (AI) machine scientist can store all space data discovered record to their machine brain every day. Hence, the human space scientist does not need to spend much time to do gathering data job. The (AI) machine space scientist can help whom to do these space data gathering job, then the human space scientist can concentrate on spendin time to do space research job in whose space science laboratory every day.

For earth science example, the (AI) machine earth scientist can help the earth scientist to go to anywhere to gather earth or ocean natural activity data in our earth every day. Then, the earth scientist only need sit in whose earth laboratory to wait the (AI) machine earth scientist to come to whose laboratory to give whose gathering every day earth or ocean natural activity data to do future research job. Hence, the earth scientist does not need to leave whose laboratory to do any data gathing jobs concern earth or ocean natural activities. The (AI) machine earth or ocean scientist had helped him/her to go to our earth or ocean anywhere to do any earth or ocean activities data gathering jobs every day. Hence, the earth or ocean scientist can concentrate on spending whose time to do any research jobs in laboratory. It means that the (AI) machine earch or ocean scientist had replaced whom to do all outdoor original gathering data jobs.

For these human or animal behavioral psychological scientist, climate scientist, chemical or drug scientist, scientist all examples, the (AI) machine human or animal behavioral psychological scientist can help them to do any data gathering job, e.g. the (AI) machine scientist can learn how to help human or animal behavioral psychological scientist to contact human or animal to observe their daily activities and record all their activities data to transfer all these daily activites data to let the human or animal behavioral psychological scientist to do psychological researching analysis only. The (AI) machine climate scientist can help the human climate scientist to arrive anywhere to observe climate changes and record climate changes daily. Then, the human climate scientist only need to wait the (AI) machine climate scientist's gathering climate change data record from whose machine brain to do climate changing predict research job in climate laboratory every day. The (AI) chemical or drug machine scientist can help the drug or chemical scientist to gather data of new drug or chemical from internet channel every day. So, the human chemical or drug scientist only need to do researching job after the (AI) machine scientist transfers all daily chemical or drug information to let them to know from internet channel. It means that the chemical or drug human scientist does not need to spend much time to gather drug or chemical new data development trend from internet. The (AI) machine chemical or drug scientist had helped them to do data gathering job every day.

Consequently, future (AI) machine men can do education and research aspects of jobs duties and their role are only human scientists or primary or secondary teachers or university lecturers whose assistants either to share scientist's data gathering job or share teachers or lecturers' teaching job.

Artificia intelligent future education market development

9.1 Artificial intelligent robots invention

negative impacts

Indeed, artificial intelligence, computing can learn to something that effectively reasons, thinks, if (AI) learns more powerful and valuable complement to human capabilities; improving medical diaguoses, weather prediction, supply-chain management, transportation and even personal choices about where to go on vacation or what to buy, how to learn teaching any subjects knowledge to assist school teachers to teach students, e.g. accounting, law, architecture, language etc. subjects knowledge It can bring benefits to human it (AI) robots can learn teacher's any subjects to teach students or skillful knowledge, e.g. driving, taking care old people at home, manufacturing vehicles in factories. Otherwise, if (AI) robots learn how to be applied to be weapons to attack enemy. It will bring harm to human's life safety. As a result, technology can assist human's development when it can be applied to do meaning jobs, but it can also damage human's development when it can be applied to do harmful human's behaviour, e.g. attacking enemy to cause robot technological war. Consequently, the negactive impact will be depending on how humans teach (AI) robots to learn when some humans intends to teach (AI) robots to attack enemy. Thus, scientists need to know to educate (AI) robots to learn positive knowledge to attribute to us to raise our standard of living, if it is a tool in the service of humans, making our lives better. Otherwise, who ought not educate (AI) robots to learn negative knowledge to attribute to influence our's life safety when they are applied to attack enemies.

However, some scientists believe (AI) robots will have negative impact to influence our economy and society. Such as (AI) robots will destroy most jobs, it will make humans foolish, due to humans depend on (AI) robots to assist us to do any jobs; it will destroy people's privacy and it will enable bias and abuse and it will eventally exterminate humanity. Hence, when time comes to develop computers really think, intelligence is the same things as consciousness, the brain is a computer. Then, scientists must have responsibilities to concern how to teach (AI) robots to learn useful or attributable human's knowledge, not harmful human's knowledge to avoid future invented (AI) robots are taught to learn how to attack ourselves.

9.2 Future AI tutoring system potential

development market

Humans can learn (AI) robots how to educate our next generation after it has be taught any knowledge from human. Thus, it brings this question: How can human teach new knowledge to (AI) robots to learn successfully? I shall indicate some scientists' evidences to explain that it is possible (AI) robots had abilities to learn human's mind, judgement and analytical abilities, reading, writing, speaking abilities in future one day absolutely. I shall indicate what is intelligent tutoring systems (ITS) example as below:

When are computer-based tutors which act as a supplement to human teachers. The major advantage of an (ITS) is, it can provide personalized instructions to students according to their cognitive abilities. In this scenario, an intelligent tutoring system can be quite relevant to solve unavailability of skilled teachers challenges. Such as India has many students who demend teachers to learn to them, but India is facing skilled teachers shortage challenge. (ITS) are computer-based tutors, which act as a supplement to human teachers. An intelligent tutoring system is educational software containing an artificial intelligence component. The software tracks students work, tailoring feedback and hints along the way. By contenting information on a particular student's performance, the software can make interences about strengths and weaknesses, and can suggest addition work.

The ITS's one of main advantages is individualized instruction delivery to the group of same learning ability of student in evey classroom, which means the system ill adapt itself to different categories of students. A real classroom is usually heterogeneous where these are different kinds of students, from slow learners to fast learners. It is not possible to provide attention to them individually. Thus, the teaching may not b beneficial to all students. An ITS can eliminate this problem, because in this virtual learning environment, the tutor and the student has a one-to-one relationship. When the school applies TIS intelligent tutoring systems to teach its students, the students can learn in whose own method. Another advantage is their using this system teaching can be accomplished with there is trained teachers. Hence, the functionalities of this intelligent tutor system can be divided into three major tasks: (1) organization of the domain knowledge, (2) keeping track of the knowledge and peformance of the learner, (3) planning the teaching strategies on the basis of the learner's knowledge state. Hence, to carry out these tasks, the ITS have different modules and interfaces for communication ( Wenger, 1987; Freedman, 2000; Chou et. at, 2003).

The criteria of robots in education include domain of the learning activity, location of the activity, the role of the robot, types of robots and types of robotic behavior. Scientists indicate that robots are primarily used to provide language, science or technology education and that a robot can take in the role of tutor, tool or peer in the learning activity. However, robotics can include these kinds, such as social robotics, pedagogy, human robot interaction and educational robotics.

Early, industrial robots are invented to apply to manufacturing industry only in 2008 beginning. robots are slowly beginning a process to everyday, lives both at home and at school (IFR, 2008). Nowadays, the most popular countries accept to apply robots to replace human's some jobs, include Japan, Korea, USA, Australia, Germany, Holland.

What is the domain of the learning activity to robots? To develop robots to education industry, the first criterion of the two main categories are robotics and computer educaion ( the awareness of technology that could be referred as technical education) and non-technical education ( science and language). The technical education means giving students the knowledge of robots and technology. For example, introducing computer science and programming and to familiarize undergraduate students with technology and schol students were gradually exposed to technical subjects using robots. A lesson plan usually involves first an initial introduction to programming the robot ( introduction phase) and then the students apply their knowledge practically by making their robots work ( intensive phase. The second observed domain in the area of robots in education are non-robots in education are non-technical subjects (such as the sciences), where schools witness the employment of robots as an intermediate tool to impact some form of education to students in classrooms, such as mathematics. The third common domain is the use of robots to teach a second language. For example, English was taught to Asia countries children by robots in by researchers from the robotics laboratory. For another example, the implication of using robots to teach a second language have been well documented by computer science researchers in Taiwan, where it is stated that children are not as hesitant to speak to robots in a foreign language as they are talking to a human instructor. So, it seems, future robots can be a language teachers to the learning foreign language students. However, the language robots require having accurage speech recognition ability and how to learn in acknowledging the use of robots for language instruction ability in future robots language education market.

9.3 How can potential social teaching robots
assist to teacher in school?

What benefits do (AI) robots provide to education industry? Artificial intelligence can indicate how to improve courses, when teachers may no always be aware of gaps in their lecturers and educational materials that can let students confused about certain concepts. Different students have different learning styles, abilities, interests and needs. For classroom situation example, on teacher in a classroom of 20 to 30 students will rarely be able to cater to each of those needs. Homework and classes could be customized based on a student profile, interests can be cultivated based enhanced by exposing students to different course and content.

Artificial intelligence can offer a way to solve that problem. For online or non-online course providers cases, will have already benefits if these apply (AI) robots to educate. When a large number of students are found to submit the wrong answer to a homework assignment, the system alerts the teacher and goves future students a customized message that offers hints to the correct answer.

This type of system helps to fill in the gaps in explanation that can occue in online or non-online courses, and helps to ensure that all students are building the same conceptual foundation. Rather than waiting to either hear back from the professor in lecturng hall or classroom from classroom education method or receive back professor's feedback message from the internet online education channel. So, classroom students or online learning students can get immediate feeback that helps them to understand a concept and remember how to do it correctly the next time around if (AI) robots can be applied to assist university professor or primary/secondary teachers to teach whose students either in classroom teaching learning environment or online teaching learning environment. Thus, (AI) educational and social robots can be attempted to accepted to teach primary school, high school or university students.

How to apply (AI) education robots to teach students in which different educational functional aspects?

Firstly, on grading system hand, artificial intelligence can automate basic activities in education, like grading. In, college, grading homework and tests for large lecture courses can be for much grading activites to secondary teachers or university lecturers to spend much time to do grading jobs for every student. Even in lower grades, teachers often find that grading takes up a significant amount of time, time that could be used to interact with students, prepare for class, or work on professional development. When, (AI) may not every be able to truly replace human grading, it's getting pretty close. it's now possible for teachers to automate grading for nearly all kinds of multiple choice and fill-in-the blank testing and automated grading of student writing may not be far behind. Today, essay grading software will improve over the coming years, allowing teachers to focus more on in class activities and student interface then grading.

Secondly, on (AI) tutor's support hand, students could get additional support from (AI) tutors. When, there are obviously things that human tutors can offer that machines can't. The future could see more students being tutored by tutors that only exist in zero and ones. Some tutoring programs based on artificial intelligence already exist and can help students through basic, mathematics, writing, accounting, law and other subjects. (AI) tutors can teach students fundamentals for helping students learn high-order thinking and creativity, something that real-world teachers are still required to facilitate. It should not rule out the possibility of (AI) tutors being able to these things in the future. With the rapid technology advancement that has marked the past few decades, advanced (AI) tutoring systems will be popular to be applied to teach students.

Thirdly, on (AI) driven program educators helpful feedback hand, (AI) can't only help teachers and students to craft courses, that are customized to their needs, but it can also provide feedback to both about the success of the course as a whole. Some schools, especially those with online offerings are using (AI) systems to monitor student progress and to alert professors when there might be an issue with student performance. These kinds of (AI) systems allow students to get the support they need and for professors to find areas where they can improve instruction for students who may be given feedback how to learn subject matter. (AI) programs at thes schools aren't just offering advice on individual courses. However, some are working to develop (AI) systems that can help students to choose majors based on areas where they succeed.

Fourthly, on (AI) changing the role of teachers hand, there will always be a role for teachers in education, but what role is and that it entails may change, due to new technology in the form of intelligent computing systems. As (AI) can take over tasks like grade, can help students improve learning, any may even be a substitute for real world tutoring. (AI) systems could be programmed to provide expertise, serving for students to ask questions, and find information or could even potentially replace the teachers for basic course materials. In most cases, however, (AI) will shift the role of the teacher to that of facilitator. Teachers will supplement (AI) lessons, assist students who and provide human interaction and hands-on experiences for students. Hence, (AI) technology has already changed in the classroom teaching method, specially in schools that are online lessons.

Fifthly, on (AI) trial-and-error learning hand, trial and error is a critical pasrt of learning, but for many students, the idea of failing, or even not knowing the answer. An intelligent computer system, designed to help students to learn, to deal with trial and error. Artificial intelligence could offer studrnts a way to experiment and learn in a relatively judgement, free environment especially when (AI) tutors can offer solutions for improvement. In fact, (AI) is the perfect format for supporting this kind of learning as (AI) systems themselves often learn by a trial and error method. Sixthly, on changing student individual learning hand, (AI) has the potential to change where students learns, who teaches them, and how they acquire base skills, using (AI) systems, software and support. Students can learn from anywhere in the world at any time and with these kinds of programs taking the place of certain types of classroom instruction. So, (AI) robots may replace teachers in some instances ( for better or worse). Educatonal programs is powered by (AI). They are already helping students to learn basic skills, but as these programs grow. Consequently, as (AI) technology can give above advantages to educational organizations. So, in the future, it seems that (AI) can be popular to be used in online or non-online classrooms to assist teacher individual job to raise educational quality.

9.4 Psychological (AI) eduational

social robots and students relationship

Whether can (AI) educational robots bild good social relationsip to students? Some scientists had attempted to do experiments to prove whether (AI) educational robots can do good or bad social relationshp to students.

For example, Knox, W.B. el. (2012) had ever attempted to do two experiments to research whether (AI) robot educational robot can build better or worse social relationship to compare human robot. Their two experiments aim to ask how differing conditions affect a human teacher's feedback frequency and the computational agent's learned performance. The first experiment considers the impact of a self-perceived teaching role in contrast to believing one is critiquing record. The second considers whether a human trainer will give more frequent feedback if the agent acts less ( i.e. choosing actions believed to be worse). When the trainer's recent feedback frequency decreases. From the results of those experiments, they draw three main conclusions that inform the design of agents. More broadly, these two studies indicate as early examples of a nascent technique of using agents as highly specifiable social entities in experiments on human behavior. Thus, it implies (AI) educational robots have ability to learn human teacher to teach students in good social relationship learning environment with students. Even, (AI) educational robots can build better learning relationship to compare human teachers between students, it seems (AI) robots have attractive ability t raise student individual learning interest, after which can applied to assist teachers to teach whose students in classrooms or lecture halls or online classroom channels.

Other scientist had ever attempted to do experiments about " reinforcement learning" (RL) to research how result of interactive supervisory input between human teacher and both robot and software agents relationship. M. Mataric (1997) attempted to do one experiment concerns that reinforcement learning is designed for interactive supervisory input from a human teacher, several works in both robot and software agents have adapted it for human input by letting a human trainer control the reward signal. He aimed to examine the assumption, namely that the human-given reward is compatible with the traditionanl RL reward signal. He described an experimental platform with a simulated RL robot and present an analysis of real time human teaching behavior found in a study in which untrained subjects taught the robot to perform a new task. For the experiment, who reported three main observations on how people administer feedback when teaching a robot a task through reinforcement learning : (a) they use the reward channnl not only for feedback, but also for future directed guideance, (b) they have a positive bias to their feedback, possibly using the signal as a motivational channel, and (c) they change their behavior as they develop a mental model of the robotic learner.

Thus, whose experiment concluded that machine learning shall play a significant role in the development of robotic assistants that operate in human learning environment. ( e.g. homes, schools, hospitals, offices). Considering the difficulty of hard-coding all information needd for the robot to play a long term role in az dynamic world, human users will need to be able to easily teach such robots. However, various works have addressed some of hard problems robots face when learning in the real-word.

In robots learning process, what difficulties which will face, the scientist indicated this question concerns how to influence robot's learning abilities: Has RL certain desirable qualifties, such as the learning abilities possibility to explore and learn from unsupervised experience? Many also queation RL as a variable technique for learning in complex real-world environments because of practical problems, such as long training time requirements, non-scaling state representations, sparse rewards ( resulting in slow utility propagation) and safe exploration strategies. As a result, reinforcement learning has been utilized for teaching robots and game characters, incorporating real-time human feedback by having a person supply reward and/or punishment as an additional input to the reward function. Consequently, the scientist discovered human's teaching method is the most important factor to influence the robot's learning abilities amonf all other environment factors. He also assumed and argued that reinforcement based learning approaches should be reformulated to move effectively incorporate a human teacher. To do this properly, the educational robot must understand the human teacher's contribution; how does the human teach? and what does the educational robot try to communicate from a robot learner? The scientist suggested human trainers ought use these methods to educate robot learners to learn more easily. His main findings indicates reward factor can influence the human robot teachers' motives to teach robot learners to learn, due to more reward can encourage the human robot trainers to teach robots to learn their knowledge and skill. He also found that robot users read the behavior of the robot machine learners and adjust the human robot trainer whose training strategies as whose mental model of the different robot human trainer education method changes. Viweing the human input as a traditional RL reward signal does not take advantage of the fact that a teacher adjusts hose training behavior to best suit the robot educational learner.

In addition to the related RL works mentioned above. Every human robot trainer needs to consider the topic of human input for machine learning systems. Personalization agents and adaptive user interfaces are examples of software that learns by observing human behaviors modeling humann preferences or activities. It empathizes how human teaches the robot learner through interaction, various works address trainable software and robotic agents, exploring explicit human input: learning classification tasks and navigation tasks via natural language, robots that learn by demonstration or/and software agents that learn or training. It seems how to design the robot machine learning software technology will also influence the robot's learning abilities. Thus, human trainer's software learning machine and how to give reward to encourage the human trainer's teaching behavior to let the robot to learn, these factors will influence the robot's learning ability to be applied to education industry to assist human teacher's teaching job successfully. Because how much knowledge and skill, the robot can learn from the human trainer, how much educational knowledge that the robot can own to prepare to teach any students to learn easily. Hence, the human trainer's knowledge and skill will quality to satisfy its primary, secondary and university students learning need.

9.5 Can robots be teachable agents to students really?

In the future, I believe robots have potential to contribute to education by acting as subordinate learners for students teaching. The benefits that the act of teaching provides for one's own learning have long been recognized, tutoring associated improvements in measures like achievement for the tutor role than the lecturer role. However, scientists had confirmed that human teaching robots has been concerned with learning for the benefit of the robot rather than that of the human.

Intelligent autonomous robots are making their way into an increasing range of application areas: Manufacturing, transportation, surveillance and monitoring, rehabilitation, agriculture, service. Hence, if teaching robots can be confirmed to assist teachers to teach students in any schools. It seems that if can be applied to teach manufacturing workers how to produce any products in manufacturing process; it can be applied to teach drivers to drive cars, or teach pilots to control plane engineering machines to fly to sky. So, it can be applied to teach any occupation learners to learn how to operate any engineering machines to replace any human machine training teachers in possible. Hence, it is possible, future one day, robots can be any mahine operating occupation job trainees' teachers ( tutors) or trainers.

In education industry aspect, for computer subject example, robotic systems have potential to increase studetn involvement and motivation and to improve the effectiveness of learning. Possible ways in which they could do so range from small personal robots used as teaching tools for areas like programming and computational thinking to move science-fictional future scenanios wirh humanoid robots supplement or even replacing teacher role in schools.

Recently " teachable agents" have been used as a tool to help students learn, e.g. s student is tasked with training a simulated computer agent on course material, which can lead to a deeper and more committed understanding on the student's own part. For example, a robotic presence in classrooms may one day help to solve problems of teacher shortage, as learning-by-teaching systems have historically been used to do. And the data about human teaching styles that will be generated by widespread use of teachable robots in classroom settings will be invaluable in developing approaches to help robots learn more effectively from humans, where it is rapid and natural instruction of a robot in a new task.

Consequently, whether future robot role is teacher role better or teacher's assistant role or tutor role better in education industry. It depends on the general student's preferable teaching choice in the school. Such as if the school's students prefer to be applied to be taught by human teachers. Then, robots can be choosed to be the school teachers' teaching assistant role to assist the school's teachers' teaching jobs only. Otherwise, if the school students prefer to be accepted to be taught by robots. Then, robots can be choosed to be the school teachers' role to replace human teachers to teach students in the school. Thus, future robots can be either teacher role or teacher's assistant or tutor role for child age, young age or adult age student learning consumers in any primary or secondary or university in future robot computer educational machine development.

Reference

Chou, C. Chan T., & Lin, C. 2003 Redefining the learning companion. The past, present and future of educational gents, computers & education, v.40 n.3, pp. 255-269. April 2003.

Freeman, R. 2000, What is an intelligent tutoring system? Published in Intelligence 11(3): pp.15-16.

IFR, Statistical Department, World Robotics Survey, 2008.

Knox, W.B. Breazeal , C.Stone,.O: Learning from feedback on actions part and intended : In proceedings of 7th ACM/ ZEEE International conference on human-robot interaction, late- breaking reports session ( HRI 2012).

M. Mataric, " Reinforcement learning in the multi- robot domain," Autonomous robots, vol.4, no 1, pp.73-83. 1997.

Wenger, E. 1987. Artificial intelligence and tutoring systems. Los altos, CA: Motgan Kaufmann.

impact to influence human's future life. Thus, (AI) scientists need to consider how to apply (AI) technology.

Finally, I recommend that as the technology of (AI) continues to develop, practitioners must ensure that (AI) enables systems are governable, that what their inventions need to be openness to let public to know clearly and understandable; that they can work effectively with people and that their operation will remain consistent with human values and aspirations. Researchers and practitioners have increased their attention to these challenges , and should continue to focus on their future any (AI) inventions.

Hence, (AI) safe system ought to be applied to solve the biggest challenges that society faces, such as mobility for the elderly and those with disabilities, smart buildings may save energy and reduce carbon emissions, precision medicine may extend life and increase quality of life, smarter government may solve citizens more quickly and precisely., better protect those at any immoral invention risk and save money.

Moreover, (AI) enhanced education may help teachers give every child on education that opens doors to a secure and fulfilling life. Thus, these are the future human's potential benefits if the (AI) technology is developed to its benefits and scientists ought avoid to manufacture (AI) tools to cause weapon risks and challenges.

Consequently, the main point is that how experts invent (AI) systems. (AI) systems ought not be advanced weapon systems, it doesn't seem to be thought similar human soldiers mind and behaviors. (AI) system ought be systems that think like humans. (e.g. cognitive architectures and neural networks), systems that act like humans ( e.g. pass the test via natural language process, knowledge representation, automated reasoning, and learning), systems that think rationally , e.g. logic solvers, inference and optimization and systems that act rationally e.g. intelligence software agents and embodies robots that achieve goals via perception, planning reasoning, learning , communicating, decision-making and acting function.

In conclusion, it is horror (AI) scientists will invent (AI) systems to be owned human's (soldier's) mind and attack strategic behavior to attack other countries easily, who must need to consider (AI) system ought be invented to own scientists' creating mind and non manual assistance functions for positive attribution to human's society. I expect that (AI) system can only be invented to create human's welfare in our future.

(AI) soldier weapon ethical, social and
economic negative impact

In the future, how human can avoid (AI) technological ethical, social and economic negative impact. Scientists need to concern these questions: how to develop of a good (AI) society, how the role and responsibility of the government, the private sector, and the reserch community( including education), in pursuing such a development, whether how the recommendation to support , such a (AI) system development may be in need of improvement.

However, none appers to deliver a comprehensive explicit vision of the role that (AI) system should play in mature information societies. Thus, (AI) 's potential contribution to social good shoud include an in-depth plan for linking in a comprehensive socio-political design questions of responsibility of the different stakeholders, of cooperation between them and of sharable values to understand of a good (AI) positive impact society, not a bad (AI) negative impact society.
Thus, the notion of mature information societies is introduced to stree the importance of addressing the current ethical challenges that (AI) poses in a comprehensive fashion.

It seems (AI) wil invention will be human's moral societal consideration issue. It concerns our (AI) scientists' moral issue, how who invent (AI) system to apply to which kind aspects. IF (AI) system was one direction on war weapon tools to similar to soldier's personal mind or attacking behavior. Then, it will bring poor social safety and poor economy growth our world, due to (AI) scientists' moral is low level.

Thus, the developed country US (AI) technological leader needs to focuse on the impacts of (AI)-driven customatin on the US job market and economy. It represents three specific policy responses to the perceived impact of (AI) on the US economy. They include these three aspects such as: How to invest in and develop (AI) for its many benefits, how to educate and train Americans for the jobs of the future and how to aid workers in the transition and empower workers to ensure broadly shared growth.

The future of (AI) influenced cyber conflicts need more than just the application of current and past solutions in order to ensure security and stability of societies, and avoid risks of escalation. To achieve this end, efforts to regulate cyber conflicts require an in-depth understanding of this new phenomenon, identify the changes brought about by cyber conflicts and the information revoluation, and defines a set of shared values that will guide the stakeholders operating to avoid the international (AI) war occurrence. This becomes clear when considering for example, cyber deterrence. Deploying conventional (cold war) strategies to deter (AI)-influenced cyber conflicts proves highly problematic and the urgent need to foster and coordinate new solutions able to account for the any kinds of conflicts of the cyber demain and of mature information societies to avoid (AI) technological war occurrence in the future.

We hope that in the on-going international conversations and reviews, the US government with further specify how " (AI) system invention law" fit into their vision of the future of society in this case the future of (AI) technological war and conflicts. Hence, (AI) scientists need to concern ethical issues related to (AI), like fairness, accountability and social justice can be addressed through increasing needs. Such as: how the creation of a new body focused on robotics and related (AI) system development to avoid to intent to apply weapon tools to provide advice on the policy, legl and consumer protection issues arising in these fields should be considered.

How to achieve ethical training of (AI) staff and ethical education of the public is certainly important responsibility for (AI) tools ethical behavior and design to the private sector and the citizens : of unique challenges that (AI) brings to society in terms in fairness, social equity and accountability are addresses. Thus, the development of the (AI) technology and defining good (AI) remains problematic. In particular, the US government's innovation driven approach to defining the potential, positive impact of (AI) shows that more could be done to ensure that the opportunities and advantages brought about by (AI) are shared by all society.

An initial on Robotics, based upon the ethical framework and guiding principles is proposed. It should be complementary to legislaton and comprise ethical codes of conduct for Robotics researchers and designers, codes for research ethics committees as well as licenses ( rights and duties) for designers and users. Thus, (AI) robotics invention of safety issues is very important considertion to any (AI) inventions or researchers. Every country's government ought have legal guiding to control their robotics' manufacturing intention. If their robotics (AI) is applied to seem to be soldiers to attack other countries to threaten their people's safety. Then, those (AI) inventors or researchers need to be punished by law.

In conclusion, I believe (AI) technology will be applied to weapon, when it's technological development is nearly mature to able to learn human's mind to do any behavior. During (AI) technology reachs thie mature stage, I predict the (AI) weapon tool , e.g. (AI) soldiers will have chance to be caused. This (AI) invention mature stage has these characteristics such as:

When (AI) invetion reachs this mature stage, computers and robots will develop conscious, intelligent, personified minds. Further, information technology devices and (AI) systems will be implanted into humans, enhancing, psychological and behavioral abilities and allowing for direct communication with artificial intelligent minds. There will be both artificial intelligence (AI) and intelligence amplification (AI) in the relatively near future stage.

During the (AI) invention reachs this mature stage, these will be an ongoing mulit-faceted integration of information technologies and human life. Humans and information technology will cooperate. Humans will increasingly immerse their lives and minds in (AI) systems of technological intelligence and virtual reality. The distinction between humanity and technology will increasingly close dependence.

During the (AI) invention mature stage reachs that the environment will be infused with information technology, becoming animated, communicative and more intelligent. The destinction between the artificial and the natural will increasing close dependence.

During the (AI) invention mature stage will expand through virtual reality, simulated and virtual reality will increasingly into normal reality, e.g. the (AI) weapons is virtual reality to seem to be soldier weapon.

Finally, during the (AI) invention mature stage is as the global expression of the evolving human-technology integration a " world brain" and " world mind" will emerge on the earth. This psychophysical (AI) weapon system will enhance and enrich the capacities of both individual and collective cogniton. This (AI) weapon system is a potential starting point toward the evolution of a cosmic brain and cosmic mind.

Thus, it is possible that the workship raw data was a unique way in which (AI) could be weaponized to cause war, during the (AI) invention stage reachs the invention mature stage. However, (AI) weapon manufacturing factory will be built possibly. In the future, how will we defins and locate (AI) weapon factories. Especially, as these factories are no longer solely buildings , but a mil of virtual and substantially different facilities, particularly as it shifts from a physical assemly and development model to a distributed and flexible network. Needing minimal raw materials to develop (AI) weapons, the phsysical location of their (AI) factories could be anywhere and their identification from the outside, nearly impossible. Given the expanding uses for intelligent and super-intelligent (AI). How will we tell the different form a location that is manufacturing (AI) for the creation of weapons versus creating (AI) for an innovative new gaming platform?

In conclusion, human needs to consider every (AI) scientist's personal ethical or moral mind and research intention and (AI) system invention of (AI) weapon factories cause. During (AI) invention reachs the mature stage if human expects to avoid (AI) technological war occurrence in future one day. The technological development on autonomous military robots, ideally among relevant social groups and actors including human-rights, activists, researchers developers, engineers, philosophers, policy-makers, military authorities, lawyers, journalists and the publis need to consider when human has effort to invent autonomous military robots successfully in the future one day. Finally, some ambitious countries or dominant global countries must like to apply (AI) autonomous military robots to be machine soldiers more than human soldiers if (AI) technology had reached the mature stage. So, future (AI) autonomous military robots will be the next choice of weapon to follow nuclear weapon. If civilians were used as a human (AI) soldiers, the weapon simply ignored them and targeted anyway. This scenario highlighted the dangers of proliferation and quick replication of autonomous weapons. Unlike nuclear weapon, a piece of code for (AI) artificial intelligent soldier could be obtained on the black market and replicated at little cost and the hardware for this type of weapon doesn't require costly or hard to obtain components and materials. Thus, (AI) artificial intelligent soldiers can be manufactured many at cheaper cost. Otherwise, manufacturing one nuclear bomb weapon will spend too much cost. Hence , it is possible that (AI) artificial intelligent soldier will be future new technological weapon to follow nuclear bomb weapon. Hence, any country government needs to legislate to control any (AI) scientists' inventions whether they are attributed benefits or welfares to human or damage human's safety.

How does (AI) robots' brain invention influence our lives?

(AI) research modeling the human brain has developed important technologies, and has overcome significant barriers. How will (AI) affect humanity in the near future? How will (AI) change our lives and our societies? Is the evolution of (AI) to humanity, or it represent a threat?

On white collar workers (AI) job replacement aspect, University of Tokyo, Institute of informatics, lecturers who had attempted to do experiments to take (AI) exams over a two year period. The (AI) achieved standard scores of around 50 in each subject, exceeding the norms for humans attempting the tests. The (AI)'s results in subjects emphasizing memorization, such as world history and Japanese history subjects were comparatively high, and the results of the study suggested that an appropriate selection of subjects would give at an 80% chance of passing the entrance exams of 80% of Japan's private universities.

So, if (AI) is applies to human white collar workers' job duties aspect, at this level, if white collar workers were replaced by (AI) in the future, around 30% of current staff would be replaced. Whatever, the outcome, large companies will be represented with two choices. One choice will be to protect their employees, but as a result lose their international competitiveness. The latter choice will enable them to reduce the cost of general duties, financial management procedures, accounting etc. general administrative job duties of cost in offices. Hence, it seems that (AI) will be possible to be invented to own human's brain ability to do some mind jobs in future on day.

However, the method called " deep learning" must be developed to cause (AI) to match human's brain ability as well as these were dramatic advances in technologies, such as image recognition and voice recognition, which form the foundation for (AI). Nowadays, this new method called" deep learning" does not reach the matured and stagnated stage. It needs to wait human to continue to invent to let (AI) to match human brain to achieve 100% owning human's mind ability. Nowadays, (AI) industry product include cleaning robots, smart TVs and future (AI) product development market. It will include self-driving vehicles, drones, and nursing robots.

On (AI) weapons applied aspect, if (AI) can be invented to own human's brain judgement and analytical abilities. Then, it is possible that it can be applied to attack enemy to cause war effect. For example, if weapons such as missiles were equipped with (AI) in the future, they would become able to decide on their own targets. Hence, human needs to apply restrictions when necessary.

On (AI) applied to analyzing information collected technological aspect, nowadays, every one will use wearable terminals to connect to the internet to obtain various types of information as well as computers will collect and analyze information on people. Our lives will probably be more reliant on these internet technologies than they are on smartphones today. When, (AI) can match human brain to own mind ability.

Then, (AI) can be applied to do any analyzing information and collection job duties aspect to raise large information restoring and remembering efficiency. For example, (AI) will be generally used and will be extremely useful in analyzing the information collected from wearable devices and stored in the cloud. (AI) will enable wearable devices to be of real assistance in our lives offering their users more intelligent support.

Rather than allowing (AI) to develop on serves, as something separate from humanity. It will be more meaningful to encourage its development via wearable devices, situating it under the control of human intelligence. The intelligence of (AI) will increase rapidly in the future. If this increase in (AI) occurs under human control, enabling humans to increase their own abilities, then surely it will be possible for us to put up a degree of resistance to the opposite scenario, the domination of (AI) over humanity. Hence, if (AI) can be invented to remember and store and make analytical judgement to collect any information from internet. Then, it will bring the effect, such as large international organizations' (AI) internet storage robots can bear in mind factors, such as competitors' privacy or business secret information, such as the loss equality between people and threats to privacy that will be stolen form the owning (AI) storing internet information remembering robots.

Consequently, what is the effect of successful invention of (AI) matching human brain's mind ability? (AI) present computers are adequately able to reproduce the emotional, conceptual and intuitive abilities of humans. Because of this, it is important that we should envision potential future problems that may manifest when we consider how to employ wearable devices. It will be essential to enhance our technologies in order to ensure that we can use (AI) under human control.

However, when a goal has been set. (AI) will implement an appropriate means for its realization. (AI) will be need as a tool by human society. If the capacities of analytical and judgement mind abilities of (AI) brain exceed those of human brain, it is difficult to imagine the type of technological, then singularity is represented by the creation of an (AI) by another (AI). It is important that we rapidly and accurately predict these developments, when image recognition and other individual technologies are functioning at a high level. There will be a considerable matter in different sectors of (AI) industry development.

Today, however, machines have become able to decide for themselves what they will learn, making it difficult to copy human's mind ability. What we must consider when machines exceed humans and (AI) surpasses human capabilities. May technologies exceed human capabilities, cars are faster than humans, planes are able to fly. Consequently, it brings a question that human needs to consider: When does (AI) brain technology be invented to reach the most reasonable stage to be accepted or stopped by humanity?

What is artificial intelligence
human brain invention?

A machine is likely to achieve the ability of a human brain. Does it a scientific story? Some scientists has predicted that a US$1,000 personal computer will match the computing speed and capacity of the human brain by around the year 2020 year. With human reverse engineering, human should have the software insights before 2030 year. it is possible that of machine intelligence and exotic new technology for faster and more powerful computational machines from cellular automata and DNA playing cheese game competition case example, it proves that (AI) had been invented to own human's analytical and judgement ability to exceed the best cheese game human player's brain analytical and judgement ability. Then, it seems that (AI) will have possible to be built machine brains to achieve the exceed level of human brain's analytical and judgement ability in the future one day.

Supposing we scan someone's brain and restate the resulting " mind file" into suitable computing medium. Will the entity that emerges from such an operation be conscious? How have advances in electronic communications changes power relationship? For electronic book publishing case example, a book that looks at the principles companies must adopt to meet the needs and desires of this new kind of client. So, such as paper book can be changed to electronic book for human to read. Why can't human brain be changed to (AI) machine brain to do human's analytical mind and behavioral mind of activities to replace to do any human's daily analytical and behavioral mind activities?

Over the next few decades, machine achieve super intelligence, human will encounter a dramatic phase. Will it be a "WALL" a barrier as conceptually the event of a black hole in space. Such as (AI) brain invention case, an " AI singularity" ruled super-intelligence AIs, or a gentler " surge" into a post human era of agelessness and super-intelligence brain. Will future technology, such as bio-engineered pathogens, self replicating nan robots, and super smart robots run and accelerate out of control, perhaps threatening the human race?

If one day, (AI) brain is invented to achieve agelessness possibility. It means human's brain will be old to lose mind and analytical ability when human's age is increasing. Otherwise, (AI) machine brain age won't lose mind and analytical ability, due to (AI) machine is no age increasing possibility. It is a machine brain. If (AI) machine brain can be built successfully. Scientists need to consider technological ethic matter, such as the challenge of guiding nanotechnology in a constructive direction, advances in nanotechnology and related advanced technologies can not be inevitable, any broad attempt to relinquish nanotechnology would interfere with the benefits. When actually making the dangers worse.

Keeping in mind that intelligence machines are already making their way into our blood stream. There are dozens of projects underway to create blood-stream based " biological micro electronic- system" (bio MES) with a wide range of diagnostic and therapeutic applications BioMEMS devices are being designed to intelligently pathogens and deliver medications in very precise ways. For example, a researcher at the University of Illinois at Chicago has created a ting capsule with pores measuring only seven nanometers. The pores let insulin out in a controlled manner, but prevent antibodies from invading the pancreatic Islet cells inside the capsule. These nano- engineered devices have cured rated with type I diabetes, and there is no reason that the same methodology would fail to work in humans. Similar systems could precisely deliver dopamine to the brain patients, provide blood-clotting factors for patients with hemophilia and deliver cancer drugs directly to tumor sites. A new design provides up to 20 substance-containing reservoirs that can release their cargo at programmed times and locations in the body.

Another brain health technological related invention case, such as Kensall Wise, a professor of electrical engineering at the University of Michigan, who has developed a tiny neural probe that can provide precise monitoring of the electrical activity of patients with neural disease. Future designs are expected to also deliver drugs to precise locations in the brain. Also, kazushi Ishiyama at Tohoku University in Japan has developed micro machines that use microscopic-cancer tumors.

A particularly innovative micro machine developed by Sandia National labs has actual micro teach with a jaw that opens and closes to trap individual cells and then implant them with substances, such as DNA, proteins or drugs. There are already at least four major scientific conferences on bio MES and other approaches to developing micro- and nano-scale machines to go into the body and bloodstream. All these inventions are related to how to apply machines to copy human's brain knowledge in order to achieve to do any human's brain functions.

Finally, for Freitas envisions micron-sized artificial platelets invention case example, who could achieve hemostasis ( bleeding control) up to 1,000 times faster than biological platelets. Freitas describes nano-robotic microbivores ( white blood cell replacement) that will download software to destroy specific infections hundreds of time faster than antibiotics, and that will be effective against all bacterial, and fungal infections with no limitations of drug resistance.

Consequently, such as above machine health scientific invention cases, there were many scientists had invented any health machines to apply drugs to transfer to human's brain to attempt to reduce human's disease causing risks, such as reducing cancer cell increasing number. Why it is no possible that scientists can attempt to invent (AI) brain which can own human's mind ability to judge or analyze any matters to give opinions in order to exceed human's judgement and analytical ability.

How can artificial intelligent brain
satisfy to human beneficial and
natural needs?

Nowadays, new scientists' most familiar form of this vision in our times is genetic engineering. Specifically, the prospect of designing better human beings by improving their biological systems of a small, serious and accomplished group of tailors in the field of artificial intelligence and robotics. Their goal is a simply new age of post-biological life, a world of intelligence without bodies, immortal identity without the limitations of disease, death and unfulfilled desire. If human can understand why this fate is presented as both necessary and desirable, human might understand modern science can help us to enter the good life and good society stage when (AI) brain is invented by scientists successfully in our future life.

How can (AI) beneficial brain satisfy to human natural need? For relatively recent example, similarly as a long term trend beginning with the first mechanical calculators, the evaluation of computing capacity increases in speed over time and decrease in cost. From biological evolution has been invented to influence human brain, an electronic chemical machine with a great, but finite number of computer neuron connections, the product of which we call mind or consciousness. As an electro-chemical machine, the brain obeys the laws of physics, all of its functions can be understood and duplicated. And since computers already operate at far faster speeds then the brain, they soon will rival or surpass the brain in their capacity to store and process information. When happens, the computer will at the vary least, be capable of responding to stimuli in ways that are indistinguishable for human responses. At that point, we would be justified in calling the machine intelligent, we would have the same evidence to call it conscious that human now have when giving such a label to any consciousness other than our own.

At the same time, the study of human brain will allow us to duplicate its functions in machine circuitry. Advances in brain imaging will allow us to " map out" brain functions, allowing individual minds to be duplicated in some combination of hardware and software. The result, will be a world that is remade and reconstructed at the atomic level through nanotechnology, a world whose organization will be shaped by an intelligence that surpasses all human comprehension.

Whether or not today's humans are willing or able to " download" their brains into machines, there will come a time when all human beings will be intelligent machines in the future. Computer hardware will continue to get faster, cheaper and more powerful computer software will increase in sophistication. Brain research will continue to explore the " mechanics" of consciousness. Nanotechnology will continue to develop.


There are powerful incentives, commercial, military, medical and intellectual that will drive many of the advances that the extinctions desire if for very different reasons. Much of the work in artificial intelligence and robotics is open to the same defense that is made on behalf of biotechnology: If we don't do it, they will and why suffer or be unhappy when some new agent or invention is available that will solve the problem.

Finally, we already accept significant artificial argumentation and replacement of natural body parts when those parts are missing or defective. Over time indistinguishable from or " superior" to their biological counterparts as they employ increasing computer processing power. There are powerful incentives, commercial, military, medical and intellectual that will drive many of the advances that the extinctions desire, if for very different reasons. Much of the work in artificial intelligence and robotics is open to the same defense that is made on behalf of biotechnological if we don't do it, they will and why suffer or be unhappy, when some new agent or invention is available. That will cure the problem.

Finally, we already accept significant artificial augmentation and replacement of natural body parts when those parts are missing or deductive. Over time, such replacements are only likely to get more useful and perhaps eventually indistinguishable from or superior to their biological counterparts, as they employ increasing computer processing power. Nor is there an obvious distinction between using manufactured chemicals to fight disease and using " smart" nanotechnology. The extinction project is begun by offering new routes to fulfilling old promises about doing good for human beings. But, it doesn't necessary end.

In connection with machine intelligence, it does not seem very promising to try to limit the power or ability of computers. The danger ( or promise) that computers might develop characteristics that lead some people to call them conscious and that this age of intelligent machines would mean our extinction seems remote when compared with their practical benefits. We already rely so heavily on computers that the incentives to make them easier to use and more powerful are very great. Computers already do a great many things better than we can, and there seems to be no natural place to enforce a stopping point to further abilities. Certainly mechanistic and reductionist assumptions about society, ethics and psychology the notion that we are atoms or animals, driven by chance or instinct, run deep in the present world.

Artificial intelligent brain future

innovation and attribution

16.1 (AI) brain invention successful factors

(AI) brain will bring what attribution to influence human's positive impact. How (AI) brain will be invented to apply to any businesses' needs. By how much the (AI) brains might exceed us remain unknown, but it could potentially be by a very significant degree. Future (AI) brain invention will have noted similar growth in everything from hard-drive storage density to the price and speed of DNA sequencing. On key feature of this technological growth that has not been adequately measured in the degree to which technology is becoming more intelligent.

16.1.1 (AI) brain test experiment

When there is an intuitive sense that (AI) programs/brains today are more capable than those of age, and that those were considerably " smarter" than the serial instructions that passed through the first supercomputers. Scientists will carry on testing (AI) to do any experiments to improve (AI) brain development. They will assess the progress of artificial (general) intelligence, but the need for intelligence tests and tests for other cognitive abilities will be tested in the forthcoming decades for bots, robots, avators, " animats" etc. and any collective system of these and biological systems ( humans and non-human animals).

When (AI) brain technology is invented successfully, the idea of a super intelligent computer means invention successfully also. Whether super intelligent computer or (AI) brain can be invented successfully. Similarly, many think that the future beyond the technological singularity is unknowable or even unimaginable. Since its conception, the idea has been examined and explored by technologists. (AI) brain invention will be mean human-equivalent (AI) vs human-level(AI). Many machine intelligence tests is the exclusive focus on identifying systems that achieve human equivalence. Considering our experience with studying non-human animal intelligence.

The need to distinguish human equivalent (AI) from human level (AI) seems critical. Perfect human intelligence , such as (AI) brain invention is likely to be extremely difficult to achieve. Potentially every aspect of the biological processes involved would need to be translated with very high fidelity.

Human level (AI), such as (AI) brain invention is another matter. Achieving capabilities that are equivalent to those of the human mind could be very feasible if it is not limited to perfectly processes involves. For instance, some pattern recognition algorithms are already superior to human abilities. This machine ability is not achieved by duplicating the processes our brains use, through some of the methods have been inspired by them. If researchers had been limited to replicating the brain's mind processes, We would still be waiting for the development of a (AI) brain machine equivalent.

Tests that would seen to have a reasonable chance of successfully testing non-human intelligence are those that use mathematics to define the value of a given challenges for (AI) brain invention. Such as Capability-test has some potential to generate meaningful data about (AI) machine intelligence, e.g. others in complexity theory test, it presents a series of abduction and prediction problems, similar to those in standard IQ tests to (AI) brain. Hence, IQ test and C-test will be potential tests to test (AI) brain ability.

(AI) brain test goals include that to identify a range of possible types of mind such as: super-fast human mind, mind with operational access to its source code, any mind capable of general intelligence and self awareness, general intelligence without self-awareness, self-awareness without general intelligence, super-logic , machine without emotion, mind capable of imaging greater mind or creating greater mind to compare human's brain abilities. Thus, any IQ or Capability test experiments aim to evaluate whether (AI) brain mind ability which can exceed to human brain mind ability. Thus, if future one day, scientists could prove (AI) brain mind ability can exceed to human brain mind ability. Then, they believe (AI) brain invention has ensured to achieve success.

16.1.2 (AI) brain invention successful factors

Hence, scientists want to invent (AI) brain successfully. They need to solve these challenges. Such as How robots and computers have progressively supplemented humans, initially only in relatively simple computational and manipulation tasks, but more recently in higher cognitive tasks that used to be the pre negative of the human brain, including language, mathematics, probabilistic reasoning and decision making.

An important challenge is how to enhance the productive interactions between humans and artificial intelligence? The important challenge include these major successful factors to invent (AI) brain technology, such as:

What is the state of the art in (AI) software and machine learning?

Can all aspects of brain function be manufactured by artificial system?

What is the proper form of mathematics that may capture the operation of minds and brains?

How to make (AI) brain feels consciousness?

What would it be taken for a machine to pose a sense of art in (AI) brain software and (AI) brain machine learning by artificial system?

Will machines soon surpass us in all domains of human competence?

What is the proper form of mathematic that may capture the operation of minds and brains?

What is consciousness to (AI) brain invention?

Could a machine be endowed with an artificial consciousness?

What would it take for a machine to posses a sense of self?

Will intelligence machines soon pose a danger to humanity of (AI) brain is invented to reach the mature stage successfully?

Is it possible to design and construct an intelligent robot with an artificial brain sense of ethics?

How can we enhance the humanitarian uses of artificial intelligence brain and owning mind ability of robotics, in particular in the field of education, health and emergencies?

Consequently, above all these challenges, I recommend scientists need to solve to achieve (AI) brain invention in order to reach (AI) brain invention mature stage easily.

16.2 Artificial intelligence brain
invention opportunities
and challenges

If scientists focus wrong direction to invent (AI) brain, it will bring wrong marketing development to attribute any benefits to human. So, they need to reduce a mismatch of timescales between the pace of commercial innovation and (AI) brain invention process, reduce an underappreciation of the fundamental unpredictability of (AI) brain autonomous systems and reduce a lack of a university agreed upon conceptual framework for any (AI) brain invention and reduce a disconnect between the (AI) brain design of any kind of (AI) autonomous robots. Thus, scientists need examine these gapes, provide a roadmap of opportunities and challenges and identify areas of any (AI) beneficial functions to be attributed to human to use for our daily life needs.

Future (AI) brain market opportunities, it is rapidly growing innovations in digital -electronic and information technology had development of new intelligence, surveillance, and reconnaissance platform and battle management capabilities, precision-strike weapons, stealth aircraft, smart weapons and sensors and tactical exploitation of space ( e.g. GPS).

Any one of these aspects will be (AI) brain future marketing development opportunities. Future (AI) brain invention of so called " narrow AI", (i.e. non-sentient artificial intelligence, whose problem-solving capability is confined to one narrow task. For example, (AI) brain needs to find the best method to win any one of chess player in any both human and (AI) robot chess playing game.

In smart weapon strategy industry, (AI) brain is needed to design how to analyze or mind to protect whose country to avoid enemy attack in any war by the best weapon protection strategy. In education industry, (AI) brain is needed to design how to analyze or mind how to assist teachers to educate whose students by the best education method. In aircraft industry, how to design (AI) brain to analyze or mind to assist pilot to make the most correct flying direction judgement to fly in the most safe way. In space exploitation industry, (AI) brain is needed to design how to find undiscovered natural resources to supply to human to use in anywhere space.

Thus, future (AI) brain invention needs have these features/characteristics to be designed. They include: (AI) learned on its own, where to find the information it needs to accomplish a specific task, (AI) can predict the immediate future from studying any matter, (AI) automatically needs to be inferred the rules that govern the behavior of individual robots within a robotic swarm simply by watching, (AI) needs to be learned how to navigation the acquired memories and experiences, much like a human brain, (AI) speech recognition needs to be reach human parity in conversational speech, (AI) communication system needs to be invented its own encryption scheme, without being taught specific cryptographic algorithms ( and without revealing to researchers how its method works, (AI) translation algorithm needs to be invented to remember fluent language to more effectively translate between any two languages ( without being taught to do so by humans), (AI) brain system interacted with its environment ( via virtual environment) to learn and solve problems in the same ways that a human child can do, (AI) based medical diagnosis system needs to be achieved 99% percent accuracy in any medical reviewing researches ( at a rate minimum 30 times faster than humans), ( AI) poker playing program brain development needs to be defeated some of the world's best human poker players during a minimum three-week-long tour, (AI) brain development needs to be effectively " read minds" of human test subjects looking at pictures of faces, via functional magnetic reasonable images of brain activity.

Consequently, (AI) future brain development needs to follow above directions to be invented to attribute to human's satisfactory needs.

16.2.1 (AI) brain legal remembering attribution

Future, (AI) robots can assist lawyers to deal any legal cases more efficient. If human understands that smart (AI) technology is not to replace human lawyers, but to make a better lawyer that forces who to use, emotional intelligence, and capital on lawyers' mind, then human lawyer have made the first step in future, proofing whose legal service business from (AI) legal robots' assistance. (AI) promises to be a real advantage for today and tomorrow' lawyers having to deal with the rate of legislative evolution and technological change.

In future, for the better with regard to technology in any law firms. According to the ALM 205 law tech. survey 95% of firm leaders and technologist respondents agreed with recent decisions by management regarding the firm's technology in legal service profession.

How can (AI) robots be applied to legal service industry by legal service firms? The (AI) reality in the legal world, it includes in relation to the four key elements of legal service provision, such as commodity, research, reasoning and judgement, exist and can support or replace certain aspects of every lawyer individual jobs both fee earning processes and business processes can be supported and/or replaced by expert systems, cognitive computing, robotics automated systems, (AI) and the machine learning, clients demand and expect more speedy, accurate, expert, creative, intuitive and accessible legal advice, it can assist young lawyers to innovate / tech. focused firm, reducing pressure both from within law firms ( or in house teams) and from clients to respond to the demand for client-designed service from (AI) robots assistance, technology related projects that are both user and client -centric need to be implemented successfully.

Due to the deployment of (AI) robots in the legal ecosystem where lawyers, firms, general counsel and clients are beginning to believe (AI) capability technologies, (AI) robots can assist lawyers to increasingly become more productive, efficient, accurate, better quality, less labor intensive and time intensive and the role of the lawyer is gradually changing. If we break down a lawyers' tasks in a legal project from beginning that can handle the majority of these four tasks far more quickly and accurately than human lawyer. (AI) robots can handle legal task in the four aspects as below:

First in legal aspect, it can be used for deep research and processing, such as extracting specific pieces of information from land registry documents, (AI) technology is placed top of a document set including client guidelines and similar forms. It searches through documents and extracts key data points to provide a report of data for improved coordination with clients.

Second on managed services technology aspect, (AI) platform which could have a huge advantage for general counsel and law departments in corporations and for clients of all company sizes.

Third on reading aspect, (AI) brain program that reads and analyzes , e.g. clauses in loan agreements. Its program helps its lawyers through transactions and points. Then toward the correct precedents of each stage of a process.

Fourth, on academics aspect, (AI) robots can assess the merits of personal injury cases. It can automatically review high volumes of contract documents to identify provisions that could potentially be impacted by contract law regulation.

Thus, all of these systems can handle large quantities of structured and unstructured data, and assist with the process management, research, and reasoning elements related to legal issues. Thus, in future, (AI) brain development can be invented to apply to knowledge research job, such as legal industry.

Artificial legal intelligence presents a thought-provoking approach to both computational models of legal reasoning and the use of evolutionary thinking about the law. The visions of computerized artificial legal intelligence , a vision of developments in both technology and legal history. A number of creative research projects have applied artificial intelligence techniques to the domain of legal reasoning.

Consequently, ( AI) brain for legal industry marketing development ought concentrate on those three fields of artificial intelligence at most relevant to work in the legal areas in order to achieve the excellent attribution. Such as case-based reasoning, expert systems and neural networks. Artificial intelligence program, such as the legal reasoning programs. Thus, (AI) brain invention needs to own these human legal concept knowledge in order to achieve (AI) legal brain program development successfully.

16.2.2 Whether human mind can create to (AI) brain mind

How can (AI) scientists build a machine that think? In fact, there was general agreement that minds can be existence on non-biological substrates and that algorithms are of central importance to the existence of minds. However, there are much debate about the raw hardware power present in organic brains, such as (AI) brain.

I think (AI) scientists need to invent powerful hand ware, e.g. commercial digital signal processing might be, giving an appearance even to digital operations, but nothing would ever make up the intellectual runaway that is the essence of the singularity.

I also think raw hardware power is not be able to organize the parts to behave in a super-human way as well as I also think powerful software complexity is the main factor to solve (AI) brain mind invention challenge.

Hence, future super-human (AI) brain invention will ought consider how to invent superhuman software more than hardware, because software can store any memory, i.e. human mind. When, (AI) brain invention which can achieve to own human mind ability. Then, (AI) scientists need to consider ethic matter: Does the future of (AI) pose an existential threat to humanity? How do we present learning algorithms from morally objectionable biases? Should autonomous (AI) be used to kill in warfare? How should (AI) systems be in our social relations? Is it permissible to fall in love with an (AI) system? What sort of ethical rules should (AI) like a self-driving car use? Can (AI) systems suffer moral harms? All those ethic matters. I think (AI) scientists need to consider after (AI) brain owns human's mind ability because it is possible that (AI) robots will harm human if they are educated to do any wrong or illegal or immoral mind ability. Thus, (AI) scientists need to consider (AI) robot's moral mind and judgement behavior.

16.4.3 Brain-inspired intelligent

robotics

How to solve fundamental problems in the areas of brain sciences and brain-inspired intelligence technology? One way scientists seek to accomplished the mission is to develop brain-inspired hardware including intelligent devices, chips, robotic systems and brain inspired computing systems. (AI) scientists ultimate goal, but since the robot's memory and learning after the human brain, there is still much to learn about neurobiology before that goal is attached.

In (AI) brain university research aspect, two schools of thought have emerged in robotics: bio-logically robots that include a body, sensor and actuators, and brain-inspired computing robot.

Robots have found increasing applications in industry, service and medicine, due in large part to advances achieved in robotics research over the past decades, such as the ability to accomplish complex manipulations that are essential for automated product assembly. However, robots still have these weaknesses which need to be solved if (AI) scientists expect (AI) brain invention can be success. Robots still lack truly flexible movement, have limited intellectual perception and control, and not yet able to carry out natural interactions with human. These deficits are especially critical in service robots. A critical concern of government, academia, and industry is how to advance research and development for the key technologies that can bring about the next generation of robots. Developing robots with more flexible manipulation, improved learning ability and increased intellectual perception will achieve the main goals for (AI) scientists' solutions.

Consequently future (AI) brain -inspired intelligent robotic invention needs have these competitive or attractive abilities or strengths to compare computer storage ability. Such as (AI) brain needs have perception to exceed computation ability, adaptation exceeds computing speed, flexibility exceeds, computer memory access speed, cognition exceeds computer memory lifetime, learning exceeds computer memory capacity and innovation exceeds computer memory storage abilities. Hence, (AI) brain innovation must need to exceed general computer storage, lifetime, capacity, learning abilities if (AI) scientists expect (AI) robots will be popular to be applied by any service, manufacturing, education industries.

16.3 How can web intelligence need (AI) brain informatics?

Brain informatics (BI) invention will be a new inter-disciplinary field that systematically studies the mechanisms of human information processing from both the macro and micro view points by combining experimental cognitive neuroscience with advanced information technology. (BI) studies human brain from the viewpoint of informatics ( i.e. human brain is an information processing system) and uses informatics ( i.e. WI centric information technology) to support brain science study. It seems that (AI) scientists need to further understand how human intelligence and brain sciences development through brain sciences fosters innovative web intelligence research and development because innovative web intelligence research will have ability to assist (AI) scientists to research how to invent (AI) brain robotic technology more easily. The synergy between ( web informatics) (WI) and ( brain informatics) (BI) advances our ways of analyzing and understanding of data, knowledge, intelligence, and wisdom, as well as their interrelationship, organizations and creation processes. Web intelligence is becoming a central field that information technologies and artificial intelligence to achieve human level web intelligence.

(WI) may be viewed as applying results from existing disciplines , e.g. artificial intelligence (AI) and information technology (IT) to a totally new domain the world wide web. (WI) may be considered as an entrancement or an extension of (AI) and (IT), (WI) introduces new problems and challenges to the established disciplines.

Thus, developing human-level web intelligence will be seem to develop brain level artificial intelligent technology. Because brain informatics (BI) is an interdisciplinary field to systematically investigate human information processing mechanisms from both macro and micro points of view by cooperatively using experimental, computational, cognitive neuroscience, and advanced (WI), centric information technology. It attempts to understand human intelligence in depth, towards a holistic view at a long term, global vision to understand the principles and mechanisms of human information processing system (HIPS).

So, I recommend (AI) scientists needs to research how to invent brain informatics technology and web informatics technology to achieve how to apply (AI) brain to analyze and judge or mind web informatics ability to compete internet ( web site) communication tool product industry. Hence, if (AI) brain can be applied to web informatics industry will increase attraction to an internet user market in the future.


16.4 Why model of sustainable development environment industry be (AI) robot attractive market

In the world scientific literature various conceptions of sustainable development are found, however, their basis are formed by three dimensions: environmental, economic and social development. The biggest attention is concerned to the environmental dimension which forms the basis of existence of social environment and economy. The reason is emphasized that each person must preserve and manage natural resources as the basic of economic and social development. However, in the sustainable development strategy, the sustainable development is understood as among environment protection, economic and social society that form the basis to achieve the universal welfare for present and future generations, it is difficult to let human to adapt to live in pollution environment. Thus, environment pollution will be human's concerning issue. It implies how to protect environment pollution which will be human's future challenge. If (AI) brain invention can be applied to how to solve environment pollution challenge. It is very attractive attribution to human (AI) brain environment protection function can be invented to include such as : how to apply (AI) brain to do mind to give opinions or do any environment protection behaviors/activities to assist human how to effective use of natural resources, how to effective use of universal economic society's welfare, do strong social guarantees during the period of strategy's implementation ( until 2020 year to achieve global (AI) robots environment protection mission from (AI) robots' behavioral assistance. Because environment pollution will influence world's climate to be worse to cause our food or vegetable can not grow up easily. Then, human will face food / vegetable shortage challenge. If (AI) brain invention can be applied to solve environment pollution aspects, then human will avoid food / vegetable shortage crisis.

16.4.1 How to invent (AI) brain's environment protection ability?

All similar problems significantly promotes scientists to research for new technologies of data extraction progressive software of data extraction operating on the basis of artificial neural networks allow finding the relations among various types of data in the huge data. Due to the data extraction technologies, it is possible to prove empiric observations to group, to process and model big amounts of data, distinguishing unknown schemes in data and using them in future activities ( Rotman M. J., 1995).

It seems scientists believe gather every country's climate environment data to predict environment climate change will have chance to avoid environment pollution challenge. In addition, the traditional statistic methods applied to process digital data of the indicators of sustainable development environmental dimension are not able to analyze the present situation of environmental dimension in the context of sustainable development to present possible reasons of change of environmental dimension problems to generate future forecasts characterized by a high level of accuracy.

Consequently, if (AI) scientists can apply (WI) web informatics technology and (BI) brain informatics technology both to apply to (AI) brain technology to gather global climate environment daily change data. Then, I believe (AI) brain invention can assist human to solve future environment pollution challenge Then, it is possible that (AI) brain can attribute to environment protection or how to give opinions or choose to do any environment pollution activities to avoid global warming crisis.

16.5 (AI) Asia market

When (AI) brain is successful invention, I think Asia will be a new market to need (AI) brain attribution for Asia consumer needs. I believe Asia people expect (AI) can attribute to let them to use. The (AI) ability includes the ability of machines and systems to acquire and apply knowledge, and to carry out intelligent behavior.

This includes a variety of cognitive tasks ( e.g. sensing, processing, oral language, reasoning, learning, making decision) and demonstrating an ability to move and manipulate objects according). So, future Asia people (AI) consumers expect (AI) robots can attribute to whose society's needs, such as: how to apply intelligent systems to use a combination of big data analytics, cloud computing, machine-to-machine communication and the internet of things ( IOT) to operate and learn. Asia people expect (AI) robots can give beneficial attribution to them, e.g. talking or playing a game for Asia young entertainment market; (AI) robots need to reflected by physical substance ( such as any talking or playing game robot player).

In this sense, (AI) is like a human brain. For Asia (AI) robot service industry need, Asia (AI) robot clients expect to use soft robotics ( robotic process automation) can be used to meet Asia (AI) robot service consumers' expectation of automation repetitive tasks and common processor needs, such as client servicing and sales without the need to transform existing IT system maps ( e.g. (AI) salespeople robots, or (AI) service robotic).

In (AI) office task aspect, Asia office consumers expect algorithmic game theory and computational social choice of (AI) robots to be attributed to replace some office human workers' tasks. Such as (AI) systems tat address the economic and social computing dimensions of (AI), such as how systems can handle potentially incentives, including self-interested human participants or firms, and the automated (AI) -based agents representing them, e.g. complex or simple office administrative tasks, e.g. typing, accounting, filing, etc. general office tasks which need human office workers who use computers to work to be replaced by (AI) robots to do. So, Asia office (AI) robots users who expect any office human administration tasks can be replaced by (AI) robots to do.

In Asia computer vision market, Asia computer vision ( image analytics), users expect (AI) robots can be replaced to human computer image workers to shorten time to work, or raise image quality to be more clear in the process of pulling relevant information from an image or sets of images to advanced classification and analysis. Such as hospital or clinic x-ray image vision (AI) robot invention, photo image (AI) robot invention etc. any Asia image industry (AI) robot market need.

In Asia collaborative systems work with human (AI) robot market, Asia autonomous systems robots users who expect (AI) robots can be applied its models and collaborative systems to help them to develop autonomous systems that can work collaboratively with other systems and with humans, e.g. car manufacturing, computer manufacturing or any machine manufacturing products. So, Asia machine related manufacturing product industry manufacturers who expect to apply (AI) robots who can assist factory human manufacturing workers to manufacture any products in the short time efficiently.

In Asia language teaching (AI) robot market, language educators expect (AI) robots own natural language processing ability, algorithms that process human language input and convert it into understanding representation, such as Asia translation education market ( PWC).

Thus, Asia language teaching businessmen expect (AI) brain invention which needs to be designed to own these above abilities of (AI) robot's manufacturers expectation to provide to them to use from any (AI) robots import.

16.6 Is the brain a good model

for machine intelligence?

The actions of a human " computer" using paper and pencil to perform a calculation ( as the world meant ), into a formalized machine, manipulating symbols on an infinite paper tape. But I believe it still has challenges to influence human brain can be invented to machine intelligence successfully.

I think (AI) scientists need to solve these further challenges to influence (AI) brain invention success. The limitations include such as below:

Computation is based on functions of integers is limited. They must let computer data can change to words, then words can change images, then any images can change to storage to let (AI) brain to remember to do any analytical and judgement able tasks to make any actions in the short time finally. It is (AI) brain behavioral and analytical mind speed limitation.

Moreover, anther limitation concerns biological systems clearly difference, they must respond to varied stimuli over long period of time, those responses any changes alter their environment and subsequent stimuli. The individual behaviors of social insects, for example, are affected by the structure of the home, they build, and the change their behaviors. Nowadays, (AI) brain invention is called computational neuro science, which have assured that the brain is a computer, it means a machine that is algorithms and architectures. Second, neuro science findings may validate the energy ability of existing algorithms being integral parts of a general (AI) system. It means (AI) robots change adapting environment limitation. It means how (AI) brains adapt to choose to make any analytical mind as well as how to be influenced by external environment to do their behaviors or actions in the efficient way, e.g. how to manufacture many cars efficiently in one factory in the short time.

Consequently, to solve these limitations, we need to know how to apply correct conceptual knowledge to let (AI) robots to learn, e.g. for example, if we know how conceptual knowledge was formed from perceptual inputs, it would crucially allow for the meaning of symbols in an artificial language system to be grounded in sensory " reality". When (AI) scientists can achieve how to solve all above limitation challenges, then (AI) brain invention will achieve more easily.

Reference

PWC, Sizing the prize, see: https:// www. pwc. com/gx/en/issues/data-and- analytics/publications/artificial-intelligence- study.html.

Rotman M. J., Data mining-a practical approach to database marketing (1995). IBM.

AI fiftth stage development

Can non-manual driving public transport tools bring global economic growth

Why MTR underground train transportation needs to know passenger behaviour

Understanding individual passenger behaviour is essential for the design MTR transportation, because who can choose to catch bus, taxi, tram, train ferry etc. different kinds of public transportation tools. Individual traveler who decides to catch which kinds of public transportation tools, it depends on whether the public transportation tool can provide real time travel information, liking link travel time schedule. So, MTR underground train needs to understand where it has terminal to give convenience to the local living areas of time travelers to choose to catch MTR easily. Although, MTR ticket fare is one factor to influence any passengers choice. But, those other factors can also influence them to choice. e.g. MTR any terminal location of convenience, short time travelling, none crowding in busy (peak) time, MTR platform waiting arrival time, none sudden MTR engineering machines broken accident events occurrence frequently etc. different factors, any one of these factors which can influence passengers who choose to catch MTR or other kinds of transportation tools.

Why route choice can influence passenger behavioural choice
Usually, the busy time passengers will regard the route choice as a coordination problem to influence them to choose to catch which kinds of transportation tools. The route choice is as an opportunity costs to influence any busy time passengers to decide to choose to catch which kind of transportation tool which is the best right choice in the right time among of them. In the short time, for example, it seems any busy time passengers will choose to catch bus to substitute MTR underground train transportation tool, due to who feels the bus can arrive any destinations to compare other kinds of transportation tools in the most short time. However even if the MTR can either charge cheaper ticket fare to sell full day or charge discount ticket fare to sell in the busy ( peak) time to compare to bus fare. It is possible that the busy time passengers will still choose to catch bus, if between the bus terminal and the another bus terminal that distance is the shorter time route to spend time to arrive destination to compare between the MTR terminal to the another MTR terminal arrival time . Also, although the busy time passengers will feel to enounter traffic jam to influence sitting or waiting bus time to be longer time in possible and who also feel MTR can avoid traffic jam problem. However, usually any busy (peak) time passengers will feel the chance of traffic jam occurrence will be less. So, the short bus route choice is more potential factor to influence the busy (peak) time passengers still to choose bus to catch.

However, if anyone wants to investigate results of day-to-day route choice which can be transferred to more realistic environment. It is necessary to explore individual behaviour in an interactive experimental set up to ensure busy (peak) time passenger transportation behavioural choice. For example, a passenger has a choice between a main road (M) and a side road (S) for travelling from (A) to (B). (M) is faster if (M) and (S) are chose by the same number of passengers. So, this method can be researched whether MTR terminal station is located at the main road (M) or the side road (S) where is more suitable to accept to passengers generally.

Why trip time reliability and
crowding factors can influence
MTR passenger choice.
Other problem is MTR busy (peak) time's crowding in public transportation occurrence of MTR underground train transportation tool is becoming a growth to concern as MTR demand growth at a busy (peak) time. To capture the MTR passengers benefits with reduced crowding from improved MTR public transport service and image. It is necessary a identify the relevant dimensions of crowding that are meaningful measures of what crowding means to MTR passengers. Two main influences on MTR model choice that are growing in relevance are trip time reliability and crowding. It represents a benefit-cost framework. In fact, MTR passengers can be willing to pay more expensive ticket fare, it MTR can avoid crowding and short and the accurate arrival trip time between terminals is reliable to occur. How to measure of MTR crowding, e.g. weighting the gap between the busy time, the standard ( i.e. objective) and the perceived ( i.e. subjective) metrics. We are not in a position to definitely map the two dimensions, which is a crucial requirement for translating objective improvements into equivalent subjective gains that then can be applied, willingness to pay estimates MTR ticket fares to obtain the additional MTR passenger benefits of MTR public transportation investment to any terminal stations. Because MTR crowding has a negative impact on passengers in terms of psychological on emotional distress. MTR passengers are willing to stand for up to 20 minutes of the service is fast and reliable. However crowding outweighed these benefits from a MTR passenger's perpective, experienced crowding leads a increased dissatisfaction. e.g. stress and less privacy during who needs to stand up in MTR. Due to there are no enough places to supply to them to stand up in MTR. If the MTR trip time was longer time between the passenger's terminals, who will feel more dissatisfaction and it will cause who feels whether who ought need to choose to catch other transportation tools to substitute MTR next time. e.g. bus, train, tram, ferry, taxi etc. So, from an operator's perspective, the MTR service frequency or MTR size is significantly influenced by the level of ridership, which sends a signal to respond if the monitored crowding level exceeds the benchmark standard in the busy time. e.g. in the morning time or at the night time, the students or employment people who need to go to schools or offices ( working places). The locations of different places between MTR terminals and crowding are regarded as

a key service attribute for MTR pubic transportation along with other factors, such as travelling time and reliability, e.g. service quality, none engineering machines are broken to cause MTR stops suddenly.

Given the increasing importance of crowding on both the disutility to existing MTR public transportation users and the influence to it. MTR passenger can choose to use either the MTR public public transportation or other public transportation. It is timely to review the MTR current measures of crowding defined by transportation authorities. MTR operators ought evaluate whether they apporpriately reflect MTR each traveler experiences and perceptions of crowding in busy (peak) time. I suggest that MTR needs to buy other underground trains to supply to the busy (peak) time passengers to let them have enough seats to sit down, so who do not need to stand up in any MTR underground trains when they catch MTR underground trains in busy time. It aims to let who are willingness to pay the estimation of reasonable ticket fares to compare the other kinds of transportation tools in the busy (peak) time.

What is the crowding difference
between train and MTR underground train.

In fact, crowding won't be happened to brother these transportation tools easily in the busy time and non busy time both. e.g. bus, taxi, train, tram, ferry. Because passengers can not choose to stand up in these transportation tools easily, due to these transportation tools have no enough areas (spaces) to let them to stand up easily . So, the crowding will be avoided to occur in these tranportation tools usually. Otherwise, MTR will have many passengers who can choose to stand up because MTR design of length is very long and it has enough areas (places) to let passengers to choose to stand up, even there have none any seats are provided to let them to sit down. So, MTR passengers will feel more dissatisfaction and crowding easily, especial in any peak ( busy) time every day.

Comparing to bus, much more diverse crowding measures are defined in the passenger rail industry. For passenger, different specifications for measuring crowding are found across countries and even within a country. For example, rail crowding measures in the UK, the passengers in excess of capacity is crowding measure that applies to all London and South east operators weekday train services at a London terminus during the morning peak from 0700 to 09: 59 , and those departing during the afternoon peak from 16:00 to 18:59 (office of rail regulation 2011 year). The overall PIXC figure is considered the planned standard class capacity of each train service as well as the actual number of standard class passengers on the service at the critical point. i.e. the location on a trains of standard class passengers that surpass the planned capacity as the difference between the number of actual passengers and the capacity of the train divided by the number of passenger is within the capacity . So, it seems train and MTR underground public transportaton tools had been encountering the crowding problems in peak time, the difference in train passengers need to wait next train or more train arrival is who doesn't plan to enter the train, when who discovers the current train has no seats to provide to them to sit down in whose trip. Otherwise, MTR passengers can choose either to stand up within the large areas (places) if who discovered there are no any seats to provide to them to sit down or who can wait the next MTR arrival in order to who can sit down. It seems MTR transportation tool crowding environment includes in waiting platform and inside of the MTR underground train. Otherwise, train transportation tool crowding environment only includes the waiting platform and the passengers will not have crowding feeling inside of the train, due to none of passengers choose to stand up inside any trains because any train inside has no enough places to let them to stand up.

How MTR can attract many passengers.
On the commuter departure time choice of any reference point researching hand, the departure time decisions of communters are of fundamental importance of peak period MTR traffic congestion. However, whether on the demand side, MTR underground train congestion relief measures, such as MTR ticket fare to every terminal station needs to be charged cheaper fare or discount fare in the peak (busy) time every day. To aim to attract many passengers to choose to catch MTR Underground train public transportation tools, substitute to choose other public transportation tools in the peak time.

Over the past decades, there have been very active research efforts in the departure time problem, both in econometric modeling and dynamic user equilibrium fields. Although, these works provide valuable insights into dynamic commuter decision making, they do not identify the commuters' response to gains and losses related to whole actual arrival time to reference points who may have relative. The appliability of the reference point hypothesis of prospect theory to the commuter's departure time decision making to obtain a better understanding of how departure time choice in MTR platform during their waiting underground train arrival time. However, every MTR underground train actual arrival time and deviation variables related to reference points ( gains and losses) are the key factors in the departure time choice model. How the MTR underground train of every commuter's daily departure time decision can be modelled when the reference point hypothesis of prospect theory. The MTR underground train's schedule delay is defined as the difference between the preferred arrival time (PAT) and the actual arrival time (AT) for a given MTR commuter. In a daily MTR commute, a commuter in the indifference band actual arrival time is an essential feature of MTR schedule study. Two reference points are the earliest acceptable arrival time and the work starting time for a given MTR platform waiting passengers. In psychological view point, prospect theory proposes that the displeasure of a loss is perceived or greater than the pleasure of a gain of the same attitude and therefore, the value function is stronger for losses than gains.

To conclude, it seems that if MTR waiting passengers need not spend long time to wait underground train arrival in platform and it can provide seats to let them to sit down in the busy (peak) crowding time. It will make them to feel pleasure, even the MTR ticket fare is not fair and reasonable to charge higher fare to compare other kinds of public transportation tools fares. So the peak waiting time factor can influence the passengers to choose other kind of transportation tools to catch easily. Moreover, MTR's two reference points are the earliest role. Similarly a loss is observed when the MTR platform waiting commuter experiences or actual arrival time which is beyond that the MTR schedule time. Due to that a MTR waiting commuter is as an early side arrival of whose actual arrival time is earlier than whose preferred arrival time.

Reference

Bailey, L., Mokhtarian, P.L. Little, A. (2008). The broader Connection Between Public Transportation, Energy Conservation And Greenhouse Gas Reduction, Report Prepared As Part Of TCRP Project J-11/Tasks Transit Cooperative Research Program, Transportation Research Board Submitted To American Public Transportation Association in http://www.apta.com/research/into/online/land_use.cfmi, accessed 17 April 2008.

The UK Standing Advisory Committee On Trunk Road Assessment ( SACTRA) (1999). Transport And The Economy ( Report To UK DETR). Retrieved From: http://webarchive.nationalarchives.gov.uk/20050301192906 ; http://dft.gov.uk/stellent/groups/dft-econappr/documents/pdf/dft_econappr_pdf_022512.pdf

Wikipedia Contributors (2008). Arterial Roads In Wikipedia, The Free Encyclopeda, http://en.wikipedia.org/w/index.php?title=Arterial_road&oldid=212832640(accessed May30,2008).

Why and how non-manual driving car owners need

raise public transport quality on travel time and fare
   aspects
   ● How non human driving behavior can be influence by non-manual driving cars

In fact, impact of automated vehicless on travel mode preference, it can bring both trip purposes and distances aim raising need to any kinds of public transport service passegners. Because of technology penetration in the transportation system, the automated vehicle is set to be a future mode of transport, it may bring negative impact to future any kinds of public transport passengers needs, in special on the potential impact of these non-manual driving automated vehicles on travel behaior negative impact to public transport passenger behavior. Automated vehicles will influence future public transportation passengers feel it can bring more short time travel distances and short trip purposes more benefit than any kinds of public transport choices, e.g. bus, taxi, ferry, train, tram, underground tram etc. road and sea public transport tools, e.g. ferry, water taxi. It means that when future any passenger feels above these any one kind of public transport tool needs to spend longer travel time on journey distance and trip to compare future automated vehicles, then they will choose to sit on automated vehicles in preference, due to automated vehicles can help global any one person needs to go to anywhere rapidly.

So, automated vehicles may replace general traditional public transport tools in possible, when they are popular accepted in societies. On the other, instead of shortening journey travel distance time, ( travel time) aspect, public transport fare, travel cost will be another influential factor to influence future public transport tool passengers to choose automated vehicles to replace to catch any kinds of public transport tools.

In fact, conventional cars and public transport s are perceivd as being the least attractive alternative in relation to in-vehicle travel time on short and long distance communting trips. So , future automated vehicle drivers ( non -human driving) behaviors will be likely changed to prefer this mode for long distance leisure trips rather than short distance commuting trips by automated vehicles.

In fact, advanced technologies have revolutionized many aspects of human life, include the automated vehicle transport system. Also, transport system is one of the essential development aspect to particular , such as non-manual driving automation , vehicle aims to make trips safer, faster , more efficient, automated vehicles passengers and drivers can feel enjoyable to do themselves leisure behavior , e.g. read books, listen, music, listen mobile, watch laptop movies when any one does not need to consider whether their cars are safe to be driven , even any one needs to drive the automated car, because robotic can help them to control how to automatic drive this car on the road safely.

Robotic will bring confidence to let them feel that themselves cars are moving safely on the roads . In recent years, the concept of automated driving has been introduced as on outstanding platform for the next generation of driving systems that is expected to improve safety, traffic flows efficiency, reducing traffic jams occurrence chance, avoiding traffic accidents occurrence chance, e.g. avoid to crash any one person when he/she is walking across road or crach any car is moving on the road easily, capacity, accessibility , and reducing congestion through the application of some technologies , such as vehicle to vehicle and vehicle to infrastructure communication.

So, future automated vechicles can have good driving facility systems to be installed in their cars, in order to raise safety, rapid driving speed level to let any one to feel , when they are sitting in their automated cars, e.g. using cameras, sensors, global positioning system adaptive cruise control, light detection and ranging, and advanced driver assistance system, automated vehicles can steer the vehicle and drive it automatically when passengers delegate control to a computer. Absolutely, ny replacing the driver role with an automated driving system , future one automated vehicle is able to totally free up passengers under automation levels.

So, unless future any kinds of public transport tools may apply automated robotic automated driven system replace the bus driver, taxi driver, train driver, tram driver, underground train driver to raise automated driving system service improvement level to let any one passengers to feel. Otherwise, when automated vehicles are popular to be accepted to buy in any one country in global. Then, global public tansport tool passegners number may be influenced to reduce when global any one family owns at least one automated vehicle at themselves homes .

In other words, automated vehicles can bring thes benefits to let global any one household family feels, future automated vehicles users , they can mostly behave like passengers inside the vehicle, which implies that they will be able to multitask and productive by allocating the travel time to do other activities, e.g. reading, eating, working, drinking, watching movies, listening musics, even sleeping. So, automated vechicles will motivate humans to change non-humanly driven behaviors from conventional humanly driven behavior. This non-humanly driven behavior may be one main factor to influence or encourage future any one kind of public transport passenger won't choose to pay fare to buy ticket to catch any one kind of public transport tool again, because non-manual driven behavior may hel many lazy people do not need to consdierate how to learn to drive cars skills to prepare pass any road test in order to earn the driving licnece to permit to drive cars forever. When automated vehiclesa re popular to be accepted to replace manual-driven cars in societies.

Hence, automated vehicles could potentially change the traditional human driven vehicle market to cause their manual driven cars sale buyers number reduces, when the automated vehicle buyers number increases, also they can chance globa public transport passengers behaviors to reduce to pay fares to catch any kinds of public transport tools when automated vechicles users may sit on themselves automated vehicles to go to anywhere in short time rapidly and safely in any countries.

On conclusion, future global public transport service competition is serious, because instead of global passengers had began to compare whether which kinds of public transport fares are cheaper, more safe, shortening journey time between leaving place and destination, more comfortable feeling, e.g. clean and comfortable chairs , mre free internet service facilities in order to make any one kind of catching public transport tool choice in preference. On the other hand, future automated vehicles number will increase when traditional manual driven car users begin to believe that automated vehicles can bring more safe , more comfortable, more fee- time using, more leisure satisfactory feeling, more than traditional manual driving cars. Then, when global any one household family had made choice to buy at least one automated vehice to replace themselves car(s) at home. When, they are habit to sit in themselves automated vehicles to go to anywhere, however, short or long trip . Consequently, global any one household family won't feel any kinds of public transport tools may bring personal economic saving cost, comfortable, enjoyable, free-time using benefit to compare themselves automated vehicles . It will cause global public transport tools passengers number will reduce , when many different kinds of home automatic vehicles are purchased to replace manual driving cars by global household automated vehicle users. So, in passegner transport tool choice psychological view, automatic vehicles will be possible to replace future public transport service tools. So, any public transport service providers can not neglect how to desing and improve their facilities , charge reasonable transport fare, provide more comfortable, and enjoyable sitting feeling , even applying automatic driving system to replace human drivers in order to attract passegners ' catching need choice more easily.

Bibliography

Backman, K., Backman, S., Uysal, M. And Sunshine, K. (1995). Event Tourism : An Examination Of Motivations And Activities. Festival Management And Event Tourism, 3(1), 15-24.

Fishbein, M., & Ajzen, Z. (1975). Belief, Attitude, Intention And Behaviour: An Introduction To Theory And Research, Boston: Addison Wesley.

Hsu, C.H.C., Cai , L.A., Li, M(2010). Expectation, Motivation And Attitude: A Tourist Behavioral Model. Journal Of Travel Research, 49(3), 282-296. http://dx.doi, org/10.1177/004728750 9349266.

ICT Information And Communication Technology Switzerland, 2005. ICT Fakten (ICT facts). Available from http://www.ictswitzerland.ch/de/ict%2fakten/factsfigures.asp(retrieved Dec.12, 2005) in German.

Lind, (2001): Befolkningen, Familjen, Livscykeln- Och Ekonomisk Tillvaxt. Institutet For Tillvaxtpo-litiska studier/ Vinnova/Nutek.

Lohmann, Martin (2001): The 31 st. Reiseanalyse-RA 2001. Tourism: vol. 49, no.1/2001;pp.65-67, Zagreb.

United Nations Population Division (2001). World Population Prospects: The 2000 year Revision, New York.

Weber E.U., & W, P.Bottom (1989). "Axiomatic Measures Of Perceived Risk: Some Tests And extensions." journal of behavioral decision making, 2 (2): 113-31.

Artificial Intelligent In Road Transportation Strategy

● How artificial intelligent vehicle may interact intelligent transportation tools

Can artificial intelligence (AI) and machine learning (ML) be used in the search for new " consumption" behavioral type variables that affect consumer individual or transportation service organization individual different transportation tools choices, such as road or sea or sky transportation tools? Can artificial intelligent vehicle may interact intelligent transportation tools market development?

Consumers usually have bargaining and on risk choice when they are already shopping, such as who need to accept to use any (AI) new technological products to replace human traditional behaviors, such as intelligent non-manual driving transportation market, e.g. cars are needed to be driven by human drivers on road, but it has bargaining and on risky choice, when non-manual (AI) vehicle buyers who need to depend on non-manual artificial intelligent (ML) system assists them to drive their cars on the roads.

So, any non-manual driving auto car buyers must need to believe (AI) non-manual driving vehicles (ML) systems can make accurate driving judgement to reduce or avoid any traffic accident occurrences more than human drivers' driving judgement when the (ML) systems are driving their cars on the roads. Then the intelligent vehicle manufacturers will have possible to sell their non-manual driving vehicles success.

This is the first reason or idea influences consumer individual choice to buy any kinds of (AI) non-manual driving vehicles, when consumers believe (ML) systems are more safe and make more accurate judgement to compare human or computer systems, when they are sitting in one non-manual auto driving vehicle on the road.

The another second reason or idea is that some common limits on driving consumer prediction might be understood as the kinds of errors made by poor implementation of machine learning.

Supposing driving consumers believe (AI) machine learning ability is worse to compare to human learning ability. It will also influence driving consumers do not accept to use any (AI) non-manual auto driving vehicles to replace every driver is essential on driving by himself/herself on the road. The third idea or reason is that it is important to influence driving customers believe how (AI) non-manual auto driving technology is used in them can both overcome and exploit human driving skill and safe limits and raise more auto driving safe judgement to compare human driving safe judgement.

However, how to predict any kinds of (AI) non-manual driving vehicles future consumption effort, due to different kinds of (AI) non-manual driving transportation vehicles which have different unique functions and designs to be used by different kinds of road transportation or driving demand of consumers. For example, lorry drivers need non-manual intelligent system can help them to drive fast, but safe to assist them to transport cargo to arrive destinations from their factories or offices. Otherwise, private car driver expects whose (AI) non-manual driving vehicle can auto drive to send to whom to arrive destination in safe way and non-too fast and non-too slow speed in order to avoid accident occurrences.

So, a different road intelligent consumer demand is to define whose individual driving behavior and driving habit and driving attitude and driving judgement and driving speed demand to decide how to design whose intelligent vehicle to satisfy those driving demand more generally, as simply being open-minded about what variables are likely to influence every consumer economic choice, when who decide either to buy any kinds of (AI) products or not to buy any kinds of (AI) products to replace the different demand of consumers their different (AI) useful demand.

Hence, for these three (AI) products group of stakeholders, such as home (AI) consumer group, firm (AI) consumer group and government (AI) consumer group . These consumer groups may consider whether different kinds of (AI) products can give what is special beneficial interest to them to use. These variables can be measurable properties of choices to influence them to choose to buy any (AI) kinds of (AI) products to use, e.g. psychophysiological, biological, social influences, consumer's wealth, moods and personality, (AI) product price etc. variable factors which will influence them to decide to attempt to buy any kinds of (AI) products to use.

If behavioral economics is as open-mindedness about what variables might predict. Then , (AI) machine learning system is a way to do behavioral economics because it can make use of a wide set of variables and select- which ones predict.

In behavioral economic view point, when general consumer overall demand to the product is much than the other similar ( AI) non auto driving vehicle products, such as any kinds of (AI) non-manual auto driving vehicles and any kinds of manual driving vehicles case, then any kinds of (AI) non-manual auto driving vehicles will be more attractive to cause many manual driving vehicle buyers choose to buy (AI) non-manual auto driving vehicles. Hence, it seems if any kinds of (AI ) non-manual auto driving vehicle products can make more attractive variable efforts to influence overall driving consumers to feel that they have more needs to drive non-manual auto vehicles to compare more than driving manual driving vehicle.

What is the main variable effort to intelligent vehicles to attract driving consumers to choose to accept to drive them ? However, I believe that (AI) machine learning system is a main factor to raise overall driving consumers' acceptances to drive it to replace manual driving vehicle. If it can persuade or prove (AI) machine learning system ability and judgement effort is more accurate than human or computer learning effort or judgement effort, then it is possible that any kinds of (AI) non-manual driving vehicle products will be accepted to drive on the road in popular.

Machine learning system is able to find prediction value in details of how the bargaining occurs. This discovery is the beginning of the next step for driving consumer individual driving behaviors or driving habits. It raises questions that include: What variables predict to influence driving consumers to change whose driving habits or driving attitudes? How can driving consumer individual emotion, face-to-face talking with whose friends when they are sitting in the non-manual driving vehicle to influence whom driving habit or driving attitude to be changed ? Do driving consumers consciously understand why those habit driving attitudes variables are important when they are sitting in one intelligent vehicle? Can (AI) driving machine learning methods capture the effects of motivated cognition to influence driving consumers decide to buy any kinds of (AI) non-manual auto vehicle products more attractively. So, it seems (AI) driving machine learning method is a main variable factor to influence driving consumers to feel who have more confidence to drive them more than any other kinds of similar manual driving vehicles on the road. Consequently, (AI) driving machine learning system will be one important psychological method to influence driving consumers to choose to buy (AI) auto driving vehicle products to replace manual driving vehicles. The reason is because human and driving machine learning system both which will have limited variable factors to influence general different countries (AI) driving consumers' need desire to be raised.

● Why can (AI) driving machine learning system main factor influence driving consumer individual desires ?
Driving consumer expectations are hard to measure or predict driving attitudes and driving behaviors in (AI) non-manual driving vehicles market. Artificial intelligence is another kind of computer science development to apply intelligent vehicle market. Why do driving consumers feel need to buy any kinds of (AI) auto driving vehicles to drive to replace manual driving vehicles on the roads? What are (AI) auto driving features different to manual driving features?

(AI) is the recreation of cognitive functions in computers; it enables machines to perform tasks like humans and perhaps even better than human. In the real world, scientists develop the technological singularity, in which a superintelligence emerges with unfold human consequences.

Professionals in many industries are intensely interested in the specifics of what (AI) can do today, and how can it helps. They are considering the impact of applied (AI), in which computers are used to address a particular problem, extracting and utilizing patterns found in large volumes of data. Of all (AI)'s subfields, machine learning is attracting the most attention. I shall explain why (AI) machine learning system is the main factor to lead consumers feel need to buy any (AI) products to use. Such as below:

For smartphone, fraud detection to medical diagnosis etc. applied (AI) technological products examples. (AI) machine learning systems can help any one of these products to do any exceed general computer learning systems which (AI) learning systems can do any skills to supply (AI) users to use to compare computer learning systems can not do any skills to supply compute users to use. It seems that (AI) machine learning system is the unique feature to attract consumer consideration in technological product market.

An term for different types of learning, and can be accomplished using different techniques. This has led to a perception that all marketing teams should have (AI) to bring a unified personalized customer experience, when consumers choose to buy any (AI) products to feel what are the different or unique characteristics to compare general computer products. Such as (AI) product has this unique machine learning characteristics, we can predict (AI) and machine learning is connected to influence consumers to feel needs.

Furthermore, over the same time period, and in contrast to predictions for roles in many industries. (AI) won't take the place of marketers and merchandisers themselves although it is already a new value to analytical and strategic marketing skills to persuade consumers to buy any (AI) products. It means different kinds of (AI) products will have different machine learning effort and unique characteristics to attract consumers to choose to buy them to use. Such as, when intelligent vehicles need have unique road driving or sea transportation or flying machine learning system when they are applied on these three kinds of transportation tool aspects. They need have good response safety driving and immediate response learning systems to avoid any boats or air planes or vehicles to crash to them to reduce accident occurrences immediately on any one of either road or sky or sea journey environment.

What is the reason why (AI) driving machine learning system can influence good at making sense to driving consumer desire? Only humans ( drivers) , preferably experienced, well informed humans can understand their driving customer needs and decide how to design or reengineer any (AI) intelligent vehicle product functions. (AI) intelligent vehicle can give these professionals the means to do this better to compare manual driving immediate response control function when any vehicles are driving or they will stop immediately to close / near to them in order to reduce crash occurrence on the road, and then maximize relevance through real-time customization of the non-manual auto vehicle driving user experience.

For example, as ever, senior decision makers need to be informed, decisive and results-oriented or risk losing out. Harvard Business Review indicated : Over the next decade, (AI) won't replace managers, but managers who use (AI) will replace those who don't. Such as intelligent vehicle won't replace drivers, but drivers who use intelligent vehicles will replace those who can not control how to drive their vehicles in the most safe way. So, (AI) driving machine learning system will have possible to do any drivers' (human's) driving judgement, driving analytical mind and driving effort to be more accurate than manual driving skills. Such as how to control to drive the intelligent vehicle in the most safe way. It is general manual driving skill can not achieve to drive in the safe way.

For another (AI) digital commerce example, (AI) and machine learning are the most exciting developments in marketing and merchandising to be applied to digital commerce, such as making better decisions through trend and cluster analysis, deploying product and content in mutually reinforcing combinations, increasing customer engagement and satisfaction in real time.

Hence, the key attraction in digital commerce circles is that machine learning is designed to be self-optimizing. Optimizing for revenue example will surface are increasingly profitably selection of products ( within the brand parameters selected).

When to apply (AI) capabilities and what value (AI) is delivering for customer and company like. Unlike any technology before it, (AI) is analytical and predictive capabilities offers the prospect for each and every individual. It can maximize real time and engagement. Effective tailored (AI) technology, such as digital experience cloud technology is available now. And once integrated, (AI) starts learning and delivering incremental value from day one. So (AI) could transform the digital experience to any business organizations.

Hence, (AI) driving machine learning system can be applied to road driving skill aspect. When intelligent vehicles are invented to own the most safe driving judgement skill and they can know when either they may auto drive fast speed, when they are feeling to know when there are not many vehicles are moving close/near to them or when they need auto drive slow speed, when they are feeling to know when there are many vehicles are moving close/ near to them. Then driving consumers will have more confidence to choose to buy any kinds of intelligent vehicles to replace manual driving vehicles to drive on the roads.

● Non-manual driving transportation tool market development

If Non-manual driving vehicle manufacturers expect their (AI) automatic vehicles can attract drivers to buy. I feel them to need to consider how (AI) driving machine learning system can achieve these requirements in order to satisfy manual driving vehicle drivers' requirement to change their traditional driving habit to choose non-manual driving needs. It means (AI) driving machine learning systems can help them to drive vehicles to replace manual driving vehicles on the road. This is the main factor to influence car buyers choose to buy intelligence driving vehicles replace to manual driving vehicles. I believe (AI) non-manual driving vehicle machine learning systems, need to be designed as below:

(1) Improving driving safety by preventing accidents from happening.

Every year, drivers are facing a large number of casualties, due to traffic accidents. The amount of killed and injured road traffic related accidents is increasing every year. The real cost of an accident can go well beyond the limits of immediate material destruction, and is impossible to evaluate.

Hence, researchers and car manufacturers are looking for solutions in order to reduce the amount of accidents. They already developed a considerable set of technologies in order to decrease the amount of casualties. Most of them ( like airbags, seat-belts, anti-lock systems, shock absorbing car bodies) are efficient in decreasing the impact of an accident, and in protecting the passengers of the cars. The technologies already saved a lot of lives, but they are rarely able to avoid accidents because they do not anticipate them. Moreover, if they are protecting in many cases, the passengers of the car, they do not prevent most traffic participants, like pedestrians on bicyclists from getting injured. it causes (AI) non-manual automatic car manufacturers need to consider how to design machine learning safety system is to prevent accident from happening instead of just reducing their impact.

This can only be possible using intelligent systems that can observe the driving environment, reason and decide if there is a danger, determine how to avoid it and act if necessary

(2) Reducing energy consumption by optimizing the driving.

Nowadays, global air pollution is serious. (AI) non-manual driving car manufacturers need to concern how to design (AI) machine learning system can reduce degree of air pollution to be the most minimum level to compare to traditional manual driving vehicles.

The reduction of energy consumption if certainly one of the main challenges. Transportation is one of the major factors in fossil energy consumption, and it is also responsible for a large amount of $CO_2$ pollution. It is difficult to ask individuals to voluntarily limit the use of their vehicle of they do not have a strong incentive to do so. Specially in regions where vehicles are needed to drive to go to work every day. It stands to reason that if it is difficult to decrease the amount of vehicles, part of the solution is to make them more energy efficient.

Hence, non-manual driving car manufacturers need to design how to improve engines, which are more optimized and need less fuel to operate, and hybrid and electric cars have been developed and are continuously being improved. But we can go beyond these solutions that do not take into account the environment in which a vehicle is driving. A growing number of scientific contributions presented intelligent systems used in order to improve energy efficiency and reduce fuel consumption, based on the optimization of the way (AI) non-manual driving (AI) vehicles are performing. Such as recharge batteries and electric engine will be predicted the popular fuel in order to limit fuel consumption to future (AI) non-manual driving vehicles. They can reduce air pollution, consume less fuel for (AI) non-manual driving vehicles.

(3) Improving comfort by anticipating (AI) non- manual driving vehicle drivers.

Finally, another application for intelligent vehicle is the improvement of driving comfort. Car industry is very competitive market. Many potentials (AI) intelligent vehicle customers need to enjoy to sit more comfortable intelligent vehicles, who will be attracted by (AI) comfortable systems improving when driving, so part of the research in intelligent systems from cars focuses on how to improve the driving experience, i.e. make it easier and more enjoyable, more comfortable to compare to traditional manual driving vehicles.

As an example, lane keeping assistant systems are technologies that actively keep the vehicle in the lane in highways of the driven drifts out of it. Automatic speed regulation keeps the car at a certain speed without requiring to touch the gas pedal. This can be really interesting for, e.g. (AI) non-manual driving truck drivers that spend a lot of time on highways. But these technologies have a limitation in the case of automatic speed regulation, this technology can not copy of a vehicle ahead drives slower than the desired speed, or if another vehicle cuts into the lane.

This case requires the driver to have a constant focus on the road. In order to achieve more comfort, it is better of the system can adapt to changes in its dynamic environment: let the (AI) intelligent vehicle adapt to the speed of the man-manual vehicle, or autonomously change lane when requires. Again, this requires knowledge about the environment, detection capabilities, reasoning and action planning. Intelligent systems can be used in order to create more attractive and more comfortable and more safe, less energy consumption and less fuel expenditure by intelligent vehicles.

Artificial intelligent public transport how influences passenger psychology

How technology influence passenger psychology

Nowadays, robotic invention can be applied to factory manufacture, hotel, restaurant, shopping center, customer service, accounting, law document draft etc. general office tasks aspect, evem hospital surgen patient medical operation health service aspects. If future robotic non -manual driving vehicles can be invented to reach the safe auto driving mature skill stage. Any one driver begins to believe robotic, driving safe level is bette than he/she drives himself/herself car. I assume that if future robotic public transport tool drivers can replace human public transport tool drivers to drive bus, taxi, train, tram, ferry, underground train, tram , even air plane ets. different kinds of public transport tools. How non maual driving public transport tools influence our social change either to improve better ot worse? How non manual driving public transport tools influence passenger psychology, e.g. increasing any kinds of public transport tools passengers safe feeling to choose to catch any kinds of public transport tools to go to anywhere or feeling more dangerous when the passenger himself/herself chooses to sit the non manual driving public transport tool.

In past, traditional public transport tools are driven by human drivers, if one day non manual driving skills are invented to reach the most safe level, when the car owner or passenger is sitting on the non manual driving vehicle or public transport tool, the artificial intelligent driver can help the driver to control the car wheel to avoid to crash any other cars or pedestrians to o to any far places on the roads easily. The public bus does not human driver to drive the bus, artificial intelligent driver won't feel tried, when it drives the bus long time, it does not need to leave the bus to go to toilet, to go to restaurant to eat, to go to rest room for rest, because it is one (AI) machine. So, the (AI) driver won't have negative emotion to feel angry when the bus passenger complaints its service is poor when he feels dissatisfactory to the (AI) driver bus service performance.

However, human bus driver may be complainted for unpolite or rude bus service attitude. It is common human bus driver will encounter any unreasonable passenger complain in general . Hence, when non manual driving technology can be invented to reach the most safe driving skill level, whether (AI) machine driver is the most suitable to replace any public transport tool drivers, such as bus, taxi, train, underground train, ferry, tram, even air plane to drive for future passengers service need.

In fact, any public transport tool drivers may cause traffic transport accidents, due to their careless driving to crash any other vehicles or pedestrians ( walling people). Consequently, any passengers may have chance to be killed by public transport tool crashing accident. So, it seems that global public transport tools are dangerous to any passengers, when they are sitting on the bus, taxi, train, tram, underground train road public transport tools, or ferry sea public transport tools, because any human public transport tool drivers will feel tried to drive any one kind of public transport tool long time, for example when the bus driver has no enough nervous to drive the bus, he wants to sleep, due to he often needs to follow night time bus timetable to drive bus long time at night. When he often want to sleep and he is driving the bus, traffic accidents will be caused easily. So, any passenger individual life is dominated by the sleeping bus driver. His bus dirving behavior is not safe to any one bus passenger when the bus passegner chooses to catch this feeling sleeping bus driver's bus to catch.

Otherwise, (AI) non manual driver must not feel tried or need sleep often. It is one automative driving mature, it can drive any kinds of public transport tools all day, because (AI) machine drivers do not need sleep, (AI) none human auto-driving driver can bring this important unique benefit to any kinds of public transport tools to compare human drivers. Instead of (AI) automatic driving tools' non need sleeping advantage, (AI) non-manual control auto-driving tools must not own sad, disappointing feeling , tried feeling, anygry emotion feeling when it needs to contact angry passengers every day. So, I mean that any traffic accident occurrence will reduce, when (AI) drivers often feel happy to drive any kinds of public transport tools. Otherwise, any human public transport drivers will be influenced to feel angry when they are complaint by angry passenger in any driving time easily. So, public transport traffic accident will be caused to occur easily. Althoug, it is not guarantee that it is obsolute none any public transport accident occurrence, due to crash to other vehicles, during the non manual driving (AI) driver drives the bus, tram, train, taxi, on the road, nut when (AI) non manual driving skill can be improved to the most safe driving level. I believe that non manual driving public transport tools ought be bring more safe to compare human public transport tools drivers to any one passenger individual life safety.

How non manual driving automated vehicle influences future mode of public transport service change? A survey distributed in the Netherlands in which respondents had to choose between conventional cars, public transportation for different travel distances and trip purposes. having collected information from 663 respondents, conducted a study on classic trip attributes ( such as travel time, car owner self driving time and non manual driving public transport tool driving time as well as travel costs, car owner car fuel purchase expenditure and general non manual driving public transport tool fare comparison), attitudinal factors and socio-economic variables to understand future non manual auto driving public transport tools choices. The repor indicates that automated driving transport service which they defined as an automatically controlles door-to-door transport service provided by a vehicle with similar features to a conventional car, albeit driveless. Results suggest that travellers' mode preferences vary significantly for different travel distances and purposes. They found that conventional cars and public transportation are perceived as being the least attraction alterrernatives in relation to vehicle travel time and short -and -long distance commuting trips respectively, preference for passegner choice is between the non-manual driving auto car and non manual driving auto public transport tool.

They indicated that future passengers will consider how non-manual driving public transport tools whether they can bring trips are safer, faster and more efficient to let them to feel as well as traveling time and time cost is also another factor to influence them to choose to catch non-manual driving public transport tool, when they feel safe to arrive the destination rapidly. Then, future many passengers will be persuaded to choose to catch non-manual auto driving public transport tools in preference. So, if future all public transport service providers can let passengers to feel fares are reasonable price, when their non-manual driving public service transport tools, bus, taxi, tram, train, underground train etc. they can let them to feel safe to arrive destinations rapidly, they won't need worry about passengers number will reduce when human drivers are replaced by AI robotic drivers.

In fact, when one needs to drive to arrive destination in long driving time. The car owner will feel tried, bored and he/she can not spend driving time to do his/her interesting activites in his/her car, e.g. reading, listening music, watching TV, playing electronic games from smartphone, phone talking etc. personal behaviors. So, it means that future long time trip passengers may be persuaded to catch non-manual auto driving public transport tools if they believe that this kind of new non manual auto driving pubic transport tools can provide more safe, efficient, rapid, comfortable feeling to them, when they are sitting on them.

All of these may be the main factors to influence them to choose to catch non -manual auto driving public transport tools. In general, these other factors may influence future passengers to choose to catch non manual auto driving public transport tools, they may include: whether automated vehicle would drive on populated streets better than conventional cars, whether an automated car would be comfortable entrusting the safety of a close family member, whether automated vehicle might produce fewer pollutant emissions. Because , when future non-manual auto driving vehicles are popular, many car owners will choose to buy automated vehicles to drive. So, future non manual auto driving public transport tool service providers , their competitors may be automated vehicles. If automated vehicles can let car owners to feel car prices are reasonable, they can provide safe, rapid speed, comfortable feeling to any one car owner, then he/she can sell his/her traditional car to change new automated car easily, when global many car owners begin to accept automated cars.

On conclusion, future passengers may be persuaded to choose to buy fares to catch any kinds of non manual auto driving public transportation tools. It depends on these factors, such as reasonable fares, safe feeling, efficient and rapid arriving to destinations short time journey, comfortable and clean seats facility, free personal behavior, e.g. quite reading , listening music, watching TV , free internet provision transport environment, when future any one passenger is sitting in the auto driving public transport vehicle. So, (AI) technology will have possible to influence our future social public transportation development may bring more significant new travelling experiences and it can let global passengers to feel indeed. Moreover, it will be future global public transportation service providers, they need to consider that they ought need to change their public transport tools services in order to satisfy future passengers transport needs more easily. I conclude that future global public transport service will be influenced to change non manual auto driving public transport services by future global passegner public transport service needs within 10 years. So, nowadays, any kinds of public transport service providers ought need to spend time to research how to design themselves traditional public transport service moods to change to non manual auto driving moods in order to satisfy future global passenger individual new public transport services needs successfully.


CAN NON-MANUAL DRIVING AIR PLANES EXCITE FUTURE AIRLINE AND AIRPORT TRANSPORT INDUSTRY DEVELOPMENT

Will airline industry's ticket price elasticity be influenced by demand and supply factor

● Boeing 747 manufacturing fuel cost strategy

- How airlines and airports implement successful netwpork strategies
- Performance measurement system strategy

What factors influence cost-related management quality ?

Can non-manual driving air planes assist future tourism industry development?

- Boeing 747 manufacturing fuel cost strategy

For Boeing 747 air plane manufacture example, how it can help airline to avoid travellers number reduces. Boeing 747 air plane manufacture company how achieve air plane manufacturing strategy to reponse airline traveller number market changing. What is fuel conservation strategy to Boeing 747 air plane manufacturing firm? The cost index, (CI) feature of the flight manufacturing computer (FMC) can help airlines significantly reduce operating cost. However, many operators do not take full advantages of this powerful tool. What does CI ratio mean? The CI is the ratio of the time-related cost of an air plane opertation and the cost of fuel. The value of the CI reflects the relative effects of the fuel cost on overall trip cost as compares to time-related direct operating costs.

The equation form, CI= time cost-$/hour / fuel cosst -cents/lb

The numerator of the Ci is often called time refrated direct operating cost ( minue the cost of fuel). Items, such as flight crew wages can have an hourly cost associated with them, or they may be a fixed cost and have n variation with flying time engines, anxiliary power units, and air planes can be leased by the hour or owned, and maintenance costs can be accounted for an air planes by the hour, by the calendar or by cycles . As a result, each of these items may have a direct hourly cost or a fixed cost over a calendar period with limited or no correlation to flying time.

What does this air plane fuel, cost strategy advantages to airline companies? In the case of high direct time costs, the airline may direct to time costs, the airline may choose to use a larger CI to minimize time and thus cost . In this case, where most costs are fixed, the CI is potentially very low because the airline is primarily trying to minimize fuel cost. Pilots can easily understand minimizing fuel consumption, but it is more difficult to understand minimizing cost when something other than fuel dominates. So, the cost of the CI ratio. Although, this seems straight toward, issues such among the operating locations, fuel tankering, and fuel hedging can make this calculation complicated. So , this fuel consumption cost, air plan manufacturing strategy can help any airlines to reduce air fuel useful cost and waste fuel.

However, CI can be an extremely useful way to manage operating costs. Because CI is a function of both fuel and non-fuel costs. It is important to use it appropriately to gain the greatest benefit. Appropriate use varies with each airline, and perhaps for each flight.

How low fuel situations can bring less fuel consumed benefit in the more environmentally friendly flight? Fuel conservation strategy can help airlines significantly reduce operating costs. However, many operators do not take full advantage of this powerful tool. Cruise flight is the phase of flight that falls the largest percentages of trip time and trip fuel are consumed typically in this phase of flight, which also impact trip time and fuel significantly can often be avoided through appropriate cruise planning. This fuel conservation strategy includes these characteristics. These objectives which depend on the perspective of the pilot , dispatcher, performance engineer, or operations planner can be groups into five categories , such as:

1. Maximize the distance traveled for a given amount of fuel ( i.e. maximum range).
2. Minimize the fuel used for given distance covered ( i.e. minimum trip fuel).
3. Minimize total trip time ( i.e. minimum time).
4. minimize total operating cost for the trip ( i.e. minimum cost, or economy speed).
5. Maintain the flight schedule . The first two objectives are essentially the same because in both cases the airplane will be flown to achieve optimum feel mileage.

In addition to one of the overall strategic objectives for cruise flight, pilots are often forced to deal with shortage term constraints that may require them to temporarily abandon their cruise strategy one or more times during a flight. These situations may include:

Flying a fixed speed that is compatible with other traffic on a specified route segment. Flying aseed calculated to achieve a required time of arrival at a fix. Flying a speed calculated to achieve minimum fuel flow when holding ( i.e. maximum endurance). And when directed to maintain a specified speed by a air traffic control. Hence, when the air plane faced with a low fuel situation at destination, many pilits will opt to fly LRC speed, thinking that it will give them , the most miles from their remaining fuel. Also, if fuel prices increase relative to other costs, a corresponding reduction in CI will maintain the most economics operation of the air plane. If however an airline experience rising hourly costs, an increase in CI will retain the most economical operation . For this reason, flight crews typically reserve a recommend CI value from their flight operations department, and it is generally not advisable to deviate from this value unless specific short term constraints demand it.

How it can help air planes to execute for maximum fuel savings efficiently? For example, but times have clearly changed. Jet fuel prices have increased over times from 1990 to 2008 year. At this time, fuel is about 40% oa a tycpical airline's total operating cost. As a results airlines are reviewing all phases of flight to determine how fuel burn savings can be gained in each phase and in totel.

Boeing 777-200 extended range and 747-400 and ( long-range, e.g. short range e.g. 717, medium range, e.g. 737-800 with winglate commercail air planes can impact fuel usage. However, flap setting must be appropriate for the situation to ensure air plane safely. Higher flap setting configurations use more fuel than low flags configurations.

The difference is small, but at today's prices the savings can be substantial especially for air planes that fly a light number of cycles each day. Hence, top fuel conservsation strategies for flight crews include: Take only the fuel you need, minimize the use of the nuxiliary power unit, taxi use efficiencly as possible, take off and climb efficiently , fly the air plane with minimum drag, choose routing carefully, strive to mantain optimum attitude, fly the proper cruise speed, descend at the appropriate point, configure in a timely manner.

Fuel conservation is a significant concern of every airline . An airline can choose an approach procedure and flap setting policy that was the least amount of fuel, but it should also consider the trade off involves with using this type of procedure.

Hence, Boeing flight crew can earn benefit, when they fly or air planes, such as to conserve fuel and reduce noise and emissions or to accommodate speed requests by air traffic control. All of these are fuel consumption reducing strategy to airlines.

● How airlines and airports implement successful netwpork strategies

What is airline network strategy ? Network management includes route planning, scheduling optimization, airline business planning and data analysis . How implement airline network strategy, airlines need to evaluate strengths and weaknesses of the current network strategy, identification of additional potential, recommendations for adjustments, network integration, due to merger or cooperation. Fleet and capacity evaluation , analysis of estimated passenger volume over time combined with option aircraft size and frequencies for current and potential future rotes / markets; competition response, modeling of results as an impact of competitor's action and reaction; establishing best practice network managment for new carriers include: route selection, network planning, scheduling, airline business planning includes forecasting of revenues and costs. Finally, any airlines need to establish network planning, such as network optimization and development, traffic and revenue forecasting, market -and -competitor analysis, scenarios for profit-optimized networks include: hub-strategies, evaluation of aliances, cooperation includes: route joint ventures in industry environment.

● Performance measurement system strategy

Any airlines may apply performance measurement methods to design the indicates of performance. Different models and frameworks are excellence models. Any airlines hope to achieve useful performance meaurement strategy. They need to answer these kep questions: Who are the key stakeholders and what they want and need? What strategy airlines have to put into place to satisfy the key stakeholders' want and need? What critical process do airlines need if they want to implement this strategy? What capability do they need to operate and improve this process? What does cost leadership strategy mean? Competitive price is decided for customer . Indoneasia, Malaysia , India and China countries are implementing, it can provide cheaper workforces and the cost of production will be covered.

In aviation service industry, cost strategy , it much relevant to be applied cost carrier, such as Vigin Blus, Ryanair Airways are implementing . However, some airlines combined the low cost carrier and full service ,which is known as low fare limited services.

Innovative marketing strategy is not only how to carry many more passengers, but also how to enable significant reduction in the costs of distribution and marketing. Such a strategy is used as a competitive strategy of low cost tariffs . For example, low cost focus strategy serve tourists or groups with certain destinations, e.g. the flights are carried out by certain airlines.

There are also tourist groups who travel to certain tourist destinations. Passengers only are transported be the tourist destinations, thus departure schedule won't be a basic need, lower price of ticket and flexible departure schedule as well as comfortable cabin are still the standard.

Another improvement of performance method, it is market orientation, it is believed to give psychological and social benefits to the employees , in the forms of greater pride and sense of belonging, as well as greater commitment to the organization.

Another strategy improves to service performance, it is distribution management strategy . Everyone can be brought benefits. It may achieve these benefits: Raising passenger numbers, as some airlines and airports are running at near full capacity. The effects of disruption are only becoming compounded. Social media can help airlines and airports to tackle dissuption management and avoid damaging their reputation with consumers.

So, when discuption is solved. It can help airports reduce discuption cost and damaging their reputation with consumers . Innovation may include: airlines attempt to develop standard procedure for common disruption situations, responding to regulations, such as the delay rule and compensation for cancellations with faster, more proactive decisions, collaborating with air traffic control facilities, airports, themselves views of resources, identify available options.

● What factors influence cost-related management quality ?

Thus, the reduction of costs lies at the core of the low cost airline model, which aims to offer lower fares, elimating some comfort and services that were traditionally quaranteed, e.g. employed to refer to low-cost flight. The use of an airline booking system, the suppression of free-in-flight catering, the use of secondary airports connected through a point-to-point network, and the use of homogeneous fleets are only a part of the innovative choices made by low-cost airlines. However, these are main important factors to influence route structure, type and characteristics of the aircraft cost of labor and management quality . For example, if the airline's pilot, airline passenger service people, cleaners , all salaries can be reduced to employ. Then, the airport or airline's labor cost will be reduced and it can influence it's air ticket price to be also decreased.

When the airline's air tickets price reduces, it will bring competitive low air ticket sale price to win other airline competitors more easily. How air ticket sale attractive prices ? The airline's air plane type and characteristics,, whether they ae comfortable to let passengers catch in their whole trip time. Whether their air planes are new or old model ? Whether their air planes' facilities can satisfy passengers entertainment need, e.g. chair can be bed to let passengers to sleep, television or movie is attractive to watch, music is soft sound etc. psychological entertainment facilities can let the airline passengers feel satisfactory or not.

Final view is management quality, such as whether the airline or airport CEO's management ability performance is either excellent or good or general or poor . They can influence whole airline or airport front -line service staffs' service performance to let passengers or airport visitors feel their services can satisfy their needs when they choose to catch the airline air planes to fly to the country to travel or work, but the country's airport staffs' service performances and airport facilities will influence their revisiting to the country's airport or rebuying the airline's air tickets again.

The final strategy is flying route strategy. I believe that whether the flying route is attractive or not, it can influence passengers to choose the airline preference. Which factors determine choice of flights on the Dhaka-London route? How could the airline be able to cope with the competitive advantages of its major rivals in this flying route? How is the competitive environment for airlines operating in this route? How could the airline sustain its competitive advantage and what can it do to gain more market share in this route?

A successful and attractive flying route design needs to amend and adjust its flying route design strategies and capabilities as the airline firm goes through its flying route design life cycle, when changes in passenger's flying route choice preferences. Hence, any airline needs to know whether what its SWOT ( strengths, weaknesses, opportunities, threats) before it decides to implement which flying route(s). Because manay flying routes choicew will be influenced by its current SWOT environment situation. For organization of new flying route to the UK airport from UK , e.g. London ciry airport, the HK airline needs to know whether what its strengths , e.g. customer loyalty, its strengths can provide the most rapid . The most cheapest, the more comfortable flying feeling from HK airport flys to UK London airport or from UK , London city airport flys to HK airport, its UK , London city flying route can provide competitive fare al promotion, extra baggage allowance, airline brand image , whether is famous to UK travellers, providing online seats reservation for any UK , London flights services, airline organization size and airline brand , whether it can get travelling passegners, loyalty or confidence either to fly to UK , London city from UK city airport, or flying to HK city from UK, London city airport, providing regular UK , London flying route schedule, network presence how much UK flying route cost cutting, Uk air planes' aircrafts facilities whether are enough to satisfy HK or UK to catching the airline's air planes flying entertainment needs. Does it one new route development opportunity of direct flughts from HK airport to UK , London city airport. HK air port has enough terminals number to let facilities to provide passengers stay in HK airport when UK , London city travellers arrive HK airport? Does HK airport lacks technologica facilities to satisfy new UK , London city airport flying route development , e.g. providing enough spaces to let air fleets stay in HK airport, aircraft manintenance whether is enough? Does HK airport encounter shortages of experienced front line counter service staffs to serve UK , London city travellers in airport? When the seasonal time is holiday travelling time for UK visitors, is UK new flying route rising fuel cost? Does new entrants to develop another UK , London city flying route from HK? Has new UK , London city from HK flying route , these weaknesses , such as lacking enough resource to develop another another new flying route or no direct flight or long hour flight after the HK airline developed the new UK , London city flying route to the HK airline from HK airport. Has new Uk, London city flying route development enountered these threats: strong competiton, high interest and UK foreign currency exchange rates, raising fuel cost, economic recession decline in the UK airline industry, environmental pollution etc. issues influence UK travellers choose to travel to HK desires. So, SWOT analysis must be needed to consider in order to develop any new flying route in order to avoid servious high cost expenditure loss to any airlines.

So, above these tangible , e.g. fuel cost control , route choice, network cooperation choice and intangible, e.g. service performance factors can influence whether the airline's network cooperation strategy, route choice strategy or low cost strategy can succeed or fail. So, airlines or airports can not neglect these factors to be revised in order to achieve any strategies in success more easily.


AI sixth stage development
How AI technology assists
    human to avoid energy and food waste
    House quality influences householder electricity
energy consumption behavior

Can apply artificial intelligence measure the house quality how influences the householder electricity energy consumption or useful activities to be more or less? In general, house owners have both intentions for whose property. One intention is living the house by householder himself or herself or householder with families themselves. Another intention is that renting to others to receive rent income ( landlord). So, in the housing market, the housing consumer includes either the property owner intents to rent to others to live for rent income aim or the property buyers intents to buy the house to be house owner to live. Does these both different property purchase intentions, which will influence the householder's attitude to use electricity energy consumption desire to be more or less, due to the householder's demand to whose house quality factor influence? This is one interesting question concerns the householder electricity energy consumption desire change to the householder, due to house's investment or house's living intention influences to house quality factor.

How does house quality factor influence to householder electricity energy consumption desire to be more or less? Has it relationship between house quality and house investment or living intention to cause house quality demand to influence the householder electricity energy consumption desire change or demand to be more or less? Has it relationship between regional housing market living or rent investment intentions, housing quality and electricity energy consumption more or less desire? I suppose that the determinants of the residential electricity energy demand form space-heating and cooking, due to the property quality demand influence and the householder's living or rent investment intention influence both, which will influence the householder's electricity energy consumption or useful behavior when he/she/they is/are living in the house.

I argue that rent properties are not only consumer goods, but it also constitute financial market assets. It is therefore reasonable to assume that rational ( rent income investment intention) investors choose to raise housing quality ( e.g. thermal insulation technological installing at home, heating or cooling technology or artificial intelligent window, lighting, door opening or closing) in order to attract many people choose to rent whose house to live. The householder's aim is to achieve an acceptable return on investment ( ROI) or raising rent income aim when he/she rents whose house to anyone, it is easy to attract many people to choose to pay higher rent his/her house to live in the property rent market. Moreover, the another important factor is that rents and future house sale prices of properties differ regionally ( or even locally), and largely depend on housing market fundamentals, such as either the house living buyer's income levels or the house rent buyer's income levels, vacancy rates, and/or householder investor's expectations.

Thus, if the householder expects to rent whose house and raises rent to attract many people choose to rent whose house to live, who will attempt to install many new technology in order to satisfy their high quality of life need when they can pay higher rent to rent to choose to rent whose houses to live. Their aim only achieves to raise housing quality, but any new technology will lead to increase electricity energy consumption or use in the house.

Hence, any high quality of houses will influence the householder to use or consume more electricity energy at home. It means that the householder will choose to consume or use more electricity energy at home, if he/she or the family householder demands to live more comfortable house and he/she/they can have high quality of living life at home. This comfortable living demand to the householder ( property renter or property buyer) view point can explain why the better quality of house factor will influence the electricity energy consumption desire to the householder also to be more daily.

I shall indicate one home electricity energy consumption experiment, it indicated that utilizing aggregate data on regional space-heating energy consumption form over 300,000 apartment buildings in 97 German planning regions. The study applies structural equation modelling to estimate the influence of housing market fundamentals on the level of housing quality, and subsequently on regional electricity energy consumption. Consequently, it suggests that housing market fundamental explain regional differences in the housing quality.

In particular, findings show that the level of per capita income, investor' expectations about future housing market development as well as vacancy all explain regional differences in housing quality has a significant impact on electricity energy consumption.

In the way, this experiment can indicate evidence that regional housing market fundamental have a substantial influence on regional levels of housing quality and energy consumption desires to the German regional householders. This Germany regional householder experiment found important implications for high or low housing quality of the regional property building and householders either property living or property rent intention of comfortable living feeling need factor which will influence the regional property householder electricity energy consumption desire to be raised or reduced. These factors will influence the consequence of electricity energy demand to be increased or decreased needs every day for the regional householder as well as the country's electricity energy supplier(s) can gather the regional properties whether they are high or low quality to predict the regional properties householders' electricity energy consumption supply budget more accurate. It implies that an important determinant of residential housing quality will have possible to influence electricity energy demand to be more or less for long term. IN particular, this Germany regional residential experiment can explain and find an important role in formulating assumptions about the quality factor has chance to influence the regional residential future levels of electricity energy efficiency and consumption in the country. Hence, housing developments and electricity energy firms can follow this regional residential housing quality factor to evaluate whether the regional housing market is the corresponding investment patterns as well as the energy researchers can follow the regional residential housing quality whether it is high or low housing quality factor to evaluate the more accurate models of regional electricity energy demand to any regional residential householders' houses in the country.

In consumer behavioral view point, it explains that if the country government expected many householders feel to need to spend much electricity energy or have much electricity energy useful demand or desire at home. The country government ought to encourage the country residential property or house developers choose to build many houses which have technological product installed to satisfy the regional householders' residential comfortable living need when they choose the regions to build the high quality houses to let them to live. Then, the regional householders will be influenced to consume or use much electricity energy at homes, due to they feel that they are living at high quality and comfortable and high building technological installed apartments in the country's regions. Then, the country's government and electricity energy provider(s) may be raise much electricity energy efficiency and supply and profit , due to the regional residential householders' electricity energy consumption or useful desire need is therefore influenced to be more by the regional high quality of residential houses factor. So, the regional high quality of residential house factor will have relationship to the regional electricity energy consumption and efficiency to the regional householder's houses.

Otherwise, if the country government felt electricity energy is shortage, it ought encourage the property developers build many low quality and low building technological houses to let householders to live themselves or rent to others to live in the country's different regional residential development market. Due to the low quality of properties and low technological installed to properties factor which will influence any these different regional residential householders to choose method to solve shortage of electricity energy challenge to the country.

In conclusion, to apply consumer behavioral economic theory to property development market, if property quality factor can really influence the householder's electricity energy consumption desire to be used more or less at home daily. The country's property developers can apply this factor to predict property consumer individual property buying consumption behaviors more accurate. For example, if the US property developer planned to build low quality and low technological design buildings and lesser comfortable residential houses in the region in US. Then, its residential householder target will be trended the less acceptable of electricity energy consumption property buyers to choose to buy these regional properties to live in the US region because they can only accept to spend less electricity energy to use when they are living in the houses in order to save money daily. SO, the low quality , less comfortable and low technological installed design residential houses will satisfy their living needs. Otherwise, if the US property developer planned to build high quality and high technological installed design buildings and more comfortable residential houses in the region in US. Then, its residential householder target will be trended to the more acceptable of electricity energy consumption property buyers to choose to buy these regional properties to live in the US region because they can accept to spend more electricity energy to use when they are living in the houses in order to improve their living of quality. So, they wont's consider to spend more expenditure to use electricity energy for any technological products are installed in their properties in order to satisfy their comfortable living needs at their homes every day.

Consequently, property developers can attempt to gather marketing research concerns whether how many people who accept to use more electricity energy or use less electricity energy in order to predict they ought build how many high quality or low quality houses number in different regions more accurate in themselves countries or overseas countries property development market.


Environmental impacts of householder greenhouse gas electricity energy consumption activities

Can artificial intelligence measure whether what environment factors influence householder electricity energy consumption activities? Has environment factor relationship to influence householder electricity energy consumption behaviors? Socially, householder electricity energy consumption provides us with sources of living satisfaction , but if any sudden environment factor changes, whether it will influence householder consume or use more or less electricity energy decision at home. However, I assume householder electricity energy consumption will have a considerable proportion of the environmental impacts be influenced by our way of life and our economic decision of electricity energy consumption behavior.

What different environmental factors will influence householder electricity energy consumption decision? The external environmental factors include, for example, the country's electricity firms or government changes to electricity energy regulations, electricity energy production technologies change and business practices and government policies changing etc. different external environmental factors will influence any country's electricity energy consumption to householders' consumption desire to be more or less. It will also require changes to influence the householders to consume which kinds of electric products which are needed to be used in different electricity energy natural manufacturing resources.

Why does these external environmental factors impact householders' any behaviors to influence them to concern to use more or less electricity energy power or which kinds of electricity energy products choice at homes. How any why environmental factors impact will influence householder activities at home, such as electricity energy consumption and choice? What are the key components of external environmental factors influence householders' electricity energy consumption behaviors. I shall explain as below:

Firstly, we need to know whether what external environments are which can influence why and how householders need to change their activities to choose more or less or which kinds of energy power to be provided to them to use at home. Who is householder? Householder is an individual, family, or group of individuals living together as unit in a home. Consumption of electricity energy at home may be cooking food needs, needing have colder feeling to turn on fan or air condition at home in summer or needing have warm feeling to turn on heater at home in winter, watching television programs or listening music , playing computer games or used computers activities , reading activities and applying artificial intelligent technological tools to help householders to open or close homes' windows, doors etc. different home equipment which need to use electricity energy provisions. SO, their home activities need to turn on lighting electric tools , televisions, music machines, radios etc. different equipment which need to use electricity energy provision at home. SO, the purpose of householder consumption means consumption by individuals living in a household and it includes consumption both in and outside the home. Why does environmental impacts link to householders' electricity energy consumption? I shall focus on discussing of greenhouse gases ( GHGS) energy product how any why it can influenced to householders to use.

The environmental impacts will influence this kind of greenhouse gases ( GHGS) energy in the product lifecycle or delivery of the service to link the householder's energy consumption at home such as these several aspects:

Extraction and greenhouse gases production ( supply number), physical distribution ( delivery far long or close near short distance between the greenhouse gases manufacturing factory and the greenhouse gases supplier), resources consumed by marketing and retail activities ( householder's needs to use the quantity of the greenhouse gases energy product), the greenhouse gases consumers search and purchasing activities( e.g. travel to shops, internet purchasing channel, , finding the which kinds of greenhouse gases products from internet, magazines, newspapers, radio advertisements etc. different medias,) , post-use greenhouse gases energy disposal ( resale, reused or rubbish). The householder's physical behavioral impact environmental factor will influence how and why he/she chooses to consume greenhouse gases energy daily , e.g. impacts of a housing development, or a wind –farm that supplies greenhouse gases with power. So, the householder's greenhouse gases energy consumption behavior which will depend upon individual personal and subjective perspectives and value.

So, the householder's useful behavior or attitude of greenhouse gases energy product which will influence how he/she/ the family use or consume greenhouse gases energy, such as the householder individual environmental protection attitude which can impact how he/she/the family spends the quantity of greenhouse gases energy every day at home, if the householder does not expect our air or water or land is polluted , due to extraction of any natural gas resources to be manufactured any kinds of greenhouse gases products. Then, this environmental pollution issue will influence some householders choose to reduce to use more quantity of greenhouse gases products every day. Another environmental factors include the bio relates the ( unsustainable ) use of resources to avoid wasting much greenhouse gases energy to cause greenhouse gases energy supply shortage, avoiding the cause negative impacts of quality life , e.g. noise causing when the extraction of any natural resource from lands to the householder's house is near to the natural resource extraction land and health impacts, e.g. when the greenhouse gas householder user who often use the kind of greenhouse gas product when it is used to cook or heat any equipment to cause they to breathe dirty air at home often. These impacts can be measured in different ways include: monetary costs or loss, physical quantities of resources used or waste or pollution produced and the burden the greenhouse gases energy place on environmental resources. All of these external environment factors will impact the householder individual attitude or behavior how to use or consume greenhouse gases energy product at home.

All these environmental factors concern householder greenhouse gases energy consumer individual consumption attitude is influenced by environment pollution, greenhouse resource supply shortage challenge, greenhouse gases influence the householder's negative quality of life, negative health impacts, noise, waste money , raising economic cost to the householder which will impact whether how the householder choose to use the quantity of greenhouse gases product or the kinds of greenhouse gases products or other kinds of electricity energy products.

However, these are other external environmental factors which can impact how the householder decides to use greenhouse gas product at home. They include: the changes of energy regulation, e.g. the country government has quota number implementation to prohibit to import above the limited quantities of any kinds of greenhouse gas products to any countries. So, when the greenhouse gas energy supplying quantity is decreased, but if the country has may householders who need to buy different kinds of greenhouse gases products to be used at home. Then, the different kinds of import greenhouse gases energy products prices will be raised in possible, due to demand is more than supply in the country's greenhouse gas energy product market. Consequently, if the greenhouse gas energy price us risen above the general social acceptable level to the home greenhouse gas energy product householder consumers. Finally, it will influence them to choose to buy other kinds of gas energy products to replace the greenhouse gas energy product to use at home.

Another side, if the country's greenhouse gas energy manufacturing supplier sudden changes its greenhouse gas energy production technologies to choose to concentrate on manufacturing other kinds of energy products. Then, the greenhouse gas energy supply quantities will be only decreased, even future one day , it will cause greenhouse gas supply shortage challenge to let the country's home greenhouse gas householder consumers who can not buy enough quantity of any kinds of greenhouse gas energy products to satisfy their electricity needs at home every day. Consequently, when future on day , the country greenhouse gas energy manufacturer has none any quantity of greenhouse energy products to supply to the country's greenhouse energy householders to use at home. The, they must only choose other kinds of new energy products to replace the traditional useful greenhouse gas energy products to be used at homes.

In conclusions, these non-controlled external environmental factors can impact and influence the country's every householder consumer individual attitude or consumption behavioral change to how any why the country's householders either choose to buy much or less quantity of greenhouse gas products to use at home.


The effect of house space occupancy
and building characteristics on
householder electricity energy use

Can artificial intelligence mesaure how much space occupancy is the suitable size as well as what the most suitable building characteristics influence each householder electricity energy useful activities or behavious? In general, society believes large space size occupancy house building characteristics factor which will influence householder use more energy at home, e.g. in summer, when the householder is living at the large space size occupancy house, who ought turn on all air conditions or fans at sleeping rooms or eating room or studying room. So, if the householder's house has two to three or more sleeping rooms. Then, he / she needs to buy more air conditions or fans in order to let all rooms' temperature to be fallen down to let he /she feel more cool comfortable feeling when the temperature is above 30 degree or more extreme hot in summer weather. Otherwise, when the temperature is low, e.g. between 0 degree to 10 degree or below 0 degree in winter weather. When the householder is living in one large space size occupancy appartment, which has thee to five sleeping rooms , even more and two studying rooms and one eating room, even more as well as every room has one heater. Then, he / she must turn on all heaters to let who to feel warm feeling when he / she is staying in the house. It brings these interesting questions.

Will large or small size space occupancy housing characteristics influence any householder often turn on heater or air condition or fan in whole house space occupancy area in order to the householder feels warmer or cooler feeling when he /she is staying in the house?

Has any space occupancy housing characteristics relationship to influence any householder to turn on heater or air condition or fan in whole house space occupancy area in order to the householder feels warmer or cooler feeling when he /she is staying in the house?

Does it bring positive relationship between turning on long time fan or air condition or heater and the house occupancy space characteristics is large or small size?

I shall attempt to give psychological evidences to explain the householder's house space occupancy area large or small size factor whether it can influence the householder choose to do long time or short time turning on heater or air condition or fan behavior in order to let he/she/the family to feel more cooler or warmer comfortable feeling when he/she/the family is staying in the house in summer or winter weather.

Does the house occupancy space size characteristics factor is the only one or important factor to influence the householder choose to turn on long or short time fan or air condition or heater in the house to let him/her/the family to feel more cooler or warmer comfortable feeling in summer or winter weather?

I feel that it is not exact right , due to the householder's house space occupancy size whether it is large or small characteristics to influence the householder choose to turn on long time or short time fan or air condition or heater time to let him /her/ the family to feel more cooler or warmer when he / she / the family is staying at home in summer or winter weather. The reason is because that the lifestyle of living quality need is different between developed countries and developing countries. The lifestyle of living quality factor will change the country's householder's expectation about the quality of living life. For example, for Africa, Korea, China , Japan, Hong Kong etc. developing countries. On the lifestyle of living quality need to these developing countries' householders aspect, that will cause a high environmental burden when they need to often turn on air conditions to satisfy more cooler feeling when they are staying at homes in summer or they need often to turn on heaters to satisfy more warmer feeling when they are staying at home in winter. Due to if their houses are large size space occupancy characteristics and they have more than at least two sleeping rooms and studying rooms and eating rooms and toilets number. Then, these householders who are developing countries' large space occupancy size characteristics houses, they won't like often turn on heaters long time to keep more warmer in their indoor whole space area in winter or they won't like often turn on air conditions or fans long time to keep more cooler in the their indoor whole space area house environment in summer .

The reason is possible because that the developing countries' householder chooses often to turn on their heaters or air conditions or fans long time in their houses when they are staying long time in their houses and their houses space occupancy sizes are very large, it will bring the electricity energy to be used more to these developing countries' householders' large space occupancy size characteristic houses. It means that the electricity fee will be also increased due to they often turn on heaters or air conditions or fans long time to keep their indoor temperature to be more cooler in summer or more warmer in winter. So, it seems that the developing countries' householders are living in the house whose space occupancy have very large size characteristics and more than two rooms house characteristics in the developing countries as above. Then, they won't often choose to turn on heaters or air conditions or fans long time to keep more cooler or warmer feeling in their house whole indoor space occupancy environment when they are often staying at home long time.

Their lifestyle of living comfortable feeling are lesser than the developed countries householders. Consequently, their lesser cooling or warming comfortable demand of living lifestyle factor will change their attitudes to use air conditions or fans or heaters turning on time in order to limit heaters or air conditions or fans turning on time to be shorter than the developed countries houeholders' heaters or air conditions or fans turning on time at homes. Due to the long time turnong on air conditions, fans , heaters at the developing countries' householders' homes, it will cause to spend much electricity energy to lead electricity fee charges to be raised to the developing countries' householders ' homes when they are often staying at homes in summer or winter weather. Hence, the house space occupancy large size characteristics ought not influence the developing countries householders choose to turn on air conditions , fans or heaters long time in order to let them to feel more cooler or warmer at homes in summer or winter weather.

So, the developing countries' house space occupancy large size characteristics householders won't be more acceptable to pay higher electricity energy fee when they are staying at homes at summer or winter weather. Due to they do not often choose to turn on heaters, air conditions or fans long time during they are staying at homes. Otherwise, the developed countries, e.g. UK, UK , France, Germany, Swiss, Singapore, Italy etc. countries. In general, these developed countries' householders' living lifestyle quality needs are higher than the developing countries. So, when the summer or winter weather is coming, if the temperature is extreme cold, e.g. below than 0 degree or it is extreme hot, e.g. higher than 30 degee.

Then these developed countries' householders will easy accept to turn on air conditions or fans or heaters long time at home in order to keep their appartment in door temperature to be more cooler in extreme hot in door environment or more warmer in extreme cold in door environment when these developed countries' householders are often staying at homes long time at night after their day time working time or schooling time. Because these developed counties' householders' quality of living lifestyle needs or demands are higher than the developing countries' householders. So, they won't consider that they will pay more electricity fee , due to they often turn on air conditions, fans or heaters long time to let them to feel more comfortable in cooler or warmer indoor large size space occupancy environment. So, it seems that the electricity energy efficiency will be raised to the developed countries' householders who are living in the house space occupancy large size characteristics and they will be possible to pay more electricity fees during they are often staying at home in extreme hot summer or extreme cold winter weather.

In conclusion, due to the living lifestyle quality need ( demand) is different between the developed countries' householders and the developing countries' householders. It will influence the householders' long time or short time spending time on air conditions or fans or heaters indoor space occupancy size characteristics environment in order to achieve more cooler or more warmer feeling in their houses. Consequently, the long or short time of turning on air conditions, fans, heaters for the developed or developing countries householders' activities factor will be more influential to compare the house space occupancy large or small size characteristics factor to influence their cooler or warmer feeling in their houses. SO, the house indoor environment electricity energy consumption efficiency degree to the developing or developed countries' every householder house in summer or winter to the developing or developed householders in summer or winter weather , which is more influenced by the living lifestyle qualty factor to the either developed countries or developing countries householders. Hence, any developed or developing countries' electricity suppliers need to consider the building areas of property development market buyers their living style quality demands ( needs) whether their living style quality demands are higher or lesser than the other building areas of property development market, they ought not consider whether the building locations of the houses' space occupation sizes whether they are large or small sizes in order to evaluate the householders will live at the building areas of property development locations ,whose electricty energy spending efficiency more accurate.

How to help low income household earners to reduce not essential electricity energy expenditure spending at homes

Can artificial intelligence measure whether which income level influences low income household accept to use the much electricity energy at home? Has it relationship between the householder income and the electricity energy needs? How to evaluate the subsidies and social tariffs to assist lower income earners to analyze household energy consumption more accurate?

Electricity energy is essential needs for every householder at home, e.g. lighting, cooking power, healthcare, sanitation, cooler or warmer temperature indoor control at home. However, for lower income household earners, it its burden when they need often to use electricity energy to supply power to any home electricity tools to do any acticities at homes. If any these countries' lower income householder earner target can not get the reasonable subsidies to assist them to solve any electricity energy tools' electricity energy poer needs. Due to their lower income leve, it is possible that to knfluence them have enough electricity supply to help them to use to cook rice and food and vegatabe to eat, boil water to drink, turning on light tools to help them to read, watch TV, listen radio, music any entertainment or essential needs at homes at night or morning afternoon time. These lower income household earners will be easy to sick , due to they have no enough electricity supply to help them to use use electric bottles to boil water or cook food to eat. Then they only drink not boiled water or not cooked food to eat at homes in possible, due to they have no enough income to pay electricity fees every month.

Hence, how to evaluate the lower income household earners' electricity fee need ( demand) level in order to provide the reasonable subsidies amount to assist every country's low income household earner to help them to pay the reasonable electricity fee which is one important issue to every country's government today. It brings this question: How to evaluate or analyze or predict every lower income household individual or family earner's every month electricity energy demand (need) more accurate?

It is one essential issue to be value to consider to every country's government. Moreover, to the estent that energy subsidies must be essential to be provided by public sources to all low income household earners or that a social tariff may be designed for improving access to energy for certain low income social earner groups. Hence, how to structure the energy subsidies between energy and income levels to be better target, such public mechanisms, and to avoid regressive subsidies unfairly. For example, India and China these both countries ' income poverty and energy poverty population are the large number. So , these both countries' governments need to focuse on more aggregated effects and analyze the effects of rural electrification at the local level on the decrease in energy poverty in rural low income poverty and energy poverty householders. Therefore, every country government needs to point regressivity of the subsidy for electricity. There is room to analyze to what extent low income household earners along the income distribution demand some forms of energy, and to suggest better and fair low income targeting household earners energy subsidies supply policies.

Each government does not only consider energy issues from a social point of view, it also needs have a manner to consider a possible link between energy, hunger reduction, and food security for each country's low income household earners group. So, every government has responsibility to calculate the determinants of different sources of energy consumption at the low income houehold earner level for urban and rural both populations in order to evaluate the electricity subsidies and to test whether every low income householder earner characteristics plays a role in determining energy consumption.

In general, in the use of energy measured as that for cooking, such as LPG reduces the exposure of households to hazardous, increases the consumption of different types of foods and medicines, improves the distribution of time between household memners, enables studys with more light, reduces the use of digital computer entertainment tools at home, and moderates the use of wood as fuel, preventing deforestation. These methods are the best suggestions to help low income householder earner groups to reduce time to use electricity at homes. When they spend less time to use electricity to do any not essential activities, e.g. watching television, playing electric games from home computers, listening music. They only use electricity to turn on light read, to turn on rice cooker to cook, when they feel hungey to eat. Then, I believe that these social low income household earner groups will reduce to pay much not essential electricity energy expenditure at homes. Hence, every country government ought need to persuade low income household earners to avoid to use electricity to do any not essential activities in order to raise electricity energy consumption in long term time.

It will bring less amount of energy subsidies expenditure benefits to every country's government. Hence, the success to persuade any countries' low income household earners to reduce to spend much time to do any electric entertainment activities of consumption behaviors at homes often. This is the most efficient and the most successful energy subsidiary method to help them to reduce electricity energy expenditure when they are staying at homes. Hence, if any country government expected the low income household earners can continue really reduce electricity energy expenditure, they need to learn to do the meaning essential activities which are needed to use electricity at home habitally. Then, they can change their electricity useful entertainment living habit, e.g. using computers to play games, listening music, watching television entertainment habits at homes to cause essential daily needs of electricity useful living habit, e.f. using cookers to cook rice or cook food to eat, turning on lights to read , turning on heaters to bath, turning on air conditions to keep cool temperature or turning on heaters to keep warm temperature at homes. Consequently, they won't need to pay much electricity expenditure at home, due to their waste useful electricity entetainment living habits have changed to do any essential useful electricity activities at homes.

Another kind of method to reduce the determinants of energy demand to the low income householder earners. The governments can persuade them to consider the variation factor can influence their electricity energy expenditure are increased or decreased at homes. It is not the electricity or gas price is increased from the electricity suppliers. It is that their bad living habits of waste electricity or gas to do any not essential activities at homes. e.g. the householder often turn on light tools to read or listen music or watch television in whole night, he/she ought need to sleep at night, but he/she does not go to bed to sleep in whole night. He/she chooses to turn on light to do these activities. Then, he/she will waste much electricity at whole night. Also, some householders like to bath more than half hour, even one hour, when it is winter, they need to turn on heaters to provide electricity to cause the bath room has warm water to provide to them to bath, Their long time bathing behaviors will be also waste electricity or gas energy from long time heating in bath rooms. So, they need to change their waste electricity consumption living behaviors at homes.

So, I suggest that some low income household earners will need to be taught to change their bad using electricity enery living habits from governments' public relation promotion in order to change the low income household earners' bad or incorrected useful electricity or gas living attitude to achieve and to avoid them often to do electricity or gas energy waste behaviors at homes. So, different countries' governments need to teach them how to do the correct or right electricty or gas useful activities ( living habits) or let them know or feel how to use their electricity or gas which can help them to reduce to waste the not essential extra electricity or gas energy. Consequently, they must reduce electricity or gas expenditure as well as electricity or gas shortage challenge won't be caused by their electricity or gas useful waste behaviors ( activities) at homes.

In conclusion, energy subsidies method is not the best solution to help low income household earners to reduce to use electricity or gas energy. Because it is only short term benefit to reduce their electricity or gas expenditure at homes. The best solution is that to let them to know or feel why and how they have responsibilities to change their incorrent or wrong electricity or gas consumption bad habits in order to avoid global electricity or gas energy is waste to be used, even it is caused shortage from householders' energy waste behaviors.

Factors influence householder energy

efficient consumption behaviors

at homes

Can artificial intelligence measure how householder uses energy level at home ? What factors can influence householders how to use energy in efficient way at homes. It depends on different countries householders' living habits to cause their choices to use energy efficiently at homes. In general, global householders energy every day consumption or use aims include cooking, heating, and cooling or warming rooms, lighting , water-boiled use and computer playing games entertainment etc. activities at homes every day. Some activities are often essential at homes, e.g. cooking, cooling or warming temperature in rooms, lighting , water-boiled use. So, their activities must not avoid to use energy at homes often. Otherwise, some activities are not essential at homes, e.g. playing entertainment games from computers, cooling rooms in summer, listening music, watching television etc. these activities. The householder can choose either to use energy to turn on these equipment tools or not to do these non essential activities at homes often. In general, householders rely on energy to make ourselves lives comfortable, productive and enjoyable. However, global householders need to learn how we can use energy resources wisely because global every householder has responsibility to manage resources includes: reducing total energy use and using energy more efficiently in order to avoid energy shortage crise occurrence. The choices are make about how we use energy, e.g. turning machines off when not in use of choosing to buy energy efficieny appliances will have increasing impacts on the quality of our environment and lives.

Energy conservation includes any behavior that results in the use of less energy. Energy efficiency involves the use of technology that requires less energy to perform the same function. For example, a compact fluorescent light buld that uses less energy to produce the same amount of light as an incandescent light buib is an example of energy efficiency. So, a householder's decision to place an incanadescent light bulb with compact fluorescent is an example of energy conservation. So, as individuals, every countries' householder choices and actions can result in a significant reduction in the amount of energy used in each sector of the economy.

So, I bring this interesting question: What factors can influence householder to choose to do any efficient energy consumption or useful behaviors at homes? I believe every countries' householders will have their different living attitudes and their living attitudes can influence their behaviors or activities to choose hoe to use energy at home. I shall indicate some countries' householders' living attitudes to explain the factors can influence them to use energy efficiency at homes as below:

● Is the low income and rising price of modern fuels both factors best to influence Nigeria householders choose to use energy efficiently?

Firstly, for Nigeria householders energy consumption habit at homes example, it is richly with natural resources, modern energy resources which provide many householders with biomass ( mostly firewood) and some other householders modern energy sources, such as kevosene, liquefied, petroleum, gas and electricity for their use. So, it is one country which can manufacture to provide energy for itself to use. It doesn't need to depend on other countries to import any kinds of energy to householders to buy to use at homes. But, it has social challenge, the poverty problem in Nigeria goes beyond low income, savings and growth rate, due to its low level of education, poor governamce, high level of unemployment factors influence.

It is important to know how Nigeria householders meet their basic energy needs between poverty and energy can bde described in terms of quality and quantity of energy used. Generally, most poor householders use biomass fuels because of affordability and they ( householders) do not have energy equipment ( such as, gas cookers, electric cookers etc.) . So, it seems Nigeria householders won't demand their living quality to be improved. It implies that they will use any kinds of energy efficiently at homes, e.g. gas, electricity, due to they find themselves in energy poverty. Although, this country has enough nature resources to manufacture energy to provide to householders to use, but due to many people are low income group, so they won't spend too much expenditure to buy much energy to use at homes. So, the rising prices of modern fuels, such as liquefied, petroleum , gas ( LPG) and electricity and their erratic supply have made many householders revert to the use of traditional fuel, such as firewood and charcoal.

It brings this questions: Is the low income and rising price of modern fuels both factors best to influence Nigeria householders choose to use energy efficiently?

The hypothes is predicated on the economic theory of consumer behavior. However, when income increases, householders not only consume more of the same goods, they also need higher quality . So, it applies economic theory to householder's energy consumption behavior at home. It explains why low living standards induce greater dependence on firewood and other biomass fuels owing to a combination of income and substitution effects, such as Nigeria low income household energy home users case. it explains why Nigeria householders can accept to use firewood and charaval traditional energy to replace liquefied, petroleum , gas (LPG) and electricity modern energy . So, economic theory explains the Nigeria household energy users why they can accept to use traditional energy to replace modern energy and their energy useful or consumption behaviors are efficient at homes. Although, Nigeria has enough natural resource to manufacture modern energy to supply to householders to use at homes. But, due to these modern energy products prices are raised to the price level of householders who can not accept. it causes to Nigeria householders only choose to buy the cheap biomass, firewoods to replace high price of modern energy products to use at home often. So, they can accept their quality of living to be fallen down. So, expensive modern energy product price is one factor to influence some countries' householders to choose to buy cheap traditional poor quality of nature energy, e.g. firewood or biomass, to use at homes. Hence, they can raise energy efficiency to use when they choose to use traditional nature energy to replace modern nature energy at homes.

● Does season factor influence New Zealand householders' energy consumption behaviors at homes

Secondly, for New Zealand householders energy consumption habits at homes , for example, their living quality needs are general comfortable need feeling. Their countries' houses of space heating was found to average 34% of total housholder energy use. The relation to space heating includes low indirect temperature are associated with persistent under-heating , whether some space heating sources tend to be higher or lower in winter indoor temperature than others and winter indoor temperatures are compared to international benchmarks and established healthy temperature ranges. So, New Zealand occupant's perceptions of winter indoor temperature conditions are presented and explored in relation to heating patterns and household energy consumption. So, it seems that NZ winter temperature is low. Moreover, it will influence householders need to turn on heaters to keep more warmer feeling indoor. Then, they will use more electricity energy. In special, if the householders' houses spaces are large sizes . Hence, their heaters need long time to keep whole houses' areas or spaces or rooms temperature to be rised up in order to let they do not feel very cold in winter. So, NZ's winter extreme cold weather will influence householders' energy use or consumption to be increased in winter.

The electricity efficiency to every NZ householder is very high in winter to compare spring, summer, autumn seasons. Hence, if NZ electricity suppliers expected to forecast electricity consumption more accurate in NZ. In order to ease the life for both electric net designers and electricity suppliers, it was decided to find out, how the NZ weather conditions and every householder's house space size factors to influence the power consumption to NZ householders. If there is a clear trend observed , then this relation can be used for power consumption forecasts to NZ householders.

Why does NZ weather condition factor and householder's house space size factor can predict householders' electricity consumption at homes. Due to geographic location on the global the lowest south sets specific conditions for weather, such as NZ's south island geographic location is near to south ocean in our earth. It is a country where average annual temperatures are well between 10 degree to below 10 degree at NZ south island special geographic location to near to the sourth ocean in our earth at the same time.

However, large part of mankind is living in the conditions where there are four different seasons in NZ geographic location, dark winter, which is cold and snowy, spring with rising temperature and high precipitation, sunny , dry and rather hot summer, and windy and wet autumn. These conditions lead to different patterns in electric appliances use in NZ householders, in special, in NZ south island householders. If trends in electric energy use have substantial correlation with weather conditions, this can help NZ electric energy suppliers and producers to forecast electricity consumption and thus organize and manage production of electric energy.

Consequently, it will lead to much more stability in energy supply to NZ every householder. For example, when the NZ energy supplier gathers data concerns every householder's house space size data, e.g. the house has how many sleeping rooms, toilets, bath rooms, eating rooms and reading rooms number, even the house has how many family members are living in every NZ geographical location. Then if it can follow different location of NZ houses spaces sizes whether they are large or small space size as well as whethe every house has how many family members are living to evaluate whether how much electricity efficiency can satisfy their comfortable living needs in winter. Then, it can evaluate whether they will use how much electricity efficiency for their needs in different seasons. If in winter, many householders are living in the large space size house in the geographic location. Then, it is possible that the geographic location is householders will use much electricity efficiency and where geographic location hosueholders who will be possible to pay the most highest electricity fee to compare the other geographic location of small space size of house householders. Hence, weather factor is the most influential to change NZ householders ' electricity energy consumption behaviors at homes.


● Urbanization level and income per capita both tangible factors as well as temperature ( weather variation factor) will have close relationship to influence China householder energy consumption or useful needs at home every day
For China householder energy consumption habit example, what factors can determine to impact this country's householders energy useful behavior at homes? Can the impacts of these factors be quntified? What are China householder energy consumption trends and characteristics? I shall explan as below:
I believe the influential factors include these three aspects to China householder energy users: Income per capita, urbanization level an annual average temperature ( weather). These factors will influence any China householder energy useful or consumption behavior at homes.
Temperature ( weather variation factor) is intangible from eastern region to western region of Chin, variances largely depend upon economic level and the provincial level. So, some regions were warmer and cooler temperature will influence the regional China householder how to use electricity. In addition, th influence of urbanization level varies according to income level as well as the urbanization level has more significant impact on the structure and efficiency

of China householder energy consumption thatn on its quantity. So, the urbanization level and income per capita both tangible factors will have close relationship to influence China householder energy consumption or useful needs at home every day. Moreover, these two tangible factors ( urbanization level and income per capita both factors) have the more influential to impact China any one of household family energy consumption or useful habit to compare temperature factor at home. Because temperature can only influence than to choose to turn on heaters to keep more cooler in summer or turn on air conditions ( fans) to keep more warmer in winter.

The electricity energy needs for these equopment tools which will be influenced less. Otherwise, the urbanization level and income per family householder how to choose to spend more or less electricity or gas etc. energy at homes. Because in behavioral economy view point, when individual householder has more income and the urban in the China geographic location is lising many high income and high household families memebrs to every house. Then, the urbanization household energy household enery useful or consumption level will be raised. Such as China household electricity users case, e.g. large cities have many high income and many houses have more than four families members to live on one house together. Then, the electricity or gas energy efficiency will be influenced to rise. The city urbanization and per capita income level is high to these large cities have high to income population, who are living in these cities in China.

Moreover, the impact of lifestyle on energy use mainly reflects types and purposes of fuels are chosen by different China households factor which will influence the urbanization level of energy choice use. China is a country with typical binary economics and social diversity and these is significant difference in the consumption pattern between urban and rural regions. Urban residents consume high-quality energy, such as electricity, natural gas , heating power, solar energy and gasoline. For rural residents, usually use coal, and bismass energy because they are cheaper price energy products which requires much time and labor and are heavy indoor pollutants . The difference in energy consumption pattern between urban and rural China residents is closely related related to living of quality needs, building structure, e.g. steel or stone etc. different materials, manufacture, easily access clean and effective feels through the electric grid, natural gas network and district heating systems.

Therefore, it explains why urbanization level is as an integrated variable reflecting social progress situation to influence urban and rural regions, such as large cities , small cities and rural countryside regions' household energy consumption or useful behaviors which have differnet kinds of fuel useful demands and energy efficiencies qualify and quantity demand, or needs at homes. Consequently, it explains, urbanization level and income per captia level both factors are more influential to China household energy consumption at home to compare temperature ( weather , seasonal) factor.

● Employment rates or gross domestic product macro economic variation factor, residential space size factor, and the government's implementation of energy labeling schemes provide significant impacts on Taiwan residential electricity consumption .

For Taiwan householder electricity consumption characteristics in the residential sector, which has different factors and pattern to compare China householder electricity householder electricity consumption habit at home. Although, they are the same Asia country. I shall explain these reasons as below:

For Taiwan electricity householder factors influence their energy useful or consumption behaviors at homes. The main factors can influence their electricity energy useful patterns include: employment rates or gross domestic product macro economic variation factor, residential space size factor, and the government's implementation of energy labeling schemes provide significant impacts on Taiwan residential electricity consumption . However, the impacts of electricity raising price and the energy supply reducing shortage efficiency standards do not significant to influence the Taiwan residential electricity consumption behavior at sources.

It means that it won't influence Taiwan householders to use electricity or gas or any kinds of energy number to be reduced, even the Taiwan government energy suppliers sudden raise, any kinds of energy price and reduce to supply energy to satisfy Taiwan householders daily essential needs at homes.

In fact, Taiwan had improved gross domestic product ( GDP) and it had raised employment rates recently. So, many Taiwanese has jobs to work, due to Taiwan economy had improved to be better. So, growth had also raised. The economy improvement causes many Taiwanese had enough jobs to work, due to new businesses are set up. Many consumers excit any kinds of businesses are invested to Taiwan from overseas or local investors. So, consumption is grown, the electricity consuming applicances are selected, as the household consumer focus grousp number if also influenced to be increased. So, Taiwan economy had improved to be better, it will encourage many electricity consuming applicances products are encouraged to excited to be selected to seel in Taiwan. Due to many different kinds of electricity consuming appliances are supplied to attract Taiwanese to choose to buy to bring to their homes for cooking, boiling water, or keeping rooms to be cooler or warmer temperature confortable feeling intention in winter or summer seasons. So, these electricity consuming appliances, e.g. rice cookers, heaters, air conditions, fans, bathing gas heaters etc. different home electricity consuming appliances will be increased to supply to satisfy Taiwan householders' needs. When they decide to buy any news electricity consuming applicances to bring to homes to use.

● Environment scientists' education message how to influence Greece householders home energy consumption behaviors from primary energy to change secondary energy

Finally , I shall indicate Greece, this western which will influence this country's householders have desires to do household energy conservation patterns or conservation energy consumption behaviors or energy conservation activities at homes. I shall explain the social economic variable, such as consumers' income and family size variation factor which can influence the different Greece family household members differences towards energy conservation preferences. IN addition, the variable, such as environmental information feedback and consciousness of energy problems are characteristics of the energy saver consumer.

Why and how can environmental pollution , environmental protection, energy conservation information message can influence Greece householders to choose to do energy use consumption conservation or less energy useful behaviors at homes. It is one interesting energy efficient use behaviors , due to Greece householders are influenced by energy conservation or environmental protection message.

In fact, scientists agree overconsumption of natural resources is a major threat to oue lives in earth. Environmental problems like greenhouse effect, ozone layer depletion, and acid rain effect are not any more problems of a specific region or environmental problem. Also, economic theory is indicated that in order to gain comfort and time households are becoming excessive energy users, neglecting the environmental impact of their choices.

Environment scientists bring these environment pollution message to influence Greeks ( Greece householders) to change their energy consumption behaviors at homes. The environment scientists' message indicate that we are facing global warmth and natural resource and energy shortage challenges. Due to our Earth have limited natural resource numbers to supply to us to manufacture energy, but global population has been increasing every year. Thus, it is possible that we have energy shortage crisis. Also, manufactures are spending too much energy to waste to manufacture any products, the energy will cause air or water pollution in manufacturing process or drivers are driving their vehicles to pollute air on the roads.

Hence, environment scientists' message influence Greece householders began to consider these questions concern to reduce fossil fuel energy. Why do we need to Safety in using fuel and handle gas leaks? Why do we feel town gas smell? How is electricity located at electric station far away from town area? How to solve problems caused by the use of fossil fuels? How to reduce the use of fossil fuels?

Greece householders consider to solve the problems, the best way is to reduce thir used of fossil fuel. This helps prevent fossil fuels form being used up too quickly. Also, it helps them to reduce environmental problems because fewer pollutants are given out when less fossil fuels are used. Can human help to reduce the use of fossil fuels? Fossil fuels are mainly in power station. Although they use some fossil fuels for our gas cooker and car, it won't make much difference if I use less. Fossil fuel is not used renew primary energy. Most of energy Greece householders use come from fossil fuels, for example, the electricity we use is generated in power stations by burning fossil fuels. The buses they ride use diesel oil. Therefore, they can help reduce the use of fossil fuels by saving energy in Greece daily lives.

The actions that Greece householders can take such as: setting the air-conditioner to a higher temperature, walking instead of using lift, taking a short shower instead of a bath. This reduces the use of the hot water and thus the energy needed to heat the water. Thus, many people can help a lot to reduce our use of fossil fuels to avoid fossil fuel shortage risk occurrence.

Greeks ( Greece householders) had been beginning to conern that they will face energy shortage challenge if they can not adopt more energy conservation actions. Because the Greece government began to bring negative environmental pollution and energy shortage challenge message if they often waste to use any kinds of energy, e.g. electricity , gas excessive number efficiency at homes. Then, they will be possible to fac energy shortage and environmental pollution challenge to their country in future one day. So, this energy shortage and environment pollution message has bring predictive negative worries to influence many Greece householder energy home users choose to reduce to avoid the waste of any kinds of energy use at homes.

So, their reducing energy use actions that had encouraged them to cause habits to avoid to waste excess energy to do any non essential electric appliances useful or consumption activities at homes often. Moreover, the environment protection and energy conservation message has changed many Greece householder to make decision and activities to change their lifestyle to b low living quality from high living quality. So, the environment protection and energy conservation message factor has much influential to change Greece household energy users' daily energy conservation or less energy use consumption activities at homes.

Greeks feel greenhouse energy can be environmental protection enegy. A greenhouse can trap heat in the sunlight and keeps the air inside the greenhouse warm enough for plants to grow. The glass roof and walls of a greenhouse let in sunlight but prevent heat from escape, this makes the greenhouse warm inside. Similarly, some gases in the Earth's atmosphere can trap heat from the sun and keep the Earth warm. This is called the greenhouse effect. The gases energy that can trap heat from the sun are called greenhouse gases. It is future one kind of potential primary energy to reduce environmental pollution new energy products for human consuming. So, environmental protection message influence them to consume greenhouse enegy at homes.

So, environment scientists' environment pollution message had influence Greece householders concern to apply seconday energy ( environment protection) to replace electricity energy to use at home. They will change energy to use at home. The scientists' messages have more influential Greece householders energy change consumption behaviors at homes. The messages are as below:

There are different forms of energy, e.g. light, heat, sound, wind, water, electrical kinetic, chemical and potential energy. Some form energy is primary energy and it can not renew to use, e.g. light, sound, wind, water, fossil fuel etc. Some form energy is secondary energy and it can renew to use in possible, e.g. nuclear, electric charge battery etc. Why does human need to concern how to manufacture secondary energy? Because it is possible that our natural resource will be consumed all, thus we will face primary energy shortage risk. If human can invent any new form of man-made secondary energy to renew to use in order to avoid primary energy shortage to supply to use to use, then human won't only depend on our Earth natural resource energy supply numbers. We can invent any new secondary energy to renew to use again either replaces primary energy or instead of primary energy limit number supply.

What is energy change? For television energy change power case. Firstly, electrical energy changes to television power to be used by television itself, then it changes to light power, next it changes to light power. How to choose fuel form to use? Due to energy can change to different form of powers to supply different form of power advantages to supply to human to use, so it is possible that we can also invent any secondary man made renew used energy to change different form powers to supply us to use, e.g. nuclear energy changes to light or sound or heat form of powers ; electrical charge batteries changes to light or sound or heat form powers to satisfy our daily life needs.

The environment scientists' energy consumption education influence Greece householders concern how to change to use secondary energy to replace primary energy at homes as below:

For primary natural resource fuel energy example, different fuel has different feature, e.g. easy to burn, safe to use, gives out a lot of energy, inexpensive, produces little air pollution, easy to transport and store. How can we use in different channels, such as heating food, hot pat, driving vehicles.

For example, although coal is not expensive to cause electricity energy for past transportation tool, e.g. traditional coal energy train or our daily home cooking, but it has negative influence to environment air pollution. Hence, we ought to follow the primary natural resource energy's feature to decide how to apply what aspects of our life needs.

For example, if the country's people hope to reduce pollution when who use any kind of energy, e.g. US , Europe energy markets. The energy entrepreneur ought concentrate on manufacturing the kind of energy which can reduce environment pollution to be the least level to supply the country people to use, e.g. electric charge battery supplies to these countries' drivers to drive their vehicles on the roads, wind energy or water energy to manufacture electricity power supply to reduce air or water pollution ; or if the country people hope to buy the inexpensive energy to use, even the energy's quality and performance is worse, e.g. China, India, Hong Kong markets. The energy entrepreneur ought concentrate on manufacturing the lowest cost and enough supply of natural resource to manufacture the kind of energy to sell cheap price to these countries to use, e.g. China, Africa can accept to use e.g. gas, coal, fuel energy to use to compare developed countries people, e.g. UK, US; or if the countries people who hope to use energy which can easy to transport and store, e.g. light coal. The energy entrepreneur can choose to concentrate on manufacturing much coal to supply to the countries people to use, e.g. China, Arica Thus, to choose to manufacture which kinds of energy supply to the countries market people to use, the energy entrepreneur how decides to manufacture which kind of energy, it depends on which kinds of fuel advantages of the countries people most concerning.

What is energy meaning? It is defined a dynamic quality, it is a fundamental entity of nature that is transferred between parts of a system in the production of physical change within the system, and it is usually regarded as the capacity for doing work, and it is usable power ( such as heat or electricity ) or the resources for producing such power.

Why does secondary energy own investment worth? Because the different forms of primary natural resource energy will have supply shortage crisis, such as natural resources coal, gas, solar, wind, water, geothermal, biomass( organic material) etc. However, human can attempt to explore any undiscovered Earth or Space resource to manufacture any kinds of secondary energies, e.g. nuclear energy, electric recharge battery energy to supply to electric vehicle or space robots transportation tools to use or satisfy our daily life needs in future one day. So any kind of undiscovered secondary man-made renewed used energy resources have potential commercial worth to any energy entrepreneurs, it is possible that they can replace traditional primary energy to supply to human to use for our different aspects of life needs. In the future, the secondary energy demand will increase, when primary energy supply number has decreased form natural exploration. So, it will cause the effect of any demand of secondary energy product to be raised and prices to be increased in possible. Due to global population has been growing up, considerably China and India both countries populations have been increasing rapidly. Scientists predict there are more than 1.2 billion people worldwide will lack access to electricity, and more than 2.5 billion still use wood, charcoal to cook and heat in the future when primary energy has no enough number to supply to us to use. Hence, the fact that demand is this much greater than supply to make energy a prime market for further growth.

Although, secondary energy will have much investment worth, but energy like all other investments will carry risks. The internal and external risk factors include such as: policy is always changing to prohibit which do energy trading more easily between the energy exporting and importing countries, the secondary energy manufacturer itself own abilities to invent and to manufacture any kinds of secondary energy, improved technology can quickly make an technology obsolete, geopolitical rifts can happen overnight, the country's energy consumer (user)'s preferable choice to use which either kinds of secondary energy or secondary energy. So, it seems that (man-made) renewed used secondary energy industry can provide above-average returns, but it can also bring high risk commercial investment.

Traditionally, energy supply companies will apply those methods to operate energy providing businesses. For Shell,. Exxon examples, which had own gas stations, explore and drill for gas on their own. Other companies specialize in a part of the energy market, e.g. leasing oil rigs for example, or operating a pipeline. Energy supplying companies can choose to manufacture any kinds of energy to supply, e.g. trade oil, gas, coal, uranium, electricity etc. Any energy price and supply is demanded on the countries energy users' which kinds of energy most choice need or certain energy commodities to be chose to use popularly. For example, if US most people prefer to use secondary man-made renew used energy more than primary energy. Then, US energy manufacturers ought concentrate on manufacturing much different kinds of secondary man-made renew used energy to prepare to supply to its domestic US market in order to raise secondary energy price to sell in its country. So, the energy manufacturer's energy manufacturing choice, it is depend on which the country's people prefer to use which kinds of energy for their daily life needs.

However, scientists predict secondary energy market will have large market share, due to primary energy will have shortage to explore to supply in our earth and future energy consumers(users) prefer to choose to use more efficiency, less energy consumption, none environment pollution cause, cost effectiveness, renew to use of any kinds of energy. For example, the electricity recharge battery secondary man-made renew used energy is one kind of reducing air pollution power to push any electric battery vehicles to be driven to compare gas energy during drivers are driving their cars on the roads. They can reduce noise and air pollution and drivers can drive safely, who only need to buy one electric recharge battery to recharge in any electric recharge battery stations on streets when the electric recharge battery has no enough power to push their cars and they need to recharge their electric recharge battery drive when they had driven between one to two days. Due to primary energy, e.g. fuel , gas, the kinds of primary energies will have shortage to supply to global drivers to drive their traditional cars. Thus, the electric recharge battery or any undiscovered secondary energy will be future driving market needs. So, man-made renew used secondary energy, e.g. biofuel, hydro-electric, nuclear, will be one kind of efficient, clean, less pollution cause, cost-effective of energy to supply to our global vehicle market, even any other undiscovered new markets. Supposing they are popular to be used for electric vehicle market globally in future one day, then their prices will be decreased and constructed to average car requires up to 1,700 gallons of oil. Also supposing that making average computer requires more than ten times or weight to fossil fuels, every calories of food eaten in the US requires roughly then calories of fossil fuels. Hence, cheap energy will be one successful factor to influence future potential energy consumer (user) individual choice needs. Conversely, ion good economic times, people are more willing to travel, to buy products, and all of which success demand and low process for energy.

In the future, secondary energy will be the best choice to food production market. The modern food production system is essentially a success of changing fossil fuels into food. So, raising energy prices are almost higher food costs and even shortage for fossil fuels energy. If one day, one kind of discovered secondary man-made renew used energy can supply to any restaurants or homes to be used to cook at the cheap price, then the profit is very high for this kind of food production energy. Thus, future food production secondary energy consumption market is large and because the primary energy inputs for agriculture are higher than the energy outputs of the food. However, future secondary man-made renew used energy for food production system is only one part of whole energy consumer in food industry. The food production is related to whole food consumption market which includes: household cooking energy market, agriculture or vegetable, rice, fruit etc. foods farming machines energy market, food manufacturing factories market, food machine package market, transportation food delivery market, supermarket or fruit/food sale stores market. They must need any energy inputs to achieve the food production or food transportation or warehouse / stores electricity supply or cooking energy needs. Hence, these food suppliers relate to any whole food factory manufacturers, food retailers, food wholesalers, farmers and home/restaurant cookers, all of them must need to use energy to carry on their food producing or food cooking or food transportation activities every day in overall food industry. Thus, it seems that undiscovered any second energy demand will be increased, when the primary energy supply number is decreasing. Also, when people can accept to use secondary energy to replace primary energy to be used for any cooking, transporting food, manufacturing food, food retail stores or warehouse food delivery energy need activities. Then, the secondary energy price will be fall down to attract many food energy consumers.

In conclusion, different countries will have different factors influence how the country's householders energy consumption behavioral changes. Hence, it seems that any country's householders' energy use of consumption behaviors will be possible influenced by extermal environment factors influence. Also, every country's energy providers can attempt to find whether the country has what kinds of unique factors to influence its householders' energy consumption efficiency to increase or decrease in order to find the methods to solve the energy efficiency demand reducing challenges successfully.

AI technology how impacts to avoid food shortage
    Why environmental factor can influence
food shortage

Can artifical intelligence measure which the lowest environment pollution level influences food consumers' good wastage behavior rises ? What are objective indicators of standard of living and quality of life? Objective circumstances refer to the economic and material conditions which are important aspects of the standard of living and quality of life. In the assessment, eight different indicators were used: CPI, GDP per capita, shopping basket, household's expenditures, GFIC basket, poverty rate, income inequality and HDI. However, these indicators is one number measure. It can't measure anyone's psychological feeling, such as health, safe emotion. The challenge concerns whether environmental pollution factor, such as air pollution, water pollution can cause human's health to be poor, even goes down human's quality of life and economy loss. I shall indicate some evidences to give reasons to support my conclusion why I believe that environment pollution is a factor to cause human quality of life to be poor , even it can also cause economy will encounter loss too.

In general, measure of quality of life need include human's psychological feeling indicator. I shall indicate, Hong Kong, China countries air and water environmental pollution challenges how to influence these two countries' people quality of life to be poor, even, it will cause their economy loss. Nowadays, China and Hong Kong and India and Afria are encountering health problems arising from damage to lungs, heart and blood vessels. Hong Kong and India and Afria and China e.g. Shanghai city pollution is a significant cause of premature death from cardiopulmonary disorders. Present level of pollution cause injury to the immature developing lings of children and adolescents. This damage will lead to life-long health problems in many and a reduction in life-expectancy. Although, there is no evidence from analyses of trends in pollutants that pollution measures in recent years have reduced pollutant concentrations in a way which will benefits public health.

There are clear indicators that for some pollutants. The problem is worsening. In fact, air and water pollution is Hong Kong and China and Afria etc. developing countries' the biggest cause of social and environmental injustice. It harms not only citizens today, but because its transquenerational effects on the urborn and youngest members of the society, it will cause its will health effects well into the later years of this century, even environmental pollution challenge will cause these countries will encounter economy loss.

Human activities have created forms of air and water pollution, such as gases from fuels, uncontrolled emissons from fossil fuels and other chemical sources have long been recognized as a cause of ill health and premature death. For example, in December, 1930 year, a dense fog affected the Meuse Valley in Belgium. Beginning on December, 3 date, the fog intensified over three days and was associated with laryngeal symptoms, chest pain, coughing, and breathlessness. Some patients showed signs of pulmonary oedema. Overall 60 deaths were attributed to the episode. After a long investigation, the cause was considered to be emissions from high sulphur fuels, including suplhur dioxide and sulphuric acid.

What is the current threat to health? the migigration of air polluton following the introduction of clear air has been followed by a period of unprecedented economic development creating new forms of pollution from the combustion of fossil fuels. For example, in constrast to the relatively large tar laden particulates from burning dirty coal which caused episodes like the London city, UK. Smog , traffic pollution now genertes fine with a different size and composition and gases,such as which may cause injury to the respiratory system and the effects of other pollutants. Such as particulates and drive the formation of the secondary pollutant ozone. The effects of pollution will therefore to some extent reflect genetic, environmental lifestyle and behavioral factors to develop these distance in a population together with the existing prevalence of diseases which may be polluted.

Hence, living in polluted urban environments is associated with increased levels of biological markers of inflammation compared with residence in a clean air environment. The damage is caused by air pollution manifests itself through a variety of common and recognized health problems, such as upper complaints heart and lung disease. Because of this, we can use statistical methods as well as clinical studies to detect the signal of changes in health problems and increased health care demands in the population. However, doctors had proved air or water pollution can cause these both curdiovscular or respiratory disease indirectly. Curdiovscular disease includes formation of arterial plaques, coronary artery, heart attacks, irregular heart rhythm, loss of heart rate variability, high blood pressure, stroke etc. disease. Respiratory disease includes inflammation of nasal, throat and tracheal airways with acute, lower respiratory tract inflammation and infection causing bronchitis, reduction long growth and function in young people. So, it seems environmental pollution can influence quality of life to human as well as environmental pollution and illness and poor health problem has close relationship.

On the other side, envionmental pollution can bring health risk, over it will influence social inequalities. Some researchers had found that the evidence has been compiled for six envionmental health challenges, such as air quality, housing and residential location, unintentional injuries in children, work related health risks, waste management and climate change. It seems human need to concern air and drinking water quality, waste management and climate change how to influence our environmental pollution challenge. Although, the evidence base on social inequalities and environmental risk is fragmented and data are often available for few countries only, it indicates that inequalities are a major challenge for environmental health policies. Irrespective of development status, environmental inequalities can be found in any country for which data are available. The valid for the exposure to environmental risk factor is also unequally distributed, and this unequal distribution is often related to social characteristics, such as income, social status, employment and education, even environment risk factor can influence human's quality of life.

(i) How environmental risk factor can influence different groups

However, human need to concern how environmental risk factor can influence inequally health outcomes to different groups. Such as, the first group is social determinants affect the environmental conditions of an individual and may contribute to the fact that specific individuals or population groups more often experience loss adequate or potentially harmful environmental conditions. The second group is the affected population groups could still be more exposed through e.g. the mechanism of education and health behavior. The third group is given socially disadvantaged groups could show more severe health effects of the social disadvantage is associated. The final group is social determinants affect health ( what remains unclear is the relative importance of socially determined exposure to environmental risk factors). Thus, human need to concern our behavior can lead environmental pollution to influence poor health to alive. Even, we can not neglect how to protect our natural environment to be clean issue. Due to environmental unhealth poor issue can lead our bodies to be unhealth and to be ill and we have no health to work to influence our job inefficiency and low productivity if we often need to see doctor to raise workload to my staffs often. Then, our employers will be probable to dismiss and many unhealth employee will lose jobs and unemployment ratio will raise and DP will reduce, it will influence our economic growth . Hence, we can not neglect environmental justice and environmental inequity issue, e.g. indoor air pollution and occupational or exposure to environmental tobacco smoke pollution exposure to high traffic roads or to industrial plants pollution in our society.

(ii) How Afria country environmental pollution influences

Surprisingly, most of above countries , among of them, although Africa is a green and natural environmental country, but Africa has encounted poor natural environmental quality to influence it has poor quality of life to its citizen and poor economy growth to its society both. Why does Africa encounter this natural environmental pollution challenge? Afican have now two potential sources of pollution: consumption and production . This looks reasonable to Africa, since maintenance is completely dedicated to improving the environment, when production generates pollution only as a " by product". Capital implies the possibility of a country being trappical in an economent poverty trap by both a bad environment and low longevity. Some countries ( or regions) may even experience other time, both environmental degradation and decay in expectancy. The fact that, in some cases, environmental degradation doesn't imply lower longevity may be due to the fact that economic growth might , at the same time, worsen environmental quality, but generate additional resources that can help increasing ( or preserving) longevity. However, these is also evidence of countries where environmental degradation is associated with a reduction in life expectancy. It seems worsen environmental quality will influence any country's economic growth and poor quality of life both. For example, McMichael et al. ( 2004) identify 40 countries that experienced a loss in longevity between 1990 year and 2001 year ( 26 between 1980 year and 2001), they also support that the resulting world divergence in terms of life expectancy might be explained by ".... ( the growing ) health risks consequent on large-scale environmental changes is caused by human pressr, by both bad environment and low longevity, biodiversity and sustainable energy".

(iii) How human adult consumption and environmental quality influences future environment for human survival probability of life expectancy.

I shall assure human adult consumption and environmental quality has relationship to influence the future environment ( green preferences ) to provide human survival probability, it depends on inherited environmental quality. Thus, human will increase or decrease in the survival probability when we need a higher or lower life expectancy. In general, we depend on these environmental conditions to live, which include quality of water, air and soils etc. and resource availability, biodiversity, forestry, fisheries etc.

It is interesting to analyze different possible strategies to escape from the environmental poverty trap as well as factors that could push some economies back to a low equilibrium characterized. To research whether environment factor has relationship to influence human quality of life. We need to give idea of explaining whether environmental care has relationship to an uncertain lifetime. However, I suppose that an environmental kind of factor can be instead of being defined in terms of GDP per capita, capital accumulation etc. economic factors. Poverty is now related to environmental quality. It should be clear, however, I focus only on one specific mechanism lying behind environmental traps. Just as under development traps may be related to a wide variety of factors, ranging from financial to technological ones, including human capital accumulation and life expectancy. So, I should use this assumption to explain why it has relatively between environmental quality and life expectancy.

This " synthetic" indicator ( YCELP, 2006) indicated environmental health is defined by child morality, indoor air polluton, drinking water, adequate sanitation and urban particulates and ecosystem vitality that includes factors like air quality, water and productive natural resources, A key ingredient of our setting is that survival until the last period is probabilistic and depends on the inherited quality of the environments. This survival probability affects the weight of the future environmental quality in human's utility function to achieve interest aim. Final stage, human will have optimal choices depend on life expectancy: in particular, a higher probability to be alive in the third period boosts investment in the environment and reduces consumption. In this case, a given country may be caught in a high morality/poo environment if low income is associated with a deteriorated environment.

John and Pecchenino (1994) were the first to introduce the possibility of multiple identifying, case for a poverty cause characteristic by poor economic performance and environmental degradation, however, life expectancy is assumed to be exogenous and plays no role in their model. Such as soils deterioration are the like, are all susceptible of increasing human morality ( thus reducing longevity). So, the existence of both environmental performance and longevity, with countries being concentrated around two levels of environmental quality and life expectancy respectively. The two-way causes are between the environment and longevity. If the causal relationship between environmental quality and life expectancy involves the existence of an environmental poverty, characterized by both bad environmental conditions and short life expectancy.

Human life stage will encounter generations of three periods to get utility from consumption and environmental quality. During adulthood, when all relevant decisions are taken, adult can work and allocate their income between consumption and investment in environmental maintenance: consumption involves deterioration of the future quality of the environment ( through pollution and/or resource depletion) when maintenance helps to improve it. The dynamics of environmental quality may also be affected by external factors on more resourced communities. The most importance, unhealthy physical environments across the region adversely affect everyone, ever though who are likely to be most concentrated in more burdened community which also have less social power to change those environments.

Why life expectancy and the environment has close relationship to influence quality of life? Life expectancy and environmental quality dynamics are jointly determined. Human may invest in environmental quality, depending on how much , we expect to live. However, environmental conditions affects life expectancy. In particular, some countries may encounter in a low life expectancy / low environmental quality. This outcome is consistent with stylized facts relating life expectancy and environmental performance measures. Some expects to live longer, who would be willing to invest more in environmental quality, because who feel which have causal link between life expectancy and environmental quality. However, environmental quality is a very important factor affecting health and morbidity: air and water pollution, depletion of natural resources and quality of life.

Reference

Cornelia, B.F. (1999) Rural development news, the North Central Regional Center For Rural Development vol. no 24 , IOWA.

David J. Nowak & Gordon M. Melsler (2016) " Air quality effects of urban trees and parks." National recreation and park association, USA.

De Hollander, A. E. M., J.M. Melse, Elebret & P. G.N. Kramers ( 1999), " An Aggregate public health indicator to represent the impact of multiple environmental exposures" Epidemiology: 606-617.

Felce, D. and Perry, J. (1995). Quality of life: A contribution to its definition and measurement, vol. 16, no.1 pp: 51-74.

Los Angeles Country Department Of public Health (2016), Country Health Ranking Model, Retrieved From www.countryhealthrankgings.org/our-approach. USA.

Melse, J.M. & A.E. M. De Hollander (2001). " Human Health And The Environment", background document for the OECD Environmental Outlook, OECD, Paris.

McGregor, S.L. T., & Goldsmith, E.B. (1998). Expanding our understanding of quality of life, standard of living and well-being. Journal of family and consumer science, 90(2), 2-6, 22.

McMichael, A.J. M. Mckee, J. Shkolnikov and T. Valkanen ( 2004), " Morality trends and setbacks, global convergence or divergence?", Lancet 363, 1155-1159.

Yale Center For Environmental Law And Policy (2006). Environmental Performance Index. Data available on-line at http://epi.yale.edu

The failure of human education method
influences consumers to change food and energy
waste behaviours

Can artificial intelligence analyze human education method is the main factor to influence consumers to change food and energy waste behaviors ? I shall indicate the reasons to explain why human education method can not persuade to change consumers and manufacturers to do food and energy waste or loss behaviour easily.

The first reason is that , food waste is a serious ethical, environmental and economic problem of excessive consumerism. Usually, food waste occurs in all phases of the food supply chain, starting with producers and ending with consumers. There are several factors which contribute to excessive qualities of wasted foods. Some are related to current quantities of wasted foods . Some are related to current production systems and product commercialization, including food quality and security norms, others are more personal like people's food habits, awareness, values and consumer attitude in regards to consumption and food waste.

So, changing consumers and manufacturers whose wrong food consumption and excessive quantities of wasted food behaviours. It is difficult to avoid from only education method , because human's eating habits or food habits are very difficult to change to reduce to do food waste behaviours in their eating processes easily. Education is only teaching how to avoid food waste knowledge promotion or persuasive method. It can not provide useful food waste methods to assist food manufacturers how to present to avoid excessive food quantities waste in food manufacturing processes , e.g. efficient operative production as well as it can not assist consumers to understand how and why the reasons for food waste, why to do the right food consumption behaviours, eating attitudes, the most reason level of consumer knowledge and opinions with respect to food waste as well as perception about the quantities of wasted food in the consumer's household. Otherwise, future (AI) technology can assist food manufacturers how to avoid and to cause food losses in food manufacturing process as well as it can assist food consumers understand they ought how to do to change wrong food habits to right food habits in order to do the right daily eating behaviours daily more influentially.

The second reason is that, in order to reduce consumer individual food waste in developed countries, there is a need to understand the factors which shape consumer behaviours. Future (AI) technology can be used to research and analysis of customer behaviours, knowledge and attitudes of people including analysis of different developed countries' consumer behaviours for food. The objective of (AI) technology research is to determine food waste attributes, daily food routines, shopping routines, planning as a predictor of food waste and policies varied in terms of household characteristics in each country.

The (AI) data gathering results can give qualitative information about food waste including data on frequency of wasting food and reasons for wasting it, which can be based on each country's consumers' food habits or eating behaviours analysis. The (AI) gathering data concerns consumers play a crucial role to combat food waste via their own households. It is important to find solutions that are relevant to food habits more easily.

Hence, (AI) technology can gather data to understand food waste problem to household level by picturing main causes affecting food waste giving deeper insight into consume behaviour throughout the every time of right food purchasing number, storage, preparation, consumption toll disposals. So, future (AI) technology ( big data gathering tool) can analyse origins and quantities of food waste quantifying the scale of the problem, including all causes and food waste influence as well as to understand the variety of factors which can cause influence food waste behaviours more accurately.

The third reason is that, future (AI) technology can also be applied to solve food loss challenge. Food loss is considered to be food that gets spilled before it reaches its final product or retail chains, which occurs at production, post-harvest, and at the food manufacturing or transportation processing stages. It is also caused by poor infrastructure , and logistics, lack of technology , inefficient skills, knowledge and management capacity , it may be accidental or intentional , ultimately leads to less food available. So, it seems that education method is difficult to avoid food losses for food manufacturers. However, future (AI) technology can be applied to assist farming agriculture to grow crops to avoid food loss easily in good climate or food growing environment.

Instead of (AI) technology can be applied to avoid humans do food waste or food loss behaviours easily. It can also be applied to avoid energy waste aspect. I shall explain why (AI) technology can change or influence householders and manufacturers' energy using or consumption behaviours to save more electricity, gas energy at homes, offices or plants or shopping centres etc. working or entertainment places more easily or effectively to compare energy waste educational or learning method. I shall indicate the reasons as below:

(AI) technology is the more quality training data to a (AI) data gathering energy sector, which has access to the better , the (AI) can be integrated into its everyday operations. The energy sector is one of those industries with a wealth of data that is for machine learning and its comparison technologies to assist humans to avoid to do energy waste behaviours.

The commercial opportunities from using (AI) in this energy industry come from the technology's real-time optimisation, predictive analysis and forecasting power. For example, a number of New Zealand energy distribution companies are working with postgraduate students in big data and machine learning to analyse information gathered from both their networks and smart meters. Between 75 and 85 per cent of data in this sector is structured and machine learning can be used to analyse it, platforms using unsupervised learning techniques are well-suited for energy detections which can boost the efficiency of utility operations to reduce optimising energy cost. Machine learning algorithms are being used to understand vast amounts of data to predict mining, drilling and power generation failures and then to recommend tailored maintenance based on the potential issues.

SO, (AI) systems can monitor the emission of nitrogen oxides from gas turbines and vary the distribution of fuel to maintain the required levels of energy generation. Scaled across the network, this is a genuine opportunity to better control energy useful cost to avoid energy waste. Moreover, machine learning can also manage energy to use within complex systems. For example, Goggle's deep mind (AI) achieved a 40 per cent reduction in energy used to cool the company's data centres, even after human engineers had supported optimised the facility's energy use. ON a bigger scale, machine learning can be employed to manage the use of resources, such as water energy in " smart cities" to avoid to cause water natural energy power is excessive waste.

On benefits to energy end-users, based on real-time usage data, machine learning algorithms and supervised learning can improve energy management , even in power plants. This in turn optimises total pricing for energy consumers and provides opportunities to offer promotions based on energy customer demographics. SO, machine learning is used in some of these products as part of a deeper effort to provide customers with more choices about how to produce, consume and store energy and encourage them to do in many energy saving right behaviours.

(AI) has the ability to transform resources distribution and compensate for drastic fluctuations in energy demand. The technology will also go beyond just pricing and distribution to respond directly to the dynamic needs for energy, water power and natural drinking water and waste management. Through an ecosystem of " smart city", water and electricity can be distributed from many small energy producers and be personalised to specific regions and time frames. Each energy producers could use techniques which are particularly suited to time series data . This network would take in and learn energy user behaviours and use the information to manage the energy or water power or drinking water supply . A energy producer can then sell excess energy capacity back to the grid, maximising efficiency and reducing energy wastage to energy consumers.

SO, in the future, the key (AI) technologies in energy sector, they will bring high impact on machine learning, natural language programming, chat bots which will be applied to machine learning technology in energy saving sector to avoid energy wastage when energy consumers use energy daily. With the high volume of data available, global energy businesses can readily continue develop their (AI) technologies. Better resource allocation, improved customer satisfaction are obvious benefits as well as the most important influence is that (AI) energy saving consumption or using technology can impact energy end –users to avoid to waste too much excessive energy when they are using energy at homes or any public places or private factories or offices manufacturing places for every day necessities. So, it can explain why (AI) technology can impact energy end-users energy saving and using behaviours to be improved.

Next, two chapter, I shall explain how (AI) technology big data gathering tool how it can help energy end-users and food consumers how to impact to avoid to do wastage behaviours more influentially every day.

How (AI) technology impacts food consumers
and food manufacturers food eating habits
or attitudes to avoid wastage

Can artificial intelligence impact food consumers' eating habits to be influenced to change ? Future (AI) technology ( big data gathering tool) can be used to gather data concerns to supervise or manage or control these below food consumers and food manufacturers' food waste or food loss behaviours in order to find the main reasons to cause their food waste or loss wastage behaviours in order to achieve to prevent or avoid their wastage behaviours occurrence again more easily. I shall explain to these aspects as below, they include :

On (AI) auto-supervised food consumer individual food waste behavioural aspect:

Firstly, for the original food wastage supervised, the original food which included food in unopened packages behaviours, which was thrown away because it passed the expiration date including products to like cheese, yogurts and other daily products, loose fruits and vegetables which became rotten and was never used. (AI) technology can follow these waste food gather number from different countries' supermarkets, food stores to gather waste food number in order to carry statistic analysis to find the reasons why these kinds of original food, cheese, yogurts, fruit and vegetable wastage number has increase to every month in order to attempt to find the reasons to cause wastage food behaviours , it is either caused by either food manufacturers' food losses ( good manufacturing process negligent factor) cause or food consumers' food eating habits cause in order to find the solution methods to change their wrong food loss or waste influential behaviours to the food manufacturers as well as the wrong food habits or attitudes to the food household consumers more easily and accurately.

Secondly, for another kind of wastage food is partly used food supervised, the food which could have been opened or started, but was never finished. (AI) data gathering tool can attempt to follow different countries' food rubbish to gather the number data to find how much rubbish number is belonged to the country's wastage food is partly used food or the food which could have been opened or started, but was never finished to find the main reasons ( factors) why they caused those kind of food wastage number is increased from food consumers; eating habits or what the reasons ( factors) caused their number is decreased the country's food consumers' eating habits in order to find the most effective or accurate methods to solve this kind of food wastage behaviours from the country's food consumers.

Thirdly, for the another kind of food wastage is leftover supervised, which consist of food left or the plates or were cooked in big amounts which ended is not being eaten. (AI) technology big data gathering tool can gather global these kind of food waste number concerns every householder's food left on plates or were cooked in big amounts to every country, but not being eaten number from their rubbish.

TO attempt to find what factors( reasons) influence their food habits to do food left on the plates or were cooked in big amounts , which are nor being eaten food habits. SO, it can follow this kind of global food wastage increasing or decreasing number every month statist data to attempt find what factors influence this kind of food wastage to household food consumers to be either decreased or what factors influence this kind of food wastage number to be increased to conclude the more accurate food behavioural wastage judgement for this kind of food wastage habits to every country household food consumers, e.g. life habitual factor, food price factor, food perishable factor, climate influence factor etc. different kinds of external factors to cause every country's household food wastage behaviours.

Finally, the final kind of food wastage is that preparation residues, ( vegetable peels, egg shells) supervised, this kind of food wastage could potentially be still used and not known away by global every householder food consumers, they can not be avoided to waste before cooling, due to the householder feels these foods are not fresh to eat. So, the householder chooses not to cook it to eat.

However, (AI) technology can gather data concerns different countries' householder fresh food purchasing habits ,e.g. per week or per twice week or per day fresh food purchasing frequently and fresh food purchasing number, such as vegetable peels etc. fresh food wastage rubbish number in order to find what the main factors ( reasons) is (are) to cause the next month fresh food wastage number to be increased or decreased to every country householders in order to conclude the more accurate or reasonable wrong fresh food wastage habitual behaviours to cause different countries' fresh food householders' fresh food wastage behaviours habitually.

In conclusion, basing on above evidences, I can explain why future (AI) technological big data gathering tool can be applied to find solutions to avoid that food consumers to do food wastage behaviours habitually. It also explain why food wastage education method is only knowledge concept to educate to let public to know. In fact, food consumers can choose either do or not do to avoid food wastage in their eating habits daily. Otherwise, (AI) big data fathering tool can be applied to attempt to gather global householders ( food consumers) their daily eating habitual data in order to achieve the more accurate and predictive householders ( food consumers) their eating habits or eating behaviours analysis and concludes the most efficient and effective solutions to avoid global food wastage number to be raised to every country household food consumers. So, (AI) technology can impact global householders (food consumers) eating or food habits to be improved more better to compare education method.

On (AI) auto-supervised food manufacturer individual food loss behavioural aspect:

How (AI) technology changes food manufacturers cause food loss in their food manufacturing processes, I shall indicate as below:

How to apply (AI) technology to avoid vegetable, fresh fruit loss to global farmers' fresh vegetable , fruit excessive wastage loss increasing number when their crops growth process in global agricultural sector? (AI) will enable significant and valuable new solutions to avoid fresh crops, fruit, rice, vegetable food loss in their growing or irrigation process.

The internet of things (OIT ) will assist (IA) in future intelligent agricultural systems, fuelled by large volumes of data acquired from images, videos and IOT sensors. For example, it be applied to water and sprays in agricultural sector. (AI) is smoothing the way for new levels of optimisation on our farms and across all horticultural activities. The automated irrigation systems are getting and the transporting of water to specific places. It is based on the real-time needs of plants. Moreover, (AI) techniques using IOT and sensors to analyse what's happening across many hectares of farming land in real time will enable improvements in predictive modelling.

Farmers will be able to check the advantages of specific phenotypes , or traits in certain growing over time . Predictive modelling will also help them forecast pest resurgences, dramatically preventing yield losses and reducing farmers' dependence on chemical pesticides to let vegetable, fruit, crop can be grown healthy.

For the university of Waikoto , NZ example, researchers are applying machine learning to near infra-red images of soil, meaning the soil does not have to be sent to the lab. This will enable farmers to apply fertilisers much more efficiently. Moreover, robots will be developed to roam between strains of needs. These bots are high enough that do not damage the soil, and because they release herbicides only onto the weeds. They are also doing more environmental protection behaviours. Many uses of (AI) in agriculture are focused on reducing the biological and ecological damage caused by inefficient use of pesticides.

For meet waste loss reducing aspect, (AI) technology can be applied to animal health monitoring . It can also be used to improve efficiencies in livestock management by optimising feeding and dispensing of medication. The (AI) technology can constantly monitor livestock , e.g. pigs, cows sheep animals movements, eating patterns and health and immediately flag animals that are showing unusual behaviour or reduced welling. They can then be treated quickly before they spread infection. So, farmers benefit is from cost reduction through more targeted use of antibiotics when also improving the treatment of livestock.

In the simplest terms, images are constantly captured and pre-processed through detection for frequency and density to livestock eating animals' eating behaviours and living and health conditions in order to provide health pork, beef, sheep meet to consumers to eat.

However, (AI) can be applied to crop seed or fruit improved growing or better irrigation aspect, for New Zealand kiwifruit agricultural irrigation case, yield well ahead of scheduled harvests. New Zealand kiwifruit growers have had to manually count fruit over certain areas and then to achieve the more accurate kiwifruit supplying number of consumers' demanding number scheduled harvests to satisfy New Zealand itself country's kiwifruit consumers' needs , even overseas kiwifruit consumers' needs. The agricultural sector would not know until the kiwifruit product hit supermarket shelves whether its spot sampling had been correct. Any miscalculation could cause kiwifruit waste loss. For example, if NZ kiwifruit farmers predict China kiwifruit consumer number will be on million kiwifruit consumer number in this year. However, they miscalculate the wrong kiwifruit supplying number either their needs are lesser. Then, it will occur shortage kiwifruit number to export to Chinese kiwifruit consumers in this year or their supplying export number are more, then it will occur excess kiwifruit number to Chinese kiwifruit consumers. So, (AI) technology can help NZ kiwifruit farmers to predict every country's kiwifruit needs in order to supply the enough kiwifruit number to every countries' supermarkets or fruit stores to sell. SO, it won't cause NZ kiwifruit excessive or shortage challenges occur more easily.

Hence, (AI) enabled technological tools can observant or supervise every countries' fruit consumers' eating habits to predict whether how many fruit number that they will need to eat every year more accurate. It is as simple as a smartphone to video along on trailers can help provide better estimates. The device videos , the orchard on –the-go and can then produce global fruit consumers' fruit eating habits to predict their fruit consumption behaviours more accurately in every year. From these video-based machine learning systems can detect more better and therefore count seeds and fruit months in advance. So, having these (AI) predictive crop or seeds growing number technology to predict whether is enough insights in advance would also enable harvesters to undertake section-based optimisation, improve food safety and direct fertiliser to specific locations. Global faming suppliers can better manage their crop , fruit seeds, vegetable seeds growing process and time to control or predict when they can be grown to sell when it can reaches the mature stage , even meat pricing and revenue forecasting with supermarkets or food or fruit store retailers accurately book shipping logistics and storage and reducing waste.

So, key (AI) technologies will extreme impact agriculture to change farmers' vegetable, fruit, tomato, potato, crop, and livestock animals feeding methods or behaviours to be improved better by machine learning, drones, computer vision. IOT robotics, satellite , data influence.

How to putting (AI) work in the vineyard for grape fruit growing better? For Lincoln Agri. Tech. a research and development company owned by Lincoln university , NZ case example, it is developing an (AI) solution which can make early season predictions of vineyard harvests. (AI) technology can help NZ grape growers and wineries to predict their grape yield each year. SO, (AI) technology can help them to do grape yield prediction work. A large number of manual workers do sample grape bunches work. SO, (AI) technology is working on creating a system that instead uses electronic sensors to accurately count grapes for NZ grape fruit farmers. The sensors will capture and analyse grape bunches within individual rows, and access the number, sizes and distribution, feeding these different kinds of grape number data into computer algorithms in order to predict grape yield at harvest time to calculate the different kinds of grape yield to different countries grape consumers' needs more accurately. New date will also be added to the (AI) computer system each year, leading to continuous improvements in the model's accuracy as more information is gathered under different conditions. Hence, (AI) system will enable NZ grape growers to accurately access differences in yield , not only between regions or vineyards, but also blocks and rows. Over the long term, site-specific grape yield prediction will help reduce costs by enabling better planning both in the vineyard and in NZ grape market, even overseas grape market both. This (AI) technology will benefit the agricultural industry by supporting better crop or seed management, smoother processing and fruit , vegetable, crop market based on capacity to supply the more accurate number.

Global farmers can apply (AI) technology to improve agricultural water system to let crop, fruit, vegetable to grow more easily to avoid waste loss number rises. Applying on farming sensors and other data to provide farmers with daily recommendations around nitrogen application and water and effluent irrigation . It can be accessed via smartphone, take the guess-work out of interpreting the large amounts of data farmers need to consider before making an irrigation decision.

(AI) agricultural water irrigation computer system can expand into nitrogen application management and water irrigation scheduling. Optimising the amount and timing of effluent are key components of ensuring farm sustainability and resource consent compliance, minimising leaching and optimising pasture growth.

Future (AI) technology just-in-time water management can provide enough water supply to satisfy agricultural fruit, vegetable, crop etc. food irrigation need to global different farming lands more easily. In farming industry, enough fresh water irrigation need is very important to influence any fruit , vegetable, crop etc. food growing process successfully. (AI) technology can be applied to this farming land water supply aspect. It can measure when the farming land needs how much water supply to provide to the farming land's fruit, vegetable, crop growth more easily. Either when the farming land does not need water supply, or when the farming land has still have enough water in the farming underground land. For example, a Florida water company is using artificial intelligence to reduce withdrawal of water from the area aquifer in Jacksonville, USA. Tis (AI) water supply measurement system can forecast water consumption , then monitors, regulates and adjusts supply in real-time , providing a just-in0time water supply to the farming underground lane. This minimises wall production during peak hours, optimises reservoir storage, and reduces the number of pump starts required, lowering energy consumption and maintenance costs.

The (AI) water supply management system can meet the fruit, vegetable crop growing demands of these food consumers needs and reduce the need to big new wells and preserve for un-predictive water supplying need for any fruit, vegetable, crop water need on any farming lands. So, every farming land will have accurate water supplying , it won't have excessive or shortage water supply challenge to every farming lands, if the country's farmers chose to use this (AI) water supply management system.

In the future, (AI) technology advantages to be applied to food waste aspect, it can influence: assisted farming to provide enough water supply to irrigate the most accurate and measured water to every farming underground lands to let any fruit, vegetable, crops' seeds to have enough water supply to grown rapidly, reducing open-sea fishing, considered use of farming land energy supply or affordable and clean energy supply.

In the future, artificial intelligence has the potential to let any fruit vegetable, crop seeds grow up successfully. It can be used to analyse seed genetic data to create crops that can thrive and adapt in any sudden changing environment or weather in order to keep the fruit, vegetable, crop seeds can still have large adapting effort to grow up in any bad conditions. It can increase effectiveness of food supply chains through the use of (AI) technologies to drive insights that improve any fruit, vegetable, crop growing up efficiency and reduce waste in any farm. It can also give recommendations to support for farmers to choose the best decision making in order to increase in productivity. Instead of these , it can reduce the food loss in growing or manufacturing process. It can also help human to choose the healthcare diets. It includes to mine health care records to improve quality of treatments and provide better and faoter health diets. It can analyse of large scale genetic data sets to help create new types of treatment and precision medicines customised for individuals, providing expert assistance in diagnosis especially in repetitive tasks, such as analysis of images and large bodies of research information, providing immediate first line consultation to improve waiting times to see a doctor, reducing pressure on frontline staff by using robots in healthcare, speeding up the development of new drugs allowing treatments to reach whose who need them more quickly, improving learning outcomes through the analysis of data about individual learning, social and learning contexts, and personal interests, providing virtual monitors for learning by integrating modelling, social simulation and knowledge representation, providing lifelong learning companions that help the learner to adapt and build new skills throughout their lifetime. Hence, it can bring the actual life knowledge to let food manufacturers, farmers and food consumers to learn how to avoid to do food wastage or food loss behaviours in our daily life experiences. Moreover, it can help farmers , food manufacturers to learn how to optimize energy generation and reduce environmental impact through analysis of operational and environment data as well as helping food consumers or food manufacturers to find the most affordable and clean energy in efficient ways od deals when they are cooking or manufacturing foods by using (AI) agents. Even, it can improve actions to fight climate change through better modelling and analysis of large or complex data sets in order to generate better insight to help to improve the sustainable management of land resources, e.g. soils, forest, biodiversity and allow greater understanding of the impact of better farming land use choices for fruit, vegetable, crops ' seeds growth through global agricultural industry development predictive analytic and machine learning.

In conclusion, based on above different (AI) technology predictive function s, so it explains why it can replace education method to avoid or reduce future good or energy shortage challenge more easily. It is due to food consumers or food manufacturers' food waste or loss behaviours which can not be controlled to avoided to cause easily in their food manufacturing or crop growing processes as well as food consumption processes. (AI) green data revolution will create a smarter, more flexible food waste controlling system as more data is created and shared between food supply chain partners and food consumers. It can bring positive benefits to agricultural industry, such as food factory manufacturing automation, intelligent food packaging, food waste or loss risk analytics in manufacturing processes , food supply chain number forecasting, food product personalisation and new avoiding food waste or loss ways of engaging with food consumers.

Finally, I shall discuss how (AI) technology can gather data to bring food waste or food loss threat consequent message to let humans ( food consumers or food manufacturers) to understand how to change their eating waste habits or food manufacturing loss behaviours more easily as below:

Future (AI) technology can gather global food manufacturing and food consumption data to conclude one accurate food system trend model to let us to influence our negative or wrong eating habits or food manufacturing behaviours to improve to more positive in order to avoid food loss or food wastage causes easily.. The food eating or manufacturing behavioural system trends model can be as below:

It includes how to consider a number of trends that will influence the food system over the coming decade, focusing on small number of " key trends" in agricultural industry and supermarket, food stores, food grocery and food manufacturing industry, fishing industry, restaurant and hotel food supply industry, these trends which focus on a number of issues across the natural environmental, social and economic influences , due to future food waste or food loss consequent causes.

Changing social norms and working practices continue to influence the frequencey and format of food consumption. Predicting how the soical media influences to the way householders make food purchasing and dining decisions to influence food consumers' food waste behaviors from online. Predicting how the widening wealth gap and aging population are altering household food shopping and eating behaviours changes. Predicting explosion of data enabled technology concerns how the amount of data is growing at an predictive rate. The offers potential for a smarter, more responsive reducing or avoiding food waste behavioral system to food consumers or food manufacturers. Predictinv how the positive changes or influences of tackling major public health and environmental challenges are caused from humans (food consumers and food manufacturers) whose food waste or food loss behaviours. Predicting food system ( big data gathering method) concerns food consumers or food manufacturers' frequency of food waste or food loss behavioural data to adatp to an uncertain operating environment as well as to find the methods how to fight the risks of climate related shocks to the food shortge challenge.

Hence, the (AI) data gathering aims to achieve these missions as below:

Learning how to grow in most technology is driven a revolution in how the entire food system operates from a better understanding of land resources to automated factories and kitchens. To bring data enabled technology that will be become cheaper and more accessible all the time, but the avoiding food waste system will fully capitalise on the benefits over tehe next ten years. This avoiding food waste system will be explored to companies, households and food waste policymakers seek to make better use of data. They include as below:

(1) Increasing number of devices connected to the internet and number of social media users to learn how to avoid to do food waste or food loss behaviours daily.

(2) Decreasing the cost of data storage technology and the size od data enabled devices.

(3) The food waste avoiding system can deal with th increasing complexity and sudden changing environmental , social and economic systems to avoid the failure to respond proactively to these new challenges to bring energy, water and food shortages consequences.

(4) Increaing efforts to fight the impacts of extreme weather events, agricultural pest ranges to the food avoiding waste system interconnectedness.

(5) Understanding how to increase soil health, crop diversity, global nutrition and quality and availability of water resources.

(6) Understanding how to fight climate change, it brings significantly affect the fruit, vegetable, crop's seed growth to the improved better in agricultural industry through its impact agricultural yields, agricultural food changing prices influence, reliability of supply, food quality and food safety for ensuring long term food security and supply chain ( food transportation process).

(7) Developments of the use of advanced monitoring agricultural systems to increase input efficiencies and anticipate food production risks, such as adverse weather.

(8) Skills for future food wastage or shortage challenges, learning how to fight climate change challenges.

In conclusion, (AI) preventive food wastage or shortage technological system will need to gather the data concerns when global climate and environement sudden change and finding anywhere the global most suitable farming lands are located in order to provide the right agricultural farming lands to let fruit, vegetable, crops' seed which can grow rapidly, finding anywhere the lands are located which have enough natural water supply in order to help us to solve food shortage challenge in order to irrigate enough water to farming lands to let agricultural foods grow more rapidly, and finding anywhere have enough weeds or food for pigs, cows, sheeps livestocks on the land to eat. So, above these are future (AI) preventive food wastage or shortage technological system will be invented to solve any one of these agricultural seeds growth and livestocks feeding challenges.

AI seventh stage development

(AI) Technological Space station market development

In the future, space intelligent control systems can be applied to space in these aspects:

Firstly, in space communication aspect, when spacecraft have a special need for (AI) system, due to communication delays and other factors, much of (AI) control system development is conducted for earth-based applications. As a result, (AI) communication system can let spacemen communicate to earth space station staffs between of them more easy and in short time communication in possible.

Additionally, ground support systems are required to facilitate communication with and command of (AI) controlled and other spacecraft. Future, numerous successful missions or which are controlled by (AI) technology have been launched. Thus, future many space missions seem to be controlled by (AI) control systems are better than none or lacked (AI) control systems assistance.

Secondly, in space exploration aspect, in the future, it is possible that fully autonomous spacecraft control has been confirmed in a limited capacity and appears to build promise for reducing any space exploration mission cost, increasing scientific returns and allowing the operation of more complex multi-vehicle missions from possible future of artificial intelligence control systems in space. Thus, future it is possible that (AI) control systems can been automated to assist any space scientists to solve problem during their each space exploration process.

Thirdly, in automated ground based testing system aspect, in the future autonomous spaceflight will start on the ground. For each spacecraft that is sent into space to explore, numerous concepts must be proposed, developed and tested on earth to ensure a successful mission.

Can future (AI) control systems be developed to be applied to automated test any spacecraft flight systems to ensure which are safe to leave earth to fly to space? Is (AI) automated system testing tool more accurate to achieve the testing safe measurement result to reduce the space flight risk than human testing effort? To answer these questions, scientists need to attempt to compare (AI) testing system and human effort to test many times of spacecraft flights in order to confirm whether (AI) testing system will be more accurate than human effort testing to every spacecraft equipment before it will fly to space.

Fourthly, in autonomous space exploration problem solution aspect, when a spacecraft encountered a somewhat unique problem as it moves further and further from earth. The increasing distance makes it take from ground-based controllers. As such, a spacecraft that will be any significant distance from earth most have some problems need to be able take collision avoidance actions and to reestablish communications with earth should they are lost.

The use of (AI) autonomous technologies started with meeting these basic requirements and performing actions ( such as docking), which required too much. However, (AI) technology will have possible to be implemented because it can make missions better and less expensive. When, (AI) technology is participated to solve any unpredictable challenges when any a spacecraft leaves our earth to fly to space for future mission plans. It is likely the future existing space site networks will expand and some of which may be in remote locations. As such, there is an even greater need for (AI) automated challenge solution technology to facilities the operation of these various space stations within budget.

Consequently, future (AI) technology can be used in space on communication, space exploration, automated ground based testing, autonomous space exploration problem solution aspects. Nowadays, human is waiting when (AI) technology can be invented to assist space development, such as the day comes.

## 1.1 Challenges to install (AI) automated system in space stations

I shall indicate one challenge case how Earth space station installs (AI) automated control system to assist future any each spacecraft to achieve lot missions more productive and efficient. Ton install (AI) automated control system, which will follow two challenge views to achieve in success as below:

The first challenge view aims to apply (AI) automated system to increase the quality and productivity of each space mission current experience. However, proponents of this challenge viewpoint believed that the (AI) expert system should be developed to allow and experienced operator to make better flight decisions. They do not believe future (AI) artificial intelligent automated control system can make more accurate decisions or operator to reduce or avoid sudden accident occurrence chances during their every space mission trip.

However, (AI) automated control system scientists believed that this could be done by developing a (AI) automated control system capable of continuously analyzing all incoming telemetry to a depth that would be impractical ever for an experienced operator. Moreover, the (AI) automated control system scientists felt that the potential dollar value of this (AI) space automated control system investment is difficult to quantify. They indicated that a single good decision which allowed completion of a several hundred million dollar mission would clearly pay for a large amount of (AI0 automated control system work, but it is difficult to say that the good decision was completely the result of the expert system.

Developing an (AI) space station expert system needs to meet this goal, such as : Reducing every time space mission flight accident occurrence as well as assisting human space flight operators to control spacecraft equipment to raise every space mission trip efficiency and productivity. So, it requires the incorporation of deep knowledge about a given spacecraft monitoring task.

Hence, the first challenge to (AI) expert system to be how to install to space stations, it is the unpredictable (AI) expert system investment amount spending and lacked depth knowledge skill of personnel each (AI) expert operator. It is only one (AI) expert system research stage to future space stations. It still needs time to research how to invent (AI) automated control system to assist space stations to implement productivity and efficiency as well as deducing space accident occurrence chances to future every space mission.

The second challenge view was how to use (AI) expert systems to allow each less experienced space flight personnel operator to perform at the level of more senior personnel. That has a measurable cost benefit that it allows shorter training times. This viewpoint requires an emphasis on breadth of knowledge as well as improved human-computer interfaces. Thus, the second challenge is unsure achievement of shorter training times measurement to assist less experienced personnel space flight operators from (AI) space expert systems assistance to achieve shorter training time productivity and efficiency aim.

Consequently, (AI) experts need to solve these two challenges to avoid space station operators feel worries whether (AI) technology can really help them to work in order to achieve (AI) automated operation control system achievement to space stations installation implementation for every space mission plan in success.

● How can artificial intelligence be applied to adapt the nature of space and satellite industry

The nature of the space and satellite industry how to require machine intelligence and assistance to launch, operate, maintain, control, repair and ensure achievement of any space missions. Some examples of potential (AI) application include as below:

(1) Remote sensing and monitoring to predict how and when space environment changes, spacecraft security, and space mission location or destination tracking.

(2) Communications between ground and space as well as from satellite-to-satellite ( in the case of multi-satellite stations), using radio frequencies, optical -laser communications, radar along with growing complexity of satellite-to-satellite handoffs between satellites in different orbits.

(3) Robotics in space transportation, include mission extension vehicles, space docking, satellite health monitoring manned space vehicles support( including health safety, medical , analytics, repair) and spacecraft to make their own decisions to explore, learn, identify, adapt during missions and carry out repairs.

(4) Data analytics include gathering large amounts of information and how that information can be used from national security , data privacy to assist space mission organizations to carry on any future space research analyses.

(5) Reusable launch and manned vehicles include sophisticated (AI) for return to Earth for completion of mission, e.g. researching asteroid mining resource includes analytics of substances discerned from asteroid examples, and remote mining of the same.

(6) Remote missions to Mars or Moon planets and beyond ( and a board variety of information transit, and return). Satellites as alternative to terrestrial-based systems include cloud computing, cross-border broadband services, and other multi-data and information transfer.

● What is (AI) unmanned aircraft system?

Unmanned aircraft system is a kind of commercial transportation tool to provide benefits in terms of safety and efficiency to carry on implementing future every flight space mission aim. Advances in (AI) and machine learning technology are allowing it to see and act like human pilots, and to process huge amounts of data in real time search and rescue missions, e.g. allowing farmers to be more efficient and environmentally friendly inspecting power lives and cell tower, survey and mapping of land or performing package delivery searching job duties. (AI) is allowing more automated , safer and efficient in Earth land, e.g. farming land and space planets' land, e.g. Mars, Moon planets. The applications of (AI) in Earth or space land data gathering industry are limited only be data analytics for space or Earth land industrial inspections to navigating, future space warehouses more efficiently to assist spacecraft operators to transport any things between space warehouses easily in any space planets ' lands.

● Real-time data analytics

(AI) can collect and process huge volumes of data in real time. For example, gathering space environment or space planet land imagery that is used to take humans' hours, days or weeks to review and analyze is being automated by (AI) that strategically determines that kind of data and images are important enough to collect, and can simulate a human looking a lot of any space images in short time clearly. For example, when one spacecraft needs to perform inspections whether its inside machines or equipment which parts to be needed to repair, due to they are sudden damaged in any non-predictive accident event during the space mission journey. Then , it can use a variety of onboard sensors ( AI cameras) to inspect track conditions and identify defects that are invisible in the spacecraft. Once detected, (AI) robots can be used to provide recommendations on what, if any maintenance may be necessary.

● (AI) robotic detective system

(AI) robotic detective system can assist spacecraft to avoid obstacles, such as space stones sudden collision in non-predictive space environment, when space stones are flying toward to it suddenly. So, (AI) robotic detective system can be capable of any flying autonomously space stones with human intention, and this will require (AI) robotic detective system to help the spacecraft to have able to sense when the space stones obstacles will fly toward to the spacecraft direction and react in time to avoid a collision. Hence, computer plus (AI) machine learning is helping any spacecraft navigate more effectively in dangerous space environment. (AI) robots are enabling the spacecraft will fly safely, even in dark, space stones obstacle filled space environment or beyond the obstacle reaches of GPS or other methods of Earth stations communication connectivity in dangerous space environment,

● The ethic challenges to (AI) space robotic learning system in space environment application

Firstly, space environment situational awareness challenge, (AI) needs have enable better space environment situational awareness and changing the ways are able to interact with things in any dangerous space environment. In the not-too distant future, (AI) technology will enable fully autonomous operations. So, (AI) space robots have ultimate safe responsibility to every spacecraft operators on aircrafts and on the human spacecraft pilots, when they depend on (AI) space robots to assist them to work in dangerous space environment.

(AI) space robots have responsibility to assist spacecraft operators and spacecraft pilots to make accurate judgement. It will raise important policy questions regarding the removal of human judgement from the (AI) space robot's judgement. Human spacecraft pilots make not only safety-related decisions, but in certain circumstances, especially emergencies, moral and ethical decisions, such as whether to crash space rocks sudden fly toward to the spacecraft. So, (AI) space robots will be needed to learn from space flight experience and use that learned knowledge to make appropriate moral and ethical judgements in dangerous space environment.

Hence, it brings this question: Do spacecraft pilots or/and spacecraft operators or (AI) space robots have legal responsibilities to maintain the security of any space mission flight as well as to ensure that its automation, space navigation, and space communication systems are good to be used?

So, (AI) space robots bring with its significant new issues involving (AI) spacecraft automated control system product liability, data privacy, intellectual property. The bound liability issues for risk and/or space mission business model includes allocation of responsibilities and costs for compliance between the spacecraft pilots/spacecraft operators and (AI) automated control system itself.

The standard allocation of known or anticipated risks between parties will separate provisions for allocating unknown rocks through on every space mission flight. Thinking these issues , through it is critical, and it may b e to (AI) space robots and spacecraft operators/ space pilots advantage to set cost and liabilities expectations, rather than to later dispute resolution. All reasonable scenarios should be approved when every space mission trip is planned to implement.

● Artificial intelligent robotic space mission or space tourism strategy

When one space exploration or space tourism organization can have one good strategic plan, it will have more accurate effort to predict either space tourism consumer individual leisure behavior or space exploration mission more successful. However to achieve artificial intelligence technology strategy to any space exploration mission or space tourism?

I believe that artificial intelligence technology strategy to natural language, transportation, client service, education etc. (AI) businesses implementation. It must be different to artificial intelligence space exploration mission or space tourism businesses implementation. However, artificial intelligence space exploration mission and space tourism both strategies, which will be similar , due to which concerns how to apply artificial intelligent technology to achieve spacecraft for space exploration mission or space tourism how to raise efficiency and productivity and avoid or reduce space stones collision accident occurrence chances in dangerous space environment.

Hence, space tourism or space exploration mission organizations need to concern above these aspects to implement their strategic plan. So, we know that the strategic plan for artificial intelligence technology to be applied to space exploration missions or space tourism missions which aims to help space operators and space pilots to make predictive judgement to reduce or avoid sudden space accident, e.g. space stones collision to the spacecraft to cause space operators or space pilots hurt, even die as well as how to apply artificial intelligence to assist space operators/ space pilots to raise efficiency, when they are carrying on any space exploration missions or space tourism leisure missions.

● How to achieve artificial intelligence technology to space exploration mission or space tourism mission more easier and more effective?

I recommend pace tourism or space exploration mission organizations can set up one examination structure of strategic council for (AI) technology to assist space exploration mission or space tourism development. When the strategic council for (AI) technology was established the (AI) research coordination council and space industry coordination council were also needed to be researched.

The (AI) coordination council needs to be progressed with giving shape to linkages in (AI) research and development carried out by there ministries. The space exploration mission or space tourism industry coordination council needs to carry out surveys and investigations on (1) establishing a roadmap for space exploration or space tourism industrialization, (2) fostering of space operators, space pilots, space tourism leisure customer service operator of human resources, (3) data maintenance/provision and open tools, (4) measures, such as for fostering start-ups and financial linkages, in aiming towards how to apply (AI) technology to assist any space exploration mission or space tourism research and development carried out more easily.

With regard to ethical aspects of (AI) technology to space tourism or space exploration mission, intellectual property right, personal information protection and promotion of open data, separate opportunities for examine whether the (AI) strategic plan is suitable to the space tourism leisure business or space exploration mission business. If the organization feel it is not suitable to the space tourism leisure or space exploration mission business. Then , it can attempt to find what weaknesses are for its (AI) strategic plan and to revise to correct its (AI) strategic plan more easier.

The (AI) strategic plan can concern how to apply (AI) industry roadmap technology to assist to related space exploration or space tourism technologies . Thus, the (AI) strategic plan includes new (AI) automated control system assistance service and space robotic products to be utilized and applied of (AI) technology to spacecraft technology. It aims to avoid or reduce sudden accident occurrences and it can predict when and how and why sudden accident occurrences as well as how to solve these challenges urgently during the spacecraft is flying on the space environment.

Hence, the strategic plan of (AI) technology must assist human space pilots/space operators to raise any space exploration missions or space tourism missions efficiency and it must assist them to predict when sudden accidents will happen, e.g. predicting when and where space rocks collision to the spacecraft in order to avoid the accident occurrences in space environment.

The (AI) strategic plan can include these three phases to any space tourism or space exploration missions in space industry. (AI) technology is simply a service to be provided to space operators/space pilots to use to assist them to make more efficient and reduce or avoid sudden accident occurrence chances in space environment. The three phases are as below:

(1) The first phase: Utilization and application of data-driven (AI) automated control systems/ space robots developed in the space environment. This phase aims to achieve space (AI) automated control systems and / space robots to raise space service efficiency.

● On real-time assessment of space

On operator operational status hand, it includes such as below:

How to use space robots on any unmanned space tasks assistance to space operators / space pilots? How to predict when failure of space equipment damage occurrence and how to solve those sudden accident occurrence challenges in spacecraft during it is flying gin space environment? How to apply (AI) -based prediction / matching of supply and demand to every spacecraft its space exploration mission or space tourism trip need, such as on-demand space (AI) robotic assistance supply service, optimization control of spacecraft energy consumption by using (AI) automated control energy management system technology, cooperative space tasks by human space operators/space pilots and space robots, implementation of space robots that simulate behavior of space pilots and space pilots, enhancement of space operators / space pilots cooperation with space (AI) robots? Which aims to create of new cooperation services with space (AI) robots / space (AI) automation control systems assistance in order to achieve any space tasks more efficient, reducing / avoiding any sudden space accident occurrence chances from manual or non-predictive space environment factors.

(2) The second phase: Space robots able to perform multiple functions and cooperate with each other to assist space operators or space pilots to do different tasks in the spacecraft, creation of diversified space robotic services.

(3) The third phase: It is the final strategic plan stage. It aims to achieve any space exploration missions / space tourism leisure missions which can achieve innovative space services and space (AI) robots are continuously developed to assist space operators / space pilots to do different tasks efficiently and effectively in shorten time, during they are carrying on any tasks in any spacecraft.

The final phase of space strategic plan objectives include as below:

● Space (AI) robots can assist to space operators / space pilots to solve manual task errors discovery in order to help human space pilots / space operators to solve challenges in any time in spacecraft.

● Space robots that can provide walking assistance, supervision, and support through conversation to cooperate with space pilots / space operators to attempt to solve challenges in any time in spacecraft.

● Space robots which can understand the space operator / space pilots intentions or needs how to finish or do every task or space mission efficiently and effectively.

● Space robots can collect of space traveler individual information and predict of when space environment will change using (AI) space robots and sensors. For example, space robots can take 3 D maps or photos , when they need to be assisted to take any space exploration mission photos on Mars or Moon etc. planet lands.

● Diversification of spatial space transportation vehicle devices, such as (AI) space vehicle drones assist space operators to deliver cargos / carriages on any planet lands.

● Diversification of space transportation devices, providing valuable space transportation service for space travelers.

Consequently, (AI) space strategic plan aims to let space operators lead space robots / space operation control systems can follow these three phases ( stages), in order to have right or reasonable direction to let space operators, space pilots, space tourism service operators, space exploration mission operators who can also follow right or reasonable direction to achieve whose space tourism leisure aim or space exploration mission more easier and efficient from (AI) technology assistance.

● (AI) robotic technology brings long term or short term benefits to space exploration industry

Whether do scientists spend time and money to invest artificial intelligent invention to space exploration industry, is it worth to earn short term benefits only ? As NASA exploration moves beyond earth's orbit, the need excites for long duration space systems that are needed to ensure events that compromise safety and performance. I believe that the (AI) technology advances in autonomy, robotic manipulators, artificial intelligent in-space vehicles can provide service possible at acceptance cost and risk.

I shall explain how to evaluate future space systems needed to support scientific observatories and human/robotic Mar, Moon planets exploration to access key structure design considerations. How can impact of in-space robots, (AI), in-space vehicles to support NASA's future long duration missions?

(1) Robotic in-space vehicle

In the future, new knowledge is gained by robotic scientific observatories like the Hubble space telescope and the Mars covers will have result to be applied to commercial and government spacecraft launch needs. Why do future space exploration missions need (AI) technological assistance? The reasons include improving aperture size, decreasing deployment risk, such as assembling systems are too large to fit into a single in-space launch vehicle and enabling repair and upgrade easily if any future space exploration missions choose to apply (AI) technological assistance. It will reduce cost and accident occurrence risks.

In this Mars exploration case, NASA's science and exploration missions of the future require spacecraft systems, both robotic and human tended, they can cooperate in deep space for extended provider of time. The Mars exploration mission could used (AI) tools to assist human space operators to test deep space habitation technologies for a Mars transport habitat. The robotic in-space vehicles could also be repurposed as an exploration platform surface exploration and provides a deep space vehicle assembly and servicing site in Mars. So, the robotic in-space vehicles can assist space operators to deliver cargos in Mars, different space sites.

Moreover, the space (AI) in-space robots have also these advantages, such as free-flyer robots which can fly to Mars different space sites to observe Mars different space sites situations to find whether what needs they need to assist, fuel storage function repair and assembly function, long reach manipulators functions power / data / mechanical joining technology function, high power, mass efficiency, large scientific observatories function.

Hence, future robotic in-space vehicle tool which can assist space operators to observe Mars different space sites easily, transport any cargos on Mars different sites between them easily. Even, it can assist space operators to attempt to find whether where has water resource, and find suitable farming land locations grow food to provide human to eat in Mars, and find alive existing locations in possible. Thus, space robotic in-space vehicle is a good example to assist space operators to achieve future Mars exploration mission for long term time.

- Robotic spacecraft

In the future, the first space robots will possible be spacecraft and probes and the main goals of these machines was the Moon or Mars surface aiming manned missions to the Earth natural satellite. The capability of these robotic spacecraft included photographing and recording the surface and sending television images to the Earth.

The first robotic spacecraft will send to the Moon had not even the capability to land in the surface. They were lunar impact robotic spacecraft. Television images were obtained and send to the Earth when falling down on the Moon. Other robotic space achievements by using probes were directed to Nenus, Mars and Mercury, Jupiter and Saturn.

However, space robotic spaceship will face these challenges to be solve if space scientists expect future space robotic spaceships can attempt to fly to Moon, even other planets to reach these any one of space planet destinations successfully.

The challenges include that it supposes the space journey needs to reach the space region with Voyager 1 and the most distant planets with Voyager 2. The robots are needed to be the appropriate machines to substitute the astronauts in risky activities in the non-predictive dangerous space environment. How to design the artificial intelligent robot will be more important and will be a key issue for the future of space exploration?

The orbital robot requires to solve these problems , such as radiation, strong temperature variation, micro-gravity environment and magnetic fields associated to some celestial bodies. Specially, for planetary exploration aspect of the robotic application includes surface exploration where the local gravity, soil characteristics, local pressure and atmosphere must be consideration. So, future space robotic tool development still needs to find solutions to deal these above challenges when it brings to space environment to be used.

- Space navigation and long distant mobility

In the future space stations need have robots to help them to repair when sudden accident occurrences to cause any one of space stations' machines to be damaged. In-space operations involve operations functionalities like in-space assembly, in-space inspection, in-space maintenance, extra-vehicular activities, and in-orbit scientific experiment. For example, planetary exploration robotic applications are remarkably operating on Mar surface.

Space scientists predict the next 10 years, it is expected that the space robots for surface exploration be not constrained anymore by, navigation and long distance mobility to access to most locations on a planetary surface will be possible. On the other side, it is expected that ground based planning and visualization tools enable scientists from space ground station to interact with the space robots in the surface of celestial bodies. However, robotic performance at the level of a space suited human scientist is and will continue to be a major challenge.

Anyway, the automation by using robots in space operation will bring the advent of space stations and orbital service. One of the most important accomplishment with an extraordinary space robot was the space shuttle series of spacecraft. Space shuttle can be considered future first space robotic shuttle can be provided series of spacecraft service. Thus, future space robotic shuttle can be considered in that if really implemented orbital servicing of grasping satellite for maintenance, and assembly for the international space station and inserting new satellite into low Earth orbit.

Consequently, robotic technology can bring long term benefits to any space exploration missions. It is absolute short term benefits. For example, robotic vehicles can assist space operators to deliver cargos in long distance between different space stations fast and easily. They don't need manual control. Such as on Mars or Moon planets which will let many robotic vehicles move on their lands to move lot of building materials to build houses or space stations or space warehouses or space laboratories on Mars between different space sites. The benefits are that they do not need space operators drive them to more on Mars lands. They can move on Mars land automatically. So, space operators can concentrate on doing their space exploration tasks.

For another example, space robots can take 3D photo images in Mars easily. They can deliver 3D photo images more clear and visible in short time from Mars on Moon planet to Earth space stations. They do not need space operators to control, they can automate to take 3D clear photo of Mars or Moon current images more clear in order to let Earth space station scientists to see these visible 3D Mars or Moon planets images to gather data to prepare to do any space exploration missions immediately as soon as possible.

For final example, future space robots can learn space operators to attempt to assist them to do any maintenance tasks. So, when the aircraft encounters sudden accident to cause machine ( equipment) damage, then the space robots can assist to the space operators to find the damaged equipment to repair immediately. So, they can share space operators' maintenance tasks risks if human operators feel difficult to find which are the damaged equipment which need to be repaired. Then, space robots has possible to find which are the damaged equipment to be repaired immediately. Hence, future space robotic technology seems to bring long term benefits to any exploration missions absolutely.

Future robots development to space tourism leisure catching passenger transport tool

How can apply (AI) technological and psychology method to attract future space tourism customer individual space tourism desire? I believe that it has relationship between the space tourism planner and the safe (AI) space tourism environment to attract every one space tourism passenger to catch the (AI) space boat to fly to space to travel as below:

Firstly,the space travelling planner individual psychology influence hand, the space travelling planner will have these both aspects of individual psychological influence, it includes these both either positive or negative psychological influence aspectsas below:

On the positive psychological influence aspect, if the space travelling planner has confidence to the space travelling leisure company can provide safe, comfortable, good quality of one space travelling trip arrangement, good taste food arrangement, reasonable space ticket price and every reasonable space trip for space hotel living arrangement and space garden and space farming land visiting journey arrangement, even, space swimming pool and space sport centre and space cinema leisure arrangement to let whom to stay on the planet at least one day trip, it means not one short time space trip, e.g. the spacecraft only flies about half hour or one half. It can not fly to the planet to arrive its space station destination to stay to let the space travelling planner to live at the space hotel at least one night. Then the space travelling planner will have more desire to choose to catch the space tourism leisure company's spacecraft to travel to space.

Secondly, on the negative psychological influence aspect, if the space travelling planner lacks confidence to the space travelling leisure company can provide safe, comfortable, good quality of one space travelling trip arrangement, good taste food arrangement, reasonable space ticket price and every reasonable space trip for space hotel living arrangement and space garden and space farming land visiting journey arrangement, even, space swimming pool and space sport centre and space cinema leisure arrangement to let whom to stay on the planet at least one day trip, it means not one short time space trip, e.g. the spacecraft only flies about half hour or one half. It can not fly to the planet to arrive its space station destination to stay to let the space travelling planner to live at the space hotel at least one night. Then the space travelling planner will have less desire to choose to catch the space tourism leisure company's spacecraft to travel to space.

Hence, it seems safe (AI) technological space boat factor and the space travelling planner's confidence factor to the space tourism leisure providers will influence the whole space travelling market whose space travelling consumer's space travelling leisure consumption desire to be more or less. So, any one space tourism provider will need to consider how to apply (AI) technology to assist space boat safety how to influence whose customer consumption desire.

● space tourism strategy

Future any space tourism leisure business needs have good business plan to outline the space tourism leisure business in these aspects , such as: different space tourism destinations of every space tourism journey, technical , financial and regulatory factors for growing space tourism leisure consumption into any one kind of unique artificial intelligent space tourism journey for identified passenger target group.

All how to design one space tourism business development plan to attempt to predict whether what trends will influence how every different kinds of identified space tourism journey in order to achieve passenger number growing aim as well as how to achieve one attractive space tourism leisure to satisfy future space tourism passenger individual space travel needs more easily.

I shall indicate what aspects to future every space tourism traveler who will consider in order to reduce the space tourism traveler personal worry to catch any pace boats to leave our Earth to fly to other planets to travel.

I recommend that any space tourism leisure organizations need to concern how to to apply (AI) technology to these aspects in their space tourism leisure business plan as below:

(1) safe space tourism journey

On first aspect concerns safe space tourism journey plan to let all space tourism travelers will considerate safe issue. They must ensure space boats that is safe to catch them to fly to planets in their space journeys. So, any space tourism leisure business will utilize previous flight rated and proven technologies to form the basis for manufacturing spacecraft vehicles, and will incorporate the latest modern avionics and flight systems for ensuring safety, reliability and economical operation in order to reduce any space tourism traveler personal worry to catch any spacecraft.

So, the space tourism safe journey plan is one very important factor to influence space tourism consumer number for them if any one of space tourism leisure business hoped they can grow the space tourism consumer number for long term. For example, the space boat flight hardware must often be maintained at the space station. It is needed to be considered by space boat experts as risky, extremely expensive and potentially sensitive. To aims to ensure spacecraft will offer an economical and safe alternative for any satellite manufacturers and other space tourism entertainment organizations have a desire or requirement for space tourism flight.

(2) reduction cost expense plan

On second aspect concerns reduction cost expense plan, any space tourism entertainment organizations need have the experience and capacity for safely launching a fully loaded , including space tourism passengers and passenger individual cargo for every spacecraft tourism journey. As a result of outsourcing the launch role to a major contractor, the space tourism pilot can concentrate on space boat crews flight training, planning space tourism passenger cargo capacity and preparing space flight manifests , and will as a result, avoid the expense of maintaining a launch operation on a daily basis.

In addition, by outsourcing the spacecraft manufacturing, it can avoid spending millions of dollar on facilities and equipment infrastructure and engineering manufacturing expertise.

(3) achieve any space tourism mission plan

On third aspect concerns how to achieve any space tourism mission. Every space tourism mission must be ensure that reliable service is provided to satisfy every space tourism passenger personal space traveler needs and let them to enjoy in their whole space tourism journey, let them to catch a big aircraft in comfortable environment of technologically sophisticated space boat, reasonable and competitive every time space tourism flight ticket price plan is developed and properly revised every time space tourism ticket price when performing their assigned every different space tourism journey mission.

Hence, the space tourism leisure company will provide one careful selected space tourism destination , e.g. Mar planet space tourism journey, Moon planet space tourism journey or no any space destination journey, it means that the space craft only needs to fly one circle around between Earth and Moon space journey etc. that are capable of meeting the requirements of travelling into Earth orbit. So, any space tourism journey must emphasize affordability, reliability, safety, customer service and responsiveness in responding to every client's space tourism journey requirements. Hence, any one of space tourism journey must have clear space journey mission and objective to satisfy any space traveler client target needs.

- Methods to raise space

traveler number

Future space tourism will be one kind of new travel leisure market for any new space travel leisure companies to enter this undiscovered market in the beginning. However, how to predict future 10 to 20 years , even more space traveler number that is one important issue to any new space tourism leisure companies.

I think that space tourism leisure companies need to apply (AI) technology to define what kinds of space travel leisure service to be provided to space travelling passengers, however, what age group of space passengers who will be their space travelling target client. For example, their space travel leisure must provide any flight operation that takes one or more passengers beyond the altitude of 100 km and thus into space to let space travelling passengers who have fun, exciting space travelling feeling.

Anyway, for any kind of space tourism ( leisure space travel) journey, space tourism leisure company needs anyone to be bring customer satisfaction, it is a plan or predictive methods to measure how to let every space travelling passenger to feel comfortable when they are catching the spacecraft ( space flying product) and they can have enjoyable and fun or exciting feeling when they have need providing any space tourism journey, services meet or surpass customer expectations.

Thus, any space tourism leisure company needs to evaluate the degree of every time space tourism journey's customer satisfaction and customer satisfaction is also always evaluated in relationship to the every time ticket price of the space tourism journey. So, the space tourism leisure company will predict the next time of what the space tourism journey of passenger number is more accurate, after it has evaluated what degree of every time space tourism journey's customer satisfaction is. It aims to gather their opinions to find which aspects that they need to revise, e.g. choosing where will be the next time space tourism journey destination, how to improve spacecraft staff's service attitude and performance to serve to their space tourism passengers when they are catching the spacecraft, to evaluate whether the spacecraft can provide comfortable and safe environment to let them to catch in order to let the next time space travelling passengers can feel satisfactory and enjoyable when they are catching the space tourism leisure's spacecraft to fly to anywhere in space.

In general, the expectation of factors space passengers include the following customer value elements, such as below:

- viewing space and the Earth.
- experiencing weightlessness and being able to float freely in zero gravity.
- experiencing pre-flight astronaut training and related sensations.
- communicating from space to significant others.
- being able to discuss the adventure in an informed way.
- having astronaut like documentation and memorabilia.

These objectives need to be combined with, sometimes conflicting constraints, such as guaranteed safe return, limited training time, reasonable comfort, and minimum medical restrictions. All these above issues which will be every space travelling passenger considerate matters before they choose the space tourism leisure company to catch its spacecraft to fly to space. So, all these factors will influence the next time space passenger number. Any space tourism leisure company can not neglect how to solve these all matters before they decide when their next time space tourism journey to be achieved.

Consequently, if the space craft tourism leisure company could revise what aspects of its last space tourism journey to find what are its wrong or weakness or unattractive challenges to cause any one space travelling passenger who feels unsatisfactory. Then, it can have more effort to concentrate on improving its next space tourism journey to raise its space tourism service performance level , e.g. people, food, leisure etc. service aspects and its space tourism product quality level, e.g. proving comfortable spacecraft facilities to let space travelling passengers to catch in whole spacecraft tourism journey. Then, it will have more confidence to achieve the raising space travelling passenger number when any space tourism entetainment providers can apply (AI) technology to assist their space tourism mission development.

● What is the prediction space travelling passenger desire method ?

The prediction space travelling passenger individual desire method can be one survey investigation method. When every time spacecraft finishes space tourism journey mission, after all space tourism passengers catch the spacecraft to arrive earth from space. When they arrive earth space station destination, then the space tourism leisure company can arrange survey investigation staffs to enquire their feeling for this time space tourism journey immediately.

The survey content can include as below:

Do you feel satisfactory or unsatisfactory to which aspects of this time space tourism journey?

(1) On service aspect questions include as below:

(a) Do you feel space food taste is good?

(b) Do you enjoy this time space tourism journey arrangement?

(c) Do you feel satisfactory to space staff

service performance?

( d) If you have unsatisfactory feeling for any one of above questions, which aspect issue cause you feel unsatisfactory to explain to let us to know in order to us to revise our service performance.

(2) On product aspect questions include as below:

(a) Do you feel comfortable when you are catching our spacecraft in whole space tourism journey?

(b) If you feel comfortable , may you explain the reasons what aspects of our spacecraft has weakness to cause you feel uncomfortable?

(c) Do you feel safe when you are catching our spacecraft in whole space tourism journey?

(d) If you feel unsafe, may you explain the reasons what aspects of our spacecraft has weakness to cause you feel unsafe?

Finally, we thank your ideas to be given to let us know how to improve our every time future space tourism journey in order to find what challenge cause our service performance and product quality which can not satisfy your needs. So, we shall improve to avoid future challenges continue occur.

Our mission is achievement of 100% satisfactory level to our every space travelling passenger individual feeling. Also, we hope that you can choose our space tourism leisure service again, when you have another time space tourism leisure desire need. However, we shall revise to improve our service performance and product quality to be better, after collecting your ideas from this time survey investigation. I think you spend time to give your ideas from this survey investigation faithfully.

So, survey investigation method will be one important idea gathering tool to help any space tourism leisure company to revise the weaknesses to raise or improve future every time space tourism journey service performance and product quality to achieve raising competitive effort in this new space tourism leisure market.

Hence, survey investigation method will be the best idea gathering method to predict how space travelling passenger emotion or desire need will change in order to achieve the objective of raising every time space tourism journey future space travelling passenger number more easily for every space tourism leisure company.

- The prediction of price factor influences space traveler number

The space tourism leisure organizations indicate the total cost of a trip into space is rapidly coming down from the initial price level of about US$600,000, it is obvious that the space travelling customer base is going to be rather small. Typical customers tend to belong to the top 1% income bracket. They also indicate that the price comes down , it is expected that new space travelling customer groups will enter the space tourism leisure market.

Typical new customers include people in other brackets with one-of-a kind incomes, such as inheritance or business sold. There are indications that those types of customers are becoming interested in spending on an once-in-a lifetime space experience. Therefore, the growth of the space tourism market is highly sensitive to customer satisfaction and how it is communicated through various media.

This will establish the status factors of space tourism and corresponding brand reputation service providers. They also suggest that any operator monitors space travelling customer satisfaction closely, as it will help developing increasingly accurate estimates of how the space tourism leisure market will develop.

Hence, it seems that every time space tourism journey price variable factor will influence the time space tourism of customer individual leisure desire and the space tourism passenger number. For example, the minimum price goal foe a variable space tourism business is currently estimate to be below US$3000-4000/kg for a round -trip depending on variable configuration and operation size. At this price, they estimate that somewhat over 1 % of the high income earners are potential customers.

However, for significant volume growth the longer term goal should be below US$2000/kg for a typical passenger, baggage and supplies. The lower price will probably open space tourism to a broader population, expanding the customer base and altering expectations. beyond this point space tourism will become into a travelling competitive leisure commodity, price competition will ensure and service providers need to rethink their space tourism marketing and branding and price strategies.

I shall also recommend how to attract the potential customers successfully. First, space operators need to pay special attention to the right level of customer services. Second, various preparatory customer operations cost, such as a travel to the launch site, space tourism destination accommodation, pre-flight training, medical check-ups and equipment my add up to between 10 to 15 % of the actual space travel cost. Thurs, solving the right balance between services offered and cost of client operation in order to earn the largest intangible benefits, such as loyalty, confidence, leisure enjoyment, comfortable space travelling journey as well as tangible benefits, such as profit, spacecraft manufacturing facilities, space stations, space hotels , space swimming pools, space gardens, space cinema etc. which are built to similar to earth building facilities to satisfy space travelers' needs.

- The influential factors persuade travelers choose space tourism

Nowadays, our earth is no longer an adventurous enough place for some experienced tourists. Space tourism will be a new sector of adventure tourism, which is in the near future will be fast becoming a new tourism leisure opportunity for experiencing the unknown. Of one day, space tourism is able to reach the mass tourism phase, due to improved safety and decreased operation costs, a future space tourist will possibly only need minimal training to cope with the zero cost.

Space tourism is quite well established with visits to space attraction and launch sites, and it is a wealthy trips to the international space station for any space tourism travelers. However, if any space tourism leisure companies can attempt to find what the most influential factors are to persuade travelers feel attraction more than travelling in our earth.

It aims to let travelers to choose space travelling more than earth travelling when they feel travelling leisure need. I shall indicate what will be the most important influential factors to persuade travelers to choose space tourism more than earth tourism as below:

Firstly, I shall argue that the majority of different new space tourism journey destinations will be needed to find to satisfy different aged space travelers and different income space tourism consumers' needs. For example, the rich people have effort to consume longer time and reach any space tourism destinations where are far away from our earth of their every space tourism journey.

Otherwise, the middle income people will choose shorter space tourism journey distance from our earth and short time space tourism journey. Also, younger space tourism clients can accept more longer journey time, exciting fast speed spacecraft flying journey. Otherwise, old space tourism clients can only accept comfortable and shorter time safe space journey. So, it seems that safety, comfortable feeling, shorter time space tourism journey won't be one important influential factor to excite any young people who choose to consume space tourism leisure. Otherwise, safety, comfortable feeling, shorter time space tourism journey will be one important influential factor to excite any old people who choose to consume space tourism leisure.

Secondly, the another most important influential factor to excite space travelers to choose space tourism , it concerns whether the space travelers will feel what tourists benefits can be earned from a substantial variety of destinations choice. In general, space tourism with those of aviation, space travelers will hope space tourism will be travelling distances by air in a very short time, safely and comfortably, to bring them to arrive any space planet destinations when spacecraft reaches any space stations to stay in any space destinations.

Hence, space destination factor will bring important influential choice to any space destination journeys. As a result of the space technological tourism boom, the number of potential different space destination, choice attractions have grown with far fewer places on earth to which human do have access yet. However, the ultimate different space destinations to which many of us dream is not on earth, but as least 100 km above us, anywhere in space any planets. If the space tourism leisure company can provide different space tourism destination choices to young or old age both space traveler target consumer groups. They will feel a real holiday when they will be able to enjoy a great image of the earth from planets. It might mean that every space tourism journey can provide different space tourism destination to let space travelers have another new travelling destinations where are far from our earth anywhere.

Hence, the different space tourism destinations will give them an unforgettable adventure. Think of how it would be to be able to check in at a " billion strategy" luxury hotel in space one planet, it means that the space planet destination can provide one luxury hotel to let space travelers to live one night or more in the space planet destination, how it would be to schedule the space traveler' vacation at one of the space tourism leisure company luxury resorts on the Moon or Mars.

This images seem from science fiction movies, but one should not forget that 100 years ago, the Wright brothers, aviation pioneers inventors and builders of the air plane, would not have imagined how, every day it is possible that future spacecraft can fly to any planets to let human have chance to stay in the space hotel one night or more.

Consequently, space destination choice and space tourism journey service performance, aviation safety, ticket price and leisure satisfactory feeling which will be important influential factors to attract future space travelers to choose space tourism leisure to replace earth tourism leisure in future one day.

● Raising space tourism leisure
consumption strategies

Although, space tourism industry is a real enjoyment and exciting travelling leisure to human. It is possible that human will choose to consume space tourism leisure to replace earth tourism leisure, if human felt that earth tourism leisure is not attractive to them to consume to go to anywhere to travel in their leisure time.

But, I believe that space tourism industry has still many factors to influence human to choose to consume space tourism leisure, even they will consider space tourism leisure consumption I is only one time space tourism in their life time. Hence, space tourism companies ought achieve this aim to persuade or attract everyone prefer to spend space tourism leisure at least one time in their life, then it can represent success. However, I think to achieve this aim, it has these challenges to influence their success, even they believe space tourism leisure business is one potential attractive travel entertainment business. These challenges include such as: expensive space tourism ticket price issue, catching spacecraft safe issue, space traveler personal body health issue, age issue, family and friend relationship influence issue, working time and holiday time arrangement issue, the space trip arrangement issue, weather issue etc. different challenges, which will have possible to influence every space tourism planner either who decide change to cancel the time space tourism plan, or forgive to choose space tourism leisure in their life forever.

Hence, how to raise space tourism leisure consumption desire will be one considerable matter for any space tourism leisure businessmen. I shall indicate my personal three aspect of strategical opinions to let them to know how to raise every space tourism planner individual space tourism leisure consumption desire to avoid every time space tourism passenger number will have decrease failure chance as below:

● (1) Strategic opinion

On the first aspect of strategic opinion, I feel that the space education tutor can teach new space knowledge to let every space traveler to learn any new space and earth knowledge during he/she is catching on the spacecraft in personal contact learning experience environment which can raise space tourism consumption desire. The reason is because the space tourism leisure traveler can raise extra space and earth learning knowledge when they can catch the spacecraft to fly and contact the space environment to learn and feel what the differences are between space and earth by himself or herself. Hence, it is very attractive to the space traveler student target group and I believe that their parents will encourage their sons or daughters to participate the time of space trip and they are more preferable to help them to buy the time space trip ticket, due to their sons and daughters can learn any space knowledge when they are studying. Moreover, every space traveler will feel surprise to learn any new space and earth knowledge from the space tutor's teaching, due to he/she is unknown that this space travel trip includes learning space and earth knowledge.

I suggest that the space tourism leisure businessmen can give learning opportunity to every travel trip space travelers to feel that this space actual environment can bring what disadvantages or advantages to influence our earth when they are catching aircraft to fly to space to travel in every space trip. The space and earth learning knowledge can include these two aspects of space learning knowledge and experience below:

On the teaching of space environment learning knowledge hand, the topics can include as below:

Firstly the space learning topic can concern how space environment influences water and hydrated minerals change , they can learn what our drinking water function how is applied to space environment. For example, in the space environment, they can learn and attempt to feel that how water can be used in protecting astronauts against harmful radiation from the sun and cosmic rays by cloaking spacecraft with a thin layer of water in the actual space environment as well as the space travelers can also feel water is same as fuel when they are catching the spacecraft, they can feel the water is heavy to transport into space when they are catching the spacecraft to fly to space during their whole space tourism journey.

Moreover, when their spacecraft reaches anyone of planets and it stays on the planet's space station, e.g. Moon space station. They can learn how to attempt to contact the hydrated minerals to learn and feel what they contained in some asteroids may be possible sources of water and fuel in the actual space environment. When they are walking in actual space environment, such as Moon planet, they can contact or touch this hydrated minerals to learn how water molecules can be extracted and separated chemically to produce hydrogen fuel knowledge in the actual space environment. This is one exciting space learning experience to the space travelling student passengers.

Secondly the space learning topic can concern how human fights space threats , even when their whole space leisure journey, the space science teacher can let the space trip student passengers to feel that they are learning new space knowledge between the space science teacher and whose space trip student passengers. Such as how to protect our earth knowledge: Teaching them to know when will be threats to our earth from space. The space science teacher can explain how this space threating environment influences our life safety and let them to feel that a mass extinction can be triggered if an asteroid 10 kilometers across hit the earth. Even being the apex species in the food chain did not space carnivorous dinosaurs from such disaster, who knows if this terrifying scene won't happen before our eyes? So, the space travelers can image and feel how the space threating environment can influence their life safety in the actual space environment as well as the space science teacher can let whose space travelers to feel and image the actual earth disaster will possible happen suddenly to let they feel afraid in the actual space environment. Also the space science teacher can teach how our earth can fright the space stones attack to let the space traveler to know, when an impactor targets an asteroid for a controlled well-times wallop. The collision will change the asteroid's momentum, deflecting it from its original orbital path which intersects with that of the earth. So, at the moment, the space travelers can image they are a larger spacecraft near an asteroid which can also change the path. Given enough time, the gravitational pull from the spacecraft will be able to steer the asteroid away from the earth. So, every space traveler will feel that they are catching the spacecraft in the safe space environment to avoid the Earth disaster from space sudden unpredictable attack.

It is more fun real space tourism knowledge learning feel to let every space traveler has chance to learn any new space science knowledge when he/she is catching the spacecraft to fly to space to travel. Hence, one successful space trip ought include trip and learning experience both contents in order to raise every the space tourism planner individual space trip consumption desire.

● (2) Strategic opinion

On the second aspect of strategic opinion, space tourism leisure companies need to let planning travelers feel that anyone of space tourism leisure is very different to general tourism leisure. In general, tourism leisure is visiting at least one night for leisure and holiday, business or other tourism purposes in Earth only. Otherwise, space tourism leisure is other kind of an unique trip leisure or entertainment method, e.g. the space traveler can catch the spacecraft to visit any planets to stay to live at the planet's space hotel at least one night, e.g. Future potential populated Moon or Mars space hotel space trip. Moreover, the space travel companies ought give chance to let them to feel what weightless feeling is in weightlessness environment when they are walking on Moon or other planets in possible. Even, they can attempt to build these entertainment facilities, instead of space hotels, such as space swimming pools, space gardens, space cinema etc. building facilities. It aims to let them to feel what the differences between Earth and space life when they are walking on the Moon, when they are swimming on the space pools, when they are living in space hotels, when they are watching movies in space cinemas, when they are seeing flowers and different species of planets and fruits. e.g. oranges, apples, bananas, and vegetable and potatoes and tomatoes in space gardens. It is very exciting and fun space trip life experience between one days to seven days. So, they believe that they must not feel these space life experience if they do not choose to participate this time space trip planning journey by the space trip company preparation.

Also, due to that the space tourism passengers need to the pre-flight checks and training before they ensure to qualify to permit to participate the space trip. So space travel companies need to concern how to take care their health check and training matter considerately. It aims to let every space traveler will feel a market segment with fitness and extreme experiences as well as he/she will become popular with a market segment passenger to the space tourism leisure company, although he/she must not guarantee to pass the space training and/or pre-flight health checks to permit to participate the space trip. However, he/she can believe that he/she is one worth space travelling passenger to the space tourism leisure company, even this time pre-flight health check or/and the short time space trip training requirements are failure. However, the space tourism leisure company must need to let all pre-flight health check and space trip training passengers to feel that it is only one space tourism which can give them and let customers view the space travel is as the ultimate showcase for health, even though a majority of the population can pass the pre-flight medical and other tests in order to raise their confidence and safety to catch the spacecraft to fly to space to travel when they are confirmed to pass these tests to permit to catch the spacecraft later.

In general, the expectations of future space passengers include the following customer value elements, such as below:
● Viewing space and the Earth.
● Experiencing weightlessness and experiencing pre-flight astronaut training and related sensations.
● Communicating from space to significant others.
● Being able to discuss the adventure in an informed way.
● Having astronaut-like documentation and memorabilia.
● Enjoying one exciting and fun space trip.

However, instead of considering these objectives need to be combined with, sometimes conflicting , constraints such as guaranteed safe, return , limited training time, reasonable comfort, and minimum medical restrictions. So, space tourism companies need to reduce every space traveler individual worries before they decide to make the time of space tourism journey. Then, it can increase their confidence to raise their space tourism consumption desire more successfully.

Consequently, instead of these consideration, a space travel operator must pay attention to the total customer experience over the entire customer process, starting from how the service is presented, proposed and sold. The service package must include training, instructions, travel to the launch site and various post. Travel activities to generate maximum customer satisfaction and brand building opportunity.

- (3) Strategic opinion

On the final aspect of strategic opinion, I think any space tourism companies space tourism companies need to consider every time space tourism ticket price and space tourism trip issues. It is important factor to influence every space traveler individual consumption desire. Due to space trip ticket price must be more expensive to compare common Earth trip travelling ticket price, so this kind of tourism leisure market target customer will be the rich and high income customer group.

On the space trip ticket challenge issue, despite that fact the total cost of a trip into space is rapidly coming down from the initial price level of about US$60,000, it is obvious that the customer base is going to be rather small and the client target customer is only high income or rich consumer group. Typical customers tend to belong to the top of the top 1% income bracket. So, ensures that space traveler number must be less than common Earth traveler number.

Also, such as the space trip ticket price, it is expected that new middle rich level or middle high level income customer target group will enter the space trip leisure market, when every space trip ticket price falls down about 1% Typical new customers include people in other income brackets with one-of-a-kind incomes, such as inheritance or business sold space traveler target group. These people will be space travel new client group, when its every space trip ticket price can be reduced to close 1 to 2 % nearly. If any space tourism leisure companies expect to attract new rich and/or high income target customer group to choose any one kind of space trip journey planning to consume.

These are indications that these types of customers are becoming interested in spending on an once-in-a-lifetime space experience. Therefore, the growth of the space tourism market is highly sensitive to customer satisfaction and how it is communicated through the various media. This will establish the status –factor of space tourism, and corresponding brand reputation of service providers. The minimum price goal for a variable space tourism business is currently estimate to be below US$3-4000/kg for a round-trip depending on vehicle configuration. So, space travel leisure companies need to concern every round space trip cost, it can depend on the space vehicle number and weight issue to influence every space trip ticket price variable to achieve how much it can earn.

On space journey design factor aspect, it includes these different facilities aspects how to design, because future space travelling consumers will concern whether the space travel company can provide special entertainment to satisfy their needs. The facilities include as below:

How to design space hotels to let them to live in comfortable space environment and eat the best taste and fresh food quality when the cookers need to cook in the space hotel in the space environment? How to design space swimming pools to let them to swim in safe space environment? How to design space sport centers to let them to run more easily in one space sport warm and safe environment? How to design one space garden to let them to see different species of Earth flowers, or plants? How to design one space farming land to let them to see different species of Earth fruits, vegetables, tomatoes, potatoes etc. fresh foods growth in warm and safe space farming land environment? How to design one space cinema to let them to watch movies in one safe and warm space cinema environment? All these facilities will be any one of future space trip's' important and attractive space trip leisure facilities to influence every space traveler to choose to buy the space tourism leisure company's space trip leisure service.

Instead of these space building entertainment facilities, they also need to concern how the space vehicle entertainment tools are provided the entertainment service to satisfy their needs. When the space travelers can sit on the space vehicles to move on any planets' lands, such as Moon. A number of space vehicle options exist in the market, mainly differing based on the seat capacity as well as the in-flight experience level offered. The typical space vehicle solution is a small, relatively light weight spacecraft taking between 2 to 10 passengers. The number of passengers depends on the service level, amenities and extra offered. The trip typically lasts about 10 hours and of which about 4 hours are spent in space. The main attraction is the weightless time after in space. The main attraction is the weightless time after re-entry has started. It is a rather low-G technology and therefore the medical requirements for participants are nor very high.

Consequently, the space vehicles, space leisure building facilities, the space trip reasonable price ticket level, every safe space trip journey arrangement, clean and fresh and good taste space food arrangement, space traveler individual real learning experience etc. these factors will be the main influential factors to raise the space tourism leisure company's competitive effort and the space traveler consumer individual consumption desire to the space tourism leisure company in the future.

● Space travel markting strategy

Any space travel organization needs have good marketing strategy to prepare how to operate its space travelling leisure business in order to attract many space travelling clients to choose its space travelling service. I shall indicate these different strategies aspects whey they are needed to be concerned as below:

(1) On concept of spacecraft design aspect

Firstly, on concept aspect, any one space travelling leisure company needs have at least one spacecraft to catch clients to fly to space to travel. So how to design the spacecraft and its quality and safety and comfortable environment spacecraft machine concept aspect issue which is one challenge to be concerned. Because many space travelling passengers ususally concern whether the spacecraft is safe, comfortable , good quality, as well as the space travelling leisure providers also need to concern whether the spacecraft is less time and energy saving efficient use, less manufactory operating cost and durable.

In general, space travelling leisure provider expects the spacecraft or spacecraft vehicle can be uesed long time. The spacecraft will be expected to utilize previous flight rated and proven technologies to from the basis for manufacturing spacecraft vehicles , and will incorporate the latest modern avionics and flight system for answering safety, reliability and economical operation.

In general, the spacecraft will be designed to carry two crew and approximately, 10,000 pounds of cargo, depending on the ultimate weight of the spacecraft. Relying on flight hardware to maintain the space station, such as Moon or Mar space station is fpr any space travelling spacecrafts to reach these space travelling destinations to stay, it is also need to consider by many space travelling experts as risky, extremely, expensive cost sensitive for any space station travelling destination design arrangement in order to future every spacecraft can fly to any planets to stay on its space station safely.

● Outsourcing spacecraft concept design strategy

As a result, outsourcing strategy is one good method to help them to reduce cost in order to achieve to let every space travel passenger has safe space journey experience and capacity for safely launching a fully loaded ( including crew and cargo). Outsourcing strategy is the launch role to a major contracor, they can concentrate on crew flight training, planning all passnegers and cargo capacoty, and preparing flight manifests, and will as a result, avoid the expense of maintaining a launch operation on a daily basis. In addition, by outsoucing the spacecraft manufacturing, the space travelling provider can avoid spending millions of dollars on facilities and equipment infrastructure and engineering manufacturing expertise.

(2) On deciding misson aspect

Secondly, on mission aspect, any space travelling journey needs have a clear mission to be planned how to achieve in order to ensure every space travelling passenger feel satisfactory in the space travelling journey. So, every whole space travelling journey arrangement, e.g. where will be the space travelling destination, how to check every space travelling planned passengers' bodies whether who are health to catch spacecraft to fly to space to travel or how to train every space travelling planned passenger to ensure whom can permit to catch spacecraft to fly to space to travel, how to arrange every space travelling journey entertainment and facilities to let either young or old age target passenger to enjoy the space trip to feel satisfactory, how to arrange different days of every space trip.

In the last few years, Virgin Galactic has been making new's headlines with its promises to provide space travel services, and announcement that it will soom offer, at quite a hefty price, trips to sub-orbit. It is generally agreed that sub-orbit exists 100 kilometres above the earth's sea-level ( Von Der Dunk, 2012). Hence, Virgin Galactic will provide travel to where customers may experience weightlessness, as well as the sight of earth's curvature. Even more interesting is that Virgin Galactic is not the only company with such a mission,there are a few more that wish to offer the same type of service. For example, some companies even aim to provide an orbital type of flight.

Orbit flight suggests that humans would venture into outer space, where they might either orbit the earth or board the international space station ( hereinafter: ISS). In addition, some envision space hotels, moon visitations and mining asteroids. Although at first such statement might seem for one must point out that a "space hotel" is already in earth's orbit and that diligent progress through flight tests is almost made the commercial aspect of regular space travel a reality; it is only the question of time and readiness for the companies to make their long-awaited and open a new industry of present day economics ( Klemm & Markkanen, 2011; Berry , 2012).

So, every space travelling mission is to ensure that reliable, technologically-sophisicated competitively-priced flight certified spacecraft are designed and properly maintained when performing their every assigned space travelling journey mission. The space traveller leisure provider will need to provide a carefully selected array of techologies that are capable of meeting the requirements of travelling into earth orbit. It will emphasize affordability, reliability, safety, customer service and responsiveness in responding to customer's space travelling requirements.

For this space tourism leisure mission example, it many include these objectives , such as below:

One trip into space, sending a space vehicle of a certain make and with a specify capacity on a space mission, provides the various grades of a core service, such as a space mission including issues such as waiting and delivery times, personal attention and advice, amenities and facilities, ensure quality assurance, it is the planned and system activities implemented in a quality system. So that quality requirements for a product or service will be fulfilled. It aims at preventing high-risk adverse events, or reducing thei impact, provides excellent customer satisfaction, it is a measure of how products and services meet the space travelling customer expectations, customer satisfaction is also always evaluated in relationship of every space travelling ticket price of the space travelling entertainment service and spacecraft product comfortable environment feeling and good leisure arrangement for every space travelling leisure journey.

## (3) On space tourism leisure organization managment aspect

Thirdly, on space tourism leisure organization management aspect, it is also important to influence efficient and excellent space service performance to be provided to satisfy every space travel organization management team needs to be consists of experienced professionals who have successfully management and operated companies specializing in the aerospace industry for a number of years.

Their knowledge and contacts within the space industry will prove invaluable in assisting the space tourism leisure provider in the achievement of its goals and objectives. In individuals on the team components that up a spacecraft tourism development organization, and have unique experience in the design, construction, operations and maintenance of the major functions will developing spacecraft for launching into orbit. Every spacecraft will be built and maintained utilizing the same high standards of quality, within budget and well within time constraints.

Hence, every space tourism provider needs have one excellent management leaders to manage every space tourism service staffs to serve passengers in order to achieve excellent service performance to let them every one to feel satisfactory, during their every space tourism journey ( trip).

## (4) On target audience prediction aspect

On target audience prediction aspect, every space trip needs have identifies target travelling passenger in order to concentrate to choose the most popular and satisfactory space travelling journey for their identified needs.

For primary audiences example, it can include space enthusiasts and educational families both. Space enthusiasts target are usually young people and they are only 20% over 65 age old people target space ethusiasts who will be the future potential space tourism target consumers as well as the educational families target who will aspect owning educational experience for children , who is the explicit reason to visit space, either he/she has interest in history of space exploration or he/she has interest in future of space exploration or he/she feels that spce trip looked like fun.

KSCVC Visots (2013) indicated that future top markets, ranked by high visitation against space enthusiasts and educational families space tourism passengers, the US cities will include: Orlando, NYC, Miami, Tampa Bay, Chicago, West plam, Philadelphia, Atlanta, Boston, Washington, DC and San Francisco cities. So, future US space travelling market will be the top one in the world.

(5) On space objective aspect

On space objective aspect, instead of any one space tourism leisure organization concerns how to achieve its mission to satisfy all space tourism passengers leisure needs. Although, it is the major missin for space tourism leisure industry. But they can not neglect what the objectives are in order to develop or achieve long term space tourism leisure missions more easily.

The objectives main open space key issues can include such as: Providing an adequate supply of land to meet the future needs of strategic opn space links, natural areas and recreational facilities on any future space tourism destinations, increasing pressure for public access to open space areas with conservation values, competing interests between adjoining land use and development on public open space and its user groups, use of public open space and recreational resources for drainage purposes, raising higher space traveller hotel residential development placing increased pressure on the demand for public open space planet land use aim and developing public open space mor intensive leisure and sport activities on any future new space tourism planet destinations.

When the space tourism leisure providers have long term objectives to attempt to solve above these any one of key issues. It will ahve a more clear objective to achieve its long term space tourism leisure business market. It's long term objectives can include such as below:

To identify existing and future active and passive recreation needs and social trends of future space tourism visitors; to provide a wide range of high quality and accessible public open space public land areas to encourage physical activity and social interaction to meet the existing and future needs of space travelling visitors; to identify existing gaps in the public open space network and develop any different kinds of space trip arrangement to satisfy the different identified target space traveller individual needs; to protect enhance and increase landcrapt values of public open space land use; to recognize the hierarchy of public open space assets; equitably distributing open space resources; access to facilities and a diverse range of opportunities to incorporate the drainage function in public open space travelling destination areas without detriment to safely, environmental, visual and recreational values.

So, these development of any space planets howo to use their lands objectives will bring long term space travelling destination beneficial advantages to raise to build the space hotels, space swimming pools, space gardens, space cinemas, space sport places to let future space travelers can stay in Mars or Moon planet destinations to enjoy these leisure facilities and they can feel which are similar to our earth leisure facilities attractively.

These space buildings are important to attract future space travellers to catch spacecraft to fly to Mars or Moon planet to travel in possible because it is fun and exciting space trip when these leisure facilities can be built on Moon or Mars to let space travellers to stay short days in either these two planets to live their space hotels. So how to build any one of these space leisure building which is another important objective for any future space tourism leisure business, instead of how to arrange any space destination trip objective. So, any space tourism leisure provider ought not neglect how to achieve these two main space tourism objectives.

However, these are key questions continually asked regarding the viability of space tourism. They concern financial, marketing and political communities. Their concerns can be best addredded in a properly, comprehensive business plan. Some questions can not be answered definitively at this time. Hoever, knowledge of the concerns and developing space businesses in any space traveling leisure planning stages and efforts to raise capital in the following questions, every spce tourism leisure business leader needs to concern this questions as below:

Can the space tourism industry into a profitable enonomic industry?

Are challenges related to financing, marketing, business methodologies or a combination of all of these facets?

Can the proponents of space tourism to be proven business tools and methodologies in their presentation of an acceptable business plan?

Can at least a cost effective, certified passenger space tourism journey to be developed for space tourism?

What effects will influence space-tourism businesses of NASA begins selling seats on the US space shuttle to civilian space tourists?

All above questions will be every new space tourism leisure businessman who needs to concern questions in order to achieve whose marketing strategy more successfully. Consequently, marketing strategy is important to be prepared in order to follow corrective steps to achieve every space tourism leisure business missions and objectives more easily.

● Space tourism leisure behavioral economic consumption model

In space tourism leisure industry, due to every time space trip needs the space travelling planner to plan how much budget to consume expensive spce ticket price. So, it seems that the target customers will be rich or high income level young people or the retirement rich old people target customer group.

So, it brings this question: How to persuade these rich or high income young people or rich retirement old people to prefer to spend spce tourism leisure at least one time in their life?

It is one valuabe research question to every future space tourism leisure provider. I shall indicate the successful factors to analyze how to persuade them to accept this kind of potential space travelling leisure in behavioral economic personal consumption view point, in order to explain the cause and effect relationship between of these factors as below:

(1) Economic environment variable factor

Firstly, it is economic environment variable factor whether it can influence to space tourism leisure consumption changing. As I discuss about economic environment variable issue will influence consumption behavior changing. For space tourism leisure case, it is not now kind of essential consumption leisure product to every one. So , even the rich or high income people who will be influences to seek this kind of leisure to play, it the economic environment is improved, it will influence they have positive attitude and interest to choose this kind of leisure consumption. However, if the economic environment is worse, it will influence they have negative attitude and no interest to choose this kind of leisure consumption, due to space travel is one kind of expensive leisure consumption to every one.

Hence, in this space tourism leisure industry, it does not ensure that the rich or high income people must be persuade to choose this kind of expensive space tourism entertainment in whose holiday or retirement time. They can have the common tourism entertainment to go to different countries to travel many times in our earth. Otherwise, space tourism leisure is more expensive to compare common earth tourism leisure , it means that the rich or high income people only spend one time spacecraft catching to fly to space to travel in their life, it is more difficult to every space traveler like to catch spacecraft to fly to space to travel more than one time, due to he/she had attempted to catch spacecraft to fly to space to travel to own space travel experience, he/she will feel enough satisfactory and enjoyment in common. Hence, it is possible that future many rich or high income people only like to spend one time space tourism leisure, then they won't continue to spend this kind of tourism entertainment again in their life.

Thus, space tourism leisure providers need to arrange any special or attractive space tourism leisure to persuade these high income or rich target clients to consume, when the economic environment will change worse. The Europen space agency (ESA), defines this phenomenon between economic environment variable and space tourism client growth or falling number relationship as: " space tourism is an execution of sub-orbital flight by privately finded and/ or privately operated vehicles and the technology development driven by space tourism market."

it seems that space vehicle is one attractive travelling desire tool will be one attractive selling point to influence space tourism leisure consumer individual entertainment choice or attitude to be changed to positive leisure consumption attitude to prefer to play this kind of space tourism activities when economic environment changes to worse. Hence, when economic environment is worse, the economic wore changing factor will influence the space travelling planner individual leisure consumption desire, even it will influence the rich or high income young people or rich retirement people target customer both groups.

As (ESA, 2008) indicated space vehicle will be one kind of attractive leisure tool for spce traveler. So, I suggest that space tourism lesiure journey arrangement needs to include that such as : the space travelers can catch space vehicle to move on Moon or Mars plants land to feel what the different feeling is between during they are catching public transportation tool, such as bus or taxi during the are catching these transportation tools on earth land and during they are catching space vehicle tools on Mars or Moon planet's lands. It is so exciting and fun catching space vehicle tool experience on these both Mars or Moon planets' lands to the young and old age space travelling passengers. Because every space vehicle's speed is not very fast and it will move on Moon or Mars planets slowly. So, any aged pace travelling passengers can attempt to play this kind of space facilities leisure after they catched spacecraft to fly to these both Mars or Moon planet to stay. They can spend half hour or one hour, even more than one hour to catch the space vehicle to go to anywhere on Mars or Moon to travel. It is possible that they can find exciting and undiscovered things on these both planets.

So, catching space vehicle to go to anywhere on either these both planets journey, it will one essential part of space travelling journey during the economic environment is changed to worse. It is extra attractive space travelling leisure journey to attract space tourism consumer individual leisure desire when economic environment is worse.

Hence, from this perspective then space tourism could be understood as a section of the tourism industry mainly based on technological development, progression and its activitity being related specifically to sub orbital flights. So, if future space tourism providers expect whether the global economic environment changing will be better or worse which won't influence space tourism leisure consumption desire to be changed. The space tourism leisure providers need to persuade the space tourism planners feel space tourism would have to be treated like an already exciting part of the tourism industry. It means that space tourism leisure is one kind of tourism leisure choice to replace common earth tourism leisure consumption. When travelers feel space tourism is another tourism leisure to replace which can replace common earth tourism leisure. It will avoid the worse economic environment changing factor to reduce the rich or high income young people or rich retirement old people whose space travelling leisure consumption desire. Consequently , the question in relation to, in what kinds of space tourism journey message do space travel providers promote behind whether space vehicle journey promotion message which is needed when economic environment will change worse. I shall be asked, as understanding the meaning in which space tourism is being marketed, communicated is seen as a factor , which can either positively contribute to future development of the tourism industry or lead into prolonging or seen stopping the space tourism industry from its progression.

(2) Space tourism leisure journey management factor

Secondly, space tourism leisure jounrey management factor, how to arrange every space tourism leisure journey which will be one important factor to influence space tourism planner individual tourism consumption desire.

In general, it can includes these several forms of space tourism leisure activities in every space lesiure trip arrangement. The following classification of space tourism include: Terrestria spce tourism ( i.e. NASA visit centre, space movies, online space experience); Atmospheric space tourism ( i.e. : MIG 31 flight, zero G. flights ) and astro ( orbita) tourism ( i.e.: trips to the international space station-beyond earth orbit) ( Cater 2010, Crouch et al. 2009).

Instead of US domestic space tourism market is potential, next country is Japan. First, the study is made by Collins et. al ( 1994, 1996) in Japan on 3030 research participants, showed that 80% of respondents under the age of 50 were willing to travel to space and out of them 20% were willing to pay year's salary for the space travel experience. Yet, it could be citicized that the Japan people age group of under 50 could be too broad, in general different generations under one groups. nest besides the willingness to go to space, the Japanese study showed respondents motivations for travelling to space, including any fun and exciting attractive space tourism journey, e.g. interest in space walk, catching space vehicle or driving space vehicle on the either Moon or Mars planets, earth view, zeo gravity experience, livin gin space hotels one night or more, watching movies in space cinemas, swimming in space pools, visiting space gardens, running in space sport centers, catching spacecrafts to view earth or Moon or Mars planets.

Hence, it seems attractive space tourism journey can persuade another country's space travelling planners, such as Japanese attempts to satisfy whose space tourism needs. So, different kinds of attractive space trip journey arrangement will be one important factor to influence young and old age travelling consumption desire. It implies that attractive space tourism journey will be one influential factor to encourage other countries tourism consumers attempt to another kind of leaving earth tourism leisure.

So, any space tourism trip destinations and leisure facilities arrangement must need to satisfy space traveler individual leisure needs and every space trip must be more fun, exciting and comfortable and enjoyable feeling to compare general tourism journey in earth. Due to general earth tourism leisure will be space tourism leisure's competitive or replaced leisure product and service. Hence, space trip destinations and leisure facilities choice will be one important factor to influence space travelling planner's consumption desire.

Every space travelling planner will compare general earth travelling leisure's destinations and leisure facilities arrangement whether the space travelling trip arrangement , leisure facilities arrangement and food arrangement, space vehicle or spacecraf leisure comfortable influence issues which will have more satisfactory enjoyable feeling to compare general earth tourism leisure and their spending expenditure to every space trip whether is value or is not value.

Consequently, economic environment changing factor and space trip and leisure facilities arrangement factor which both will influence any space tourism planner individual consumption desire mainly. So, space tourism businessmen ought concern these two aspects of factors how and when will change to adapt any country's potential space traveler's space tourism changing taste and needs in order to follow the new space tourism changing needs easily.

● Space tourism market moral ethic risk
threats

What are space tourism moral ethic risk during the space businessmen operate this businesses as well as what market threats who will encounter to face difficulties ? I shall give actul cases to explain how and why these challenges will cause to influence any new space tourism businesses development successfully.

(1) Potential accidents aspect

Firstly, space travelers will concern that public reactions to potential accidents aspect during they are catching spacecrafts to travel to space. In fact, it is moral ethic responsibility to any space tourism leisure providers to provide safe, comfortable and non accident occurrence in their whole space trip. Because once time accident will cause any one of space passenger hurt or death. So , it must be any space tourism businessmen responsibilities to concern whether they have enough confidence to ensure none any accident occurrences in every space tourism trip.

Hence, in space tourism industry, government needs have public policy to threaten or prohibit any space tourism leisure providers neglect to often check and ensure any spacecraft machines or equipments are regular opeations, as well as often renew new spacecraft machines when they are old to be used. The policy is a force effort to need them to abide every space tourism leisure safe responsibility to ensure or guarantee any one of spacecraft won't have accident occurrences during it has left earth to fly to space in whole space trip journey from the beginning to the end till to the spacecraft come to earth safely.

Hence, this policy forces any space tourism leisure providers concern to put a monetary value on increased or reduced risk of death, the " value of statistical live", used to characterize when the benefit of safety regulation is worth the cost such regulation improves. So, the country government and the country's space tourism leisure providers both have responsibilities to guarantee all space tourism passengers' life safety. It must not allow any death or hurt occurrences during every space tourism trip.

Even, the country government can have legal action to publish any space tourism leisure providers, when their every space tourism trip has occurred accidents, e.g. fire accident occurrence in spacecarft or spacecrat machines are broken to be damaged and need to be repaired during the space tourism trip. It will threaten to reduce trip accident occurrence, such as this cases. The commercial space ventures may present risk to property as well, such as a fire starting on the ground by launch-related material or problems presented by space debris.

In principle, liability law can provide incentive to deter carelessness that could lead to the destruction of property, although statutory ( rather than common law) assignments of liability for commercial launches are somewhat problematic.

Consequently, if the space tourism leisure provider expected to grow space tourism passenger number in long -term time, it must need to ensure none any accidents can occur during any space trip. Otherwise, the space tourism passengers can choose another space tourism leisure provider to replace its spce tourism leisure easily.

(2) Space tourism destinations and space tourism entertainment facilities safe arrangement challenges aspect

Secondly, it is space tourism destinations and space tourism entertainment facilities safe arrangement challenges. Nowadays, commercial space travel is looking more like a real possibility than science fiction. The usual ethical issues related to the safety of the space destination choices and the space tourism entertainment facilities, e.g. space vehicles, space hotels, space swimming pools, space sport centers, space cinemas, space gardens, space farming lands. In this strange space environment and safety concerns are just the beginning as there are othe interesting questions, such as below:

What likely would be a fair process for commercializing or claiming property in any space planets? Such as Moon or mars, when any future space tourism leisure providers who need to build above these any one of space entertainment facilities on these planets to provide to their space travelling customers to play.

How to distribute and manage these any lands ownership to these future space tourism providers fairly and legally?

How likely would a separatist movement be among space settlements to want to be free and independent states?

How to ensure above future space entertainment facilities and space entertainment places are in the safe space environment to be provided to any space travelers to play in any planets, e.g. Moon or Mars etc. planets.

So, concerning how to arrange space entertainment facilities to provide to space tourism clients to play in any safe space environment issue, it will be another concerning question to every space tourism leisure providers. When they decide to choose anywhere to the space hotels, space swimming pools, space gardens, space cinemas or space farming lands or space sport centers. These space buildings will need to be built in the safe, on stable stone lands environment and none any natural distaster, such as large wind or space underground water etc. unpredictable space natural distasterr attack to these space buildings suddenly. Because it has responsibility to any space tourism leisure providers to guarantee any one of these space buildings are safe to be built in the planet's safe land environment. It aims to achieve none any accident occurrences during their space tourism clients are staying to enter these any one of space buildings to visit or play any space entertainment facilities safely, e.g. space vehicle.

So, they must need to ceck anywhere the space planet's places to be ensured safe to build any buildings. Then, they can choose the suitable locations to build space entertainment facilities or buildings more confidently.

In fact, any space entertainment facilities, e.g. space hotels, space farming lands as well as space transportation tools, e.g. spce vehicle, spacecraft , these things will be value to be concerned to any space tourism leisure providers and it is business moral ethic responsibility to every one of them, when they plan to develop their space tourism business in any planets.

(3) Space tourism market competition challenge aspect

Thirdly, any provate space tourism development leisure businesses will face market competitive challenge, such as large spacefaring countries, e.g. US, UK have possible to dominate future space tourism leisure business ( government can own space tourism leisure business). They will be main actors in space were nation-states. Large spacefaring counties can build the space vehicles, that can take people and cargo into orbit and to the Moon, or Mars crafted international space law and shaped the main investments in space tourism leisure technology.

So, it is possible that the own space technological developed countries, such as US, UK, these countries governemts will have possible to operate public fund to support space tourism leisure business. It implies that private space tourism leisure businesses will face public space tourism leisure business and themselve private space tourism leisure business market competition in space tourism leisure industry.

If these two countries governments also participate this private space tourism leisure market. It will raise market threats to any private space tourism organizations.

Whether will developed countries governments participate private space tourism market? It is possible that new commercial actors began to enter the space tourism leisure industry, looking to disrupt both space launch services ans use space in new exotic ways. For example, the US government also moved its purposeful degradatoin of the global positioning system ( GPS), so US government will have effort to dominate GPS global positioning system communication business also. As this GPS communication business case, future US government has possible to decide to participate space tourism leisure business also.

However, in the future, space tourism leisure industry may contribute even more the developed countries, e.g. American, England economy. Space tourism and resource recovery, e.g. mining on planet, Moons and asteroids in particular may become large parts of that space tourism industry if these countries governments participated to this space tourism industry development. Of course, their viability rests on a range of factors, including costs , future regulation, international market competivitive problems and assumption about space technological development. However, these is increasing optimism in these areas of economic production to bring human space tourism leisure enjoyment and space mining resource development benefits. But the space economy is not just about what happens in orbits or how that alters life on the ground. The growth of this economy can also contribite to new innovations across all future possible unpredictable or undiscovered technological development, instead of space tourism leisure or space mining resource exploitation development.

Consequently, any space development technological governments will have possible to bring economic benefits from either only private space tourism leisure organizations or governments and private space tourism leisure both organizations cooperate to participate to achieve space tourism misson to contribute to global economic development and create new jobs to be employed in space labor supply market. The most important successful factor is that space tourism entertainment companies need to learn how to apply (AI) technology to assist any space boats and space journeys and any space entertainment facilities in order to raise safe and efficient service performance to let space passengers to feel satisfactory and confident in every space tourism (AI) design journey.


Reference

Cater Iain , Carl 2010, " Steps to space: Opportunities for astro tourism development, tourism management 31 ( 2010); pp. 838-845; Elsevier Ltd, DOI: 10:1016/j.tourman. 2009.09.001

Collins Patric, Iwasaki Yoichi, Kanayama Hideki, Ohnuki Misuzo 1994, comercial implications of market research on space tourism. journal space technology and sciences , vol. 10 no 2, 94 Autumn, pp.3-11. copyright: Japanese rocket society; available at: www.spacefuture.com/archive/commercial-implications-of market-research-on-space-tourism.shtml.

Collins Patric, Marita M; Stockmans R. and Kobayahi S. 1996. "Demand for space tourism in America and Japan and its implications for future space activities ". sixth international space conference of Pacific basic societies; Marina del rey; California: Advantages in the Astronautica science ( AAS paper no AAS 95-605) vol. 91. pp. 601-610. Available at: http://m.internationalaerospaceconsulting.org/upload/space % 20Future%20-%20Demand% 20for%20space% 20Tourism%20in% 20America%20Japan.pdf

ESA 2008, " Richard Garriott, millionaire American space tourist. blasks off of international space station". published in 12.11.2008. Huffington post, seen on i01.04.2015; available at: http://www.huffington.com/2008/10/ 12/richard-garriott-milliona-n-1333940.html.

Klemm, G., & Markkanen, S. (2011). IN A Papathanassis ( ed.) The long Tai , tourism ( pp.95-103). Weisbaden, Germany : Gabler Verlag; Springer Fachmedien Weiesbaden GmbH.

KSCVC Visitors, 2013; MRI 2013 Market by Market

Von Der Dunk , F. (2012). The integrated approach. Regulating private human spaceflight as space activity, aircraft operation, and high-risk adventure tourism. Acta Astronautica, 92(2), 199-208.

AI brings positive or negative impact to our societies

The relationship between robotic invention and economic growth

Can robotic invention bring social economic growth? If it is possible that why and how robotic invention can assist any countries social economic growth? First, we need to know whether what economic growth means in order to answer this question. In macro economic view, economic growth may mean that GDP growth, employment ratio growth, job creating growth, unemployment ratio reduces, consumption growth, productive industries growth, service provision and service needs increases. So, it seems that any countries' social development , when it can have positive impact growth for any one of these issue. It implies that the country's economic is growing.

To discuss AI and economic growth issue, it may being these two main questions. Whether robotic invention may have direct or indirect relationship to influence any countries' economic growth? What are the main factors to cause robotics invention to bring the country's economic growth in its society?

On the first hand, some researchers believe that robotic invention may influence any countries' economic growth. However, otherwise, other many researchers find large and robust negative effects robots on employment and wages. They estimate what are more robot per thousand workers reduces the employment -to-population rate by between 0.18 and 0.34 percentage points, and is associated with a wage decline of between 0.25 and 0.5 percentage. Because many jobs can be replaced by robotics. So, it seems that robotics invention can reduce workers number and increasing wage level, even robotic increasing number, it can influence unemployment ratio increases to the country. But, some lecturers estimate that it can impact economic growth. They estimate that AI may deliver an additional economic output of around US$13 trillion by 2030 year, increasing global GDP by about 1.2% annually. This will mainly come from substitution of labour by automation and increased innovation in products and services.

What is the impact of robots on society? They may include this spillover, one robot per thousand workers has slightly less of an impact on the population as a whole, leading to an overall 0.2% point reduction in the employment-to-population ratio, and reducing wages by 0.42%. Thus, adding one robot reduces employment nationwide by 3.3 workers. So, it seems that robotic number increases may influence global workers number increases many influence global workers number reduces as the same time. It can cause unemployment workers number increases in societies. On the other side, robotic invention can increase productivity, because robots increase productivity, which means that fewer human hours are needed to produce a given output. But, higher productivity also reduces production costs and output prices. Consequently, robotic production anticipation , it can increase the quality demanded by consumers, and firms hire workers in this increased demand.

But, when robotic production participation to any industrial manufacture, it can also bring negating effects, when they are entering the workforce, e.g. higher maintenance and installation costs to factories, enhanced risk of data breach and other cybersecurity issues, reduced flexibility, anxiety and insecurity regarding the future social development. So, it seems that robotic invention can bring the future of workplace automation may reduce workers number of factories, loss of jobs and reduced employment opportunities to global future workers, potential job lose, initial investment costs to employers. However, AI may also bring harmful to our future societies because if AI surpasses humanity in general intelligence and becomes " superintelligent" , then it could become difficult or impossible for humans to control. Moreover, a second source concern is that a sudden and unexpected " intelligence explosion" might take an unprepared human race by surprise. Hence, robotic invention may become human enemy or soldier, if human applies it to social damage aspect.

However, robotic invention can also bring advantages to our future societies in possible, robotic automation may bring advantages to employers : cost effectiveness, improved quality assurance, increased productivity, avoiding workers need to work in hazardous environments. But,. AI can also bring positive impact our daily lives, such as artificial intelligence can dramatically improve the efficiencies of our workplace and can argument the work humans can do. When AI takes over repetitive or dangerous tasks, it frees up the human workforce to do work, they are better equipped for, tasks that involve creativity and empathy among others.

However, robotic invention may bring organizational benefits in our societies. Robots will have a profound effect on the workplace of the future. They will become capable of taking on multiple roles in organizations, e.g bookkeeping or writing law draft simple clerical tasks. So, it's time for us to start thinking about the way we shall interact with our new coworkers. To be more precise, robots are expected to take over half of all low-skilled jobs in our societies, e.g. cleaning , warehouse picking up delivering tasks, restaurant cooking, hotel front line customer service, hotel room food delivery tasks etc.

So, when robots would be used in many fields all over the world. However, robots can not totally rule over the workplace by replacing all humans at jobs to keep economy afloat. Hence, robots in the workplace may bring advantages, they will not have bad emotion problems, such as safety of utilizing robotics to work in dangerous workplace. Robots do not get distracted or need to take breaks robots never need to sleep or they need to divide their attention between a multitude of things, perfection, let employees to feel happier and safe to work when robotics can help workers to work in dangerous warehouses or factories, workers can be replaced from robotics, increases productivities and job creation , even raises efficiencies in any workplaces.

In our future, whether robots won't destroy humans. It depends on how humans choose to apply this technological worker tool. A robot may not injure a human being or, though inaction, allow a human bring to come to harm. A robot must obey the orders given it by human beings, expect where such orders would conflict with the first law. A robot must protect its existence as long as such protection does not conflict with the first or second laws. How AI technology affects us in the future. They are concerned that we will see increases in stress, anxiety, and depression as digital lives expand. Meanwhile , we shall need to adapt our future digital living, there will be less face-to-face interaction , increased inactivity, poor in-person communication skills and an overall distrust among people. For robotic invention case example, future robotics can may focus on developing these five major field: Human -robotic interface, mobility, manipulation, programming, sensors and their importance to robotics development.

Robots can be applied to educational aspect. Robots can be used to bring students into the classroom that otherwise might not be able to attend. Robots such as the one mentioned are able to bring school to student who can not present physically. Simulators-high school sees the strongest example of stimulators within drivers' education courses, e.g. Google's worker robots. Google is planning to produce worker robots with personalities. So, robots can help teachers automate key classroom processes, integrate advanced technologies, acclimate students to technological change and helping identify personalized learning potential and discovering key learning trends.

● How manufacturers to raise efficiency in order to improve economic growth in possible to apply manufacturing robots to improve help economic growth

(AI) can be defined as the capability of a machine to imitate intelligent human behavior. If AI can imitate any talent humans to learn how to improve their behaviors, e.g. manufacturing industry worker behavior improves to raise GDP growth or productive number grows rapidly . Can artificial intelligence being labour to become automated? Allows an ever-in-to increasing number of tasks previously performed by human labour to become automated in the ordinary production of goods and services process.

Can (AI) create new ideas and technologies to help businessmen to solve complex problems and improve automation in the production of goods and services. The question concerns how (AI) manufacturing technology can impact economic growth?

(AI)'s new form of automation live, self-driving cars, or they may bring high levels of skill,such as legal services, radiology, and some forms of scientific lab-based research. Can it allow our societies be impacted on economic growth, due to automation to disicipline our modeling of AI.

In fact, AI automation to production of new ideas to future any industrial manufacturing process. It can influence to market structure, organization restructure, reallocation and wage inequality. So, discovering to AI automation can help future any organizations need to change existing task or discovering new tasks that can be used in production a reflects the fraction of tasks that have been automated.

It brings automating old tasks when automated could be constant, leading a stable , capital share and a stable growth rate. Hence, in long term, AI automation manufacturing technology may help businessmen to reduce large manufacturing cost in order to achieve stable micro economic benefit to them when they can apply AI automation manufacturing technologies to improve their productivities in efficient manufacturing method. For Coca Cora soft drink example, if Coca Cora applies AI automation to raise its productive soft drinks number. Then , it can manufacture many soft drinks number in short time per day. Then, its sale number can be increased, when it has enough number to supply to global to sell its soft drink. Consequently, soft drink sale number must grow rapidly significantly.

The fact that automated goods are produced with cheap capital , but it can also help business to raise production number significantly . How this superintelligence affect the economy? It seems physical tasks are essential to producing output, but when the manufacturer applies robotics to help production . Then, employees number may reduce, due to robotic participation, wages expenditure may reduce , but production number may increase .

So, AI increases the motivation at physical tasks. Hence, AI must may bring production growth innovation incentives to any manufacturers. Finally, with imitation and learning being performed mainly by super machine in developed economies. Then , research labor would become devoted to product innovations increasing product variety or inventing new products ( new product lines), to replace existing products.

It is one good example to explain that how AI can bring long term social economic benefit to any manufacturers, when any old ( existing) product lines are improved to innovative new product line by robotic manufacturing participation. Moreover, AI can change market structure to be reduced competition. When be escape competition effect tends to dominate at low discouragement effect may dominate for higher levels of competition or in less advanced economies. Hence , AI can also affect innovation and growth through potential effects, it might have on product market competition. So, it seems that (AI) can respond on helping social economic growth principle.

On conclusion, although robotic invention may bring disadvantages to raise unemployment ratio in possible when there are many low-skilled jobs are replaced by robotics, but at the same time, robotic may be applied to education aspect, when we are experiencing digital knowledge social development stage. Hence, smart robots ought may help humans to raise economic long term growth in possible when robotics can help the developing countries to develop more rapid to be developed countries as well as they can help the developed countries to develop more advanced both in global whole one economic developed societies. So, when robots can assist any countries to cooperate together, any countries are not independent, we need robots to assist develop between countries. Then, robots can assist any countries to develop economic growth in possible in future this day comes.

How robotic helps to solve recession

Is the use of robots to job market increasing during the Great Recession?

Some researchers constructed a measure of the use of robots—commonly referred to as "robot intensity"—to estimate trends in robot exposure across more than 250 metropolitan areas and over time, finding that: During the Great Recession, robot intensity plummeted. But since 2009, robot intensity has sharply increased nationwide. They felt that robots may influence recession in possible. How are robots going to affect our jobs? Most analysis tends to be prospective in nature, and estimates of future impacts on employment vary widely, with some studies predicting that as many as 50 percent of all workers are at risk of losing their jobs to automation. Even less is understood about the actual impacts of robots on jobs, wages, and workers today. If there are many low skillful jobs e.g. cleaner, warehouse deliver, cooker, restaurant waitor etc., even high skillful jobs, e.g. accountant, lawyer, doctors etc. occupations are replaced by robotics. Then, our societies will increase unemployed people number. Consequently, great recession will be caused by robotic workers because when our societies have many people lose jobs, then many people loss income, then our consumption desires may be influenced to reduce. Consequently, many businesses may lose many customers. Low consumption desires may bring serious recession to any countries, due to robotic workers number increases to replace human workers in global societies.

The reason is that new technologies of the period have enabled people to be very productive while working part-time. Businesses do not need large numbers of employees, so individuals can devote most of their waking hours to hobbies, volunteering, and community service. In conjunction with periodic work stints, they have time to pursue new skills and personal identities that are independent of their jobs. Developed countries may be on the verge of a similar transition. Robotics and machine learning have improved productivity and enhanced the economies of many nations. Artificial intelligence (AI) has advanced into finance, transportation, defense, and energy management. The internet of things (IoT) is facilitated by high-speed networks and remote sensors to connect people and businesses. In all of this, there is a possibility of a new robotic society that could improve the lives of many people, but it also encourage future businesses apply robotics to replace human workers to do many jobs in finance, transportation, defense and energy management, medical health, hotel, tourism, cinema, theatre etc. entertainment service fields. So, robotics may bring reducing cost benefits to businessmen, but they can also increase workers losing jobs number in our societies if future many businesses make decision to apply robotics to replace human workers to do any simple ot complex jobs in global job market.

A McKinsey Global Institute analysis of 750 jobs concluded that "45% of paid activities could be automated using 'currently demonstrated technologies' and . . . 60% of occupations could have 30% or more of their processes automated."[6] A more recent McKinsey report, "Jobs Lost, Jobs Gained," found that 30 percent of "work activities"

could be automated by 2030 and up to 375 million workers worldwide could be affected by emerging technologies. Researchers at the Organization for Economic Cooperation and Development (OECD) focused on "tasks" as opposed to "jobs" and found fewer job losses. Using task-related data from 32 OECD countries, they estimated that 14 percent of jobs are highly automatable and another 32 have a significant risk of automation. Although their job loss estimates are below those of other experts, they concluded that "low qualified workers are likely to bear the brunt of the adjustment costs as the automatibility of their jobs is higher compared to highly qualified workers."
reference

James Manyika, Susan Lund, Michael Chui, Macques Bughin, Jonathan Woetzel, Parul Batra, Ryan Ko, and Saurabh Sanghui, "Jobs Lost, Jobs Gained: Workforce Transitions in a Time of Automation," McKinsey Global Institute, December, 2017.

Melanie Arntz, Terry Gregory, and Ulrich Zierahn, "The Risk of Automation for Jobs in OECD Countries," Organization for Economic Cooperation and Development, Working Paper 189, 2016.

However, some economists felt opposite opinions, they believes that future many businesses won't choose whole applying robotics to replace human workers in global job market. So, they are only human workers assistant role. Economists have, on the whole, been fairly discuss about the impact of robots and AI on workers. History is strewn with incorrect predictions of the looming irrelevance of human labour. The economic statistics have yet to signal the arrival of a robot-powered job apocalypse. Outside of slumps, firms remain keen to hire humans, for example. Growth in productivity—which ought to be surging if machines are helping fewer workers produce more output—has been unimpressive. A look beneath the aggregate numbers, though, reveals that change is indeed afoot. They believes that an AI-induced change in the mix of jobs need not translate into less hiring overall. If new technologies largely assist current workers or boost productivity by enough to spark expansion, then more AI might well go hand-in-hand with more employment. This does not appear to be happening. Instead the authors find that firms with more AI-vulnerable jobs have done much less hiring on net; that was especially the case in 2014-18, when AI-related vacancies in the database surged. But the relationship between greater use of AI and reduced hiring that is present at the firm level does not show up in aggregate data, the authors note. Machines are not yet depressing labour demand across the economy as a whole. As machines become cleverer, however, that could change.

Take work by Daron Acemoglu and David Autor of the Massachusetts Institute of Technology, Jonathon Hazell of Princeton University and Pascual Restrepo of Boston University, which was presented at the recent meeting of the American Economic Association (AEA). The authors use rich data provided by Burning Glass Technologies, a software company that maintains and analyses fine-grained job information gleaned from 40,000 firms. They identify tasks and jobs in the dataset that could be done by AI today (and are therefore vulnerable to displacement). Unsurprisingly, the researchers find that businesses that are well-suited to the adoption of AI are indeed hiring people with AI expertise. Since 2010 there has been substantial growth in the number of AI-related job vacancies advertised by firms with lots of AI-vulnerable jobs. At the same time, there has been a sharp decline in these firms' demand for capabilities that compete with those of existing AI.

An AI-induced change in the mix of jobs need not translate into less hiring overall. If new technologies largely assist current workers or boost productivity by enough to spark expansion, then more AI might well go hand-in-hand with more employment. This does not appear to be happening. Instead the authors find that firms with more AI-vulnerable jobs have done much less hiring on net; that was especially the case in 2014-18, when AI-related vacancies in the database surged. But the relationship between greater use of AI and reduced hiring that is present at the firm level does not show up in aggregate data, the authors note. Machines are not yet depressing labour demand across the economy as a whole. As machines become cleverer, however, that could change.

Evidence that AI affects labour markets primarily by taking over human tasks is at odds with some earlier studies of how firms use the technology. A paper from 2019 by Timothy Bresnahan of Stanford University argues that the most valuable applications of AI have nothing to do with displacing humans. Rather, they are examples of "capital deepening", or the accumulation of more and better capital per worker, in very specific contexts, such as the matching algorithms used by Amazon and Google to offer better product recommendations and ads to users. To the extent that AI leads to disruption, it is at a "system level", says Mr Bresnahan—as Amazon's sales displace those of other firms, say.

New work by Ajay Agrawal, Joshua Gans and Avi Goldfarb of the University of Toronto suggests that this state of affairs may not persist for long, though. As the quality of AI predictions improves, they write, it becomes increasingly attractive for AI-using firms to restructure in more radical ways. At some level of accuracy, for example, Amazon's ability to predict consumers' desires could encourage the firm to adjust its business model—by pre-emptively shipping goods to consumers before they ever go searching at Amazon in the first place—in ways that are likely to change how many workers and of what sort the firm requires. In that event, the influence of AI on the economy could change dramatically. So, such as Amazon case, it applies robotics are only concentrated on predicting consumer prediction aspect, AI is only assistant role to Amazon human market researchers. They help Amazon market researchers to gather consumer behavior data , but Amazon human market researchers need to do marketing analysis tasks by themselves. So, Amazon can not employ human market reseachers jobs position in itself company. Amazon needs robotic and human market researchers to do market research tasks in order to achieve how to predict consumer behavior more accurately. Thus, it seems that future large enterprises won't fire any professional staffs more easily because some complex tasks, e.g. analysis tasks, they believe that human's analysis can make more accurate judgement to compare AI's analysis.

However, some scientists believe that some skilful professional occupations , they have possible be replaced by robotic. Will Architects and Engineers be Replaced by Robots? It's not uncommon for people to think they may be replaced by a robot in the workplace. After all, it's happened plenty of times before. For example, the rise of the mechanical assembly line saw machines replace people in the early $20^{th}$ century. With recent advances in artificial intelligence (A.I.), it's entirely possible that more jobs are at risk. Even skilled workers, such as architects, programmers and engineers may be at risk. One day, an A.I. software developer may be able to do everything that a human programmer can do.

Recent reports have not abated this thought process. In fact, the 2016 Economic Report of the President seemed to suggest that artificial intelligence is playing an increasingly important role in the engineering industry. Just think about the software you use in your work. Many software packages can handle a lot of the complex calculations for you. Yes, this cuts down on the amount of work you do. However, this automation may also present a threat to your job. What if the future sees these same software packages handling data input, as well as processing.

Automation is important. The use of artificial intelligence, alongside various other technologies, has always improved production. More work gets done, which means that businesses make more money. Architects and engineers constantly look for ways to speed up their work. The desire for automation has informed many recent software innovations. Furthermore, project methodologies, like Building Information Modelling, place automation at the fore. That's great for speed and efficiency, but what does it mean for architects and engineers? History has shown that automation has a very human effect. People lose their jobs because machines can do them faster. Just think of it from a business viewpoint. Do you want to pay 10 or more employees, or invest in one machine? More often than not, the machine will cost less than the employees, even if you factor maintenance into the equation. It's a simplification, but not an invalid one. Businesses make these sorts of decisions all the time. By pushing for automation, architects and engineers may be slowly working themselves out of their own jobs.

Several studies have also suggested that artificial intelligence may cause job losses. One recent example comes from the University of Oxford. The study found that over 700 types of jobs are at risk of technological disruption. All told, this means that about 47% percent of jobs are at risk because of artificial intelligence. That is a huge amount of people who may find themselves obsolete due to advancing technology. The same study also mentioned a concept called the "technological bottleneck". The researchers used this to determine how "at risk" a job was of displacement. The bottleneck takes three factors into account:

•How much creative intelligence the role needs

•If manual manipulation and perception is required

•The role of social intelligence in the role

If a role requires a high degree of any of those three things, it's less likely that it's at risk from artificial intelligence. Architects and engineers are a good example. These professionals require a great deal of creative intelligence.

Artificial intelligence and robots may not be able to emulate that creative intelligence. As a result, it's unlikely that architects and engineers need to worry about losing their jobs. Right now, at least. The study concluded with a cautionary note. It said that just because automation enhances an architect and engineer's work right now, it doesn't mean that automation won't replace that role in the future. So, if future human architects or engineers jobs can be replaced to do by robotics. Then, robotic architects can help the architectural firms to design more attractive architectural plans to satisfy construction firms clients needs in short time or robotic engineers can help the engineering design firms to design more attractive machines to satisfy any engineeing customers needs in short time , when robotics can be invented to own excellent creative ability to compare human architects or engineers. So, these two professional occupations will be lost when robotic architects and robotic engineers can be invented to own excellent creative ability to compare human architects or engineer in future one day in possible.

However, robotic architects or engineers may bring advantages and disadvantages both aspects:

On advantages aspect:

Artificial intelligence allows us to do all of the following:

•The completion of mundane tasks that would otherwise take a lot of labour hours. Automating such tasks frees up skilled workers to work on more important tasks.

•A.I. is not as prone to making errors as a person. As long as the A.I.'s programming is good enough, you should find that calculating errors and similar issues become problems of the past.

•Speed is a key feature of artificial intelligence. Huge datasets no longer provide any problems to businesses, as automation allows for much faster processing. This means that a business can spend money elsewhere.

•The most complex A.I.s reduce the amount of risk attached to the decision-making process. The "Curiosity" Mars rover is a good example. It's programmed to choose the best course of action depending on its position.

On disadvantages aspect:

It's not all good, unfortunately. The following are some of the bad points of artificial intelligence:

•The previously mentioned job losses can cause all sorts of problems for staff morale.

•Some believe that artificial intelligence gets rid of the human element. The nightmare scenarios in films like "The Terminator" may seem far-flung, but that doesn't mean there isn't a risk in letting machines make all the decisions.

•A.I. relies on pre-existing knowledge, which means it lacks creativity. Attempting to use it for creative endeavours may result in failure.

•Algorithms may not be able to make judgement calls in disaster situations. Again, the A.I. may not take the human element into account, no matter what's actually happening on the ground.

Oe people management aspect, what do you think would be the reaction to a robot attempting to manage people? It's likely that a lot of people won't take to kindly to artificial intelligence telling them what to do. Many underestimate the importance of people skills in the architecture and engineering profession. Architects and engineers must be able to organise workloads and manage individuals. It is sure, A.I. device could handle the former. Scheduling is a task that many already automate. However, A.I. will fall down when it comes to the human relationships that are so vital in a team environment. An A.I. won't understand when somebody is demotivated, or why. It won't make allowances for the human issues that affect every problem. This makes skilled team members even more valuable. As A.I. takes an increasing role in the workplace, the need for people management will become more important. Architects and engineers with those skills may even find they make more money to employ them. Although, it is possible that AI can replace human architects or engineers to do their tasks to be better , it can help any one architectural or enginering firms to improve design performace in order to satisfy customers design demand, but our societies will increase unemployment ratio to engineers and architects number, even universities will reduce architect and engineering students number. Our traditional professional knowledge will be felt to be rubbish when these professional subjects won't be useful to help us to find jobs easily. So, AI invention will influence many students won't choose to study these two subjects. Our societies will be influence to experience knowledge recession when knowledge will become rubbish because robotics invention , it can do many human professional jobs to do. " knowledge recession" will be important factor to bring economic recession, because human can not be encouraged to learn any new knowledge to prepare to enter job market due to robotic invention can replace us to do more complex tasks in our future societies.

Can robotic leadership be good solution method when recession had come to the country?

On leadership management aspect, can robots become clever leadership to any organizations? As artificial intelligence becomes further embedded into our everyday working lives, we are already seeing the footprint of machine learning, automation, algorithms and robots in many of our professions and sectors. However, when we look at the upper levels of business management and leadership, these technological shifts are less evident, with C level Executives continuing to lead and strategise as they have done before. In Ireland, there are more than 500 CEOs. The question is, when will we start to see machines and robots play a more central role in the CEO sphere, and is a 'Robot CEO' realistic in the short to medium term?

New research shows that 24% of people aged 25-29 would replace their boss with a robot, demonstrating an interesting trend among Generation Z. However, these data sets are perhaps less founded in AI and robotics and more in current employee engagement. It's telling that the 20-30% of people who would willingly replace their human boss with a robot is about the same percentage of people who are consistently classified as "actively disengaged" at work. In addition, research from analytics giant Gallup demonstrates that 70% of how we feel about work, namely our emotional commitment, is driven by who our manager is, again underlining the centrality of human behavioural traits when making decisions on leadership.

As the Irish economy moves forward, values will define how we use and leverage the potential of AI. Tomo Noda of the Harvard Business Review believes that we will need more focus on leadership with humanity, ethics and integrity, stating "only good people can create good AI."With many roadblocks and challenges for the economy looming, primarily in the shape of Brexit and trade tariffs, it is a sound integration of both human and tech which will provide the leadership required to ensure our economy remains robust. Human leaders have played a central role in helping to steer us out of the 2008 recession, and with diplomacy and relationship building key to our post-Brexit future, humans will undoubtedly be the key influencers within the C level for decades to come. Hence, it seems that future organizations ought choose to apply robots to assist leaders to do make important decision, if robotics can assist leaders to make any important decision to conclude the best results to improve any companies performance. Then, GDP may be influenced to increase or grow rapidly, when recession had come to the country. So, robotic leadership may be one solution to solve recession method in possible.

The Recession Cometh and Robots are Ready

In economic demand vs. supply theory indicates that consumer appetites for customized product and their expectations for ever-lowering costs. So the current tug-of-war over if, when, and where a recession will hit is not unchartered territory. For manufacturers, though, the uncertainty is particularly challenging, as the flexibility that allows operations to reflect the pace of the economy simply isn't there. The economy has been growing. Unemployment is down. Last year's Christmas sales were better than they've been in a long time. All good and logical reasons for manufacturers to hire.

Recently though, there have been signs that instability is coming. The US stock market experienced extreme volatility as 2018 came to a close. The Federal Reserve raised interest rates and laid down some pretty clear language that more was to come. Consumer confidence fell. In the UK, a deal on Brexit that would allow British manufacturers to continue to do business with the EU seemed elusive at best. The Chinese government announced that growth in its economy has slowed. And the "R" word started to appear with more frequency. These are not signs that inspire confidence. So, manufacturers once again find themselves in a place they know so well. The rock: the need to hire workers to keep ahead of demand. Compounding this challenge is that unemployment is low and it is very hard to find people with the skills needed to take a job in manufacturing and be ready to work on day one. The hard place: overwhelmingly, today's automation is fixed, expensive, and able to perform only a single task.

As one supply chain executive of a global automotive firm shared recently, "In a downturn...it is about flexibility. All of the automation we have cost too much and it is too complicated to change what it does. What we need is flexible automation that can respond when and how we need it to."Can robotics be applied to manufacturing industry to avoid cost reduces to manufacturers when consumption number reduces or recession is coming? So what makes the most sense? Hire, hoping that if and when recession comes, it will be short-lived and you won't have to lay folks off? Or try to invest in reconfiguring existing automation?

Hence, cobots give manufacturers the flexibility they need to thrive in good times and not-so-good times. Advances in robotic technology make it possible to put cobots to work
•at lower costs
•on more than a single task
•in the same amount of time, it takes to train a person – or even less
With collaborative robots, manufacturers can build the operations they need to compete and thrive regardless of the economic climate, where manufacturing robotic participation can help organizations to reduce labours number on strategic tasks and flexibility is part of the organizational human resource cost reducing strategy. It seems that robotic manufacturing workers can help organizations to reduce manufacturing cost when recession is coming. Consequently, these applying manufacturing robotic businesses may prolong business life time in possible. So, it seems that manufacturing robots may help organizations to reduce manufacturing cost to keep life when recession is coming.

# CHAPTER IV

# Robots how impact organization and social behavioral change

Human Behavioral network job brings social economic benefits

What does human network job mean ? Why may human network job be popular? Why human network job behavior may influence economy ?

Nowadays internet is popular to use. We can apply internet to find data , search any new things, even earn money. Why does internet

may become huma network job source. For example, e-publish may be one kind of new human network job. Any authors may apply internet

channel to help them to sell electronic or paper books from e-publisher web store. They may apply facebook, you tub etc. any online

channel to promote themselves new books to let new readers to know whether when they may buy themselves favourable new topic books to read

from electronic publisher web store.

Thus, future electronic publisher industry may help any authors to build internet network platform to help them to sell and promote

ot advertise their any one new electronic or paper book topic to let global any one reader to choose to buy their any new topic books from electronic publisher web store easily and conveniently. However, it implies that electronic network platform author may be one kind of future new human network job in our societies.

How electronic network platform author job may bring economy benefit in macro economy view? A person can have few friends, contacts and still be very influential if these few

friends and contacts are themselves highly influential, e.g. one author must not need to know any one reader in global society. When they like to choose any electronic books from electronic internet network platform. They may become the author's any one topic book buyer, when they feel the author's any one topic book is fun and attract they make decision to buth the strange author whose the topic book from electronic book publisher's platform web store conventiently in short time. Although, they are strangers, they do not know themselves , but the reader can understand what it way that made Google from writing platofrm to create new creative mind and typing network job method to replace traditional hand writing book method for global authors. It will be one kind of new human network writing job.

Hence, global any one reader can apply an innovative search engine , such as google.com to find whether whom author personal new topic books are value to read from internet.

Then, the electroniuc publisher's web store may be new book store platform sale network to help the author to sell many electronic or paper books from electronic network platform

in short time. So, internet may be future new network plaform to help global any one author to create network writing job absolutely. Furthermore, internet may be popular social media

to help any one author to build goold relationship between his/her readers. It is one kind of new network, human network job. New authors do not need to buy many paper books to prepare to put in any one book shop warehouse. Their every book can print on demand to reduce out of book stock in any one book shop. They may choose to sell either electronic books or paper books both from any one book publisher web store. So, electronic network platform may be one kind of good writing channel to help human authors to create income and it can also help authors to bring new creative mind and new topic fun content books to let readers to know and buy to read from electronic publisher network platform.

Why does human behavior may be one kind of new human network job to bring global economic advantages. ALthough, it may be free income or without inocme, but the person does the network behavior, his/her behavior may be bring advantages to influence many other people's health. For this case, when a worker in a coffee shop in an airport gets a vaccination aganinst the flu, it does not only helps him or her stay healthy, but also helps the many travellers who might otherwise have been inflected if that workers caught the flu. So, the externality , the result implies the vaccination of even a part of a community conveys benefits to the whole community. For example, governments pay special attention to the vaccinations of school children, teachers, health mothers, and the elderly, categories of people particularly susceptible not only to catching, but also to transmitting a disease.

It is not accidential that governments are heavily involved with vaccination . When there are externalities, free market, fail to persuade individual incentives with society's

their the worker's decision of whether to get a vaccine ends up attracting whether other people get sick. The workers might not fully take all these other people's potential suffering into account when making her or his vaccination decision.

As Stanford University does many suggestions, understand this and tries to help them make the right decisions and so providers free flu vaccines for its staff and students.

Small pockets of unvaccinated individuals can allow a disease to gain a spread more widely well-being. For example, parent weighing the costs and benefits of a vaccine for their child is not always thinking of the consequences of that vaccination to other people. THese are markets in which subsidizing or regulating behavior can make everyone better off. Because the reason for requiring that a child be vaccinated before enrolling in school is not just to protect that child, because each child's vaccination affects others via potential contagions.

Robots take our jobs behavioral and economy influences

Robot job behavior brings economy influences

If one day robots can replace human to do simple, even complex jobs. They will bring what influences to our global societial economy.The popular economic refrain declares that the

global middle class is dying and robots will soon take our jobs, e.g. shopping center customer service jobs, library service jobs, cinema ticket sale jobs, restaurant kitchen cooker jobs,

even, bus drivers, taxi drivers etc. public transport driving jobs, accountant, doctors etc. professional jobs. Whether it is beautiful or petty matter if our future societies have many human jobs can be replaced to do from robots. Businessman must may reduce to employ employees and reduce to pay salary or wage, when robots can be replaced to do their employees tasks. But, societies must bring unemployement rate rises , due to societies will have many people loss jobs when their employers choose to buy robots to serve their clients or do any office tasks or customer service or cleaning etc. tasks.

In micro economy view, employers may save money in long term, but in macro economy view, it will cause unemployment ratio rises , even crime rate rises when there are many people lose

jobs in societies. These models of doom, though, fail to account for the hundreds of businesses riding the waves of change in their industries when robots may be invented to replace human to do many simple , even complex tasks in our future societies.

WE may image that one small factory needs to manufacture fishes canes to sell to supermarket, the small , cheaper stuff and higher margin parts of the fishes manufacture industry. Before, this factory needs to employe many human factory workers need to help every fresh customer makeing the perfect fishing gear, designed for performance, durability, and cost in order to achieve to manufacture every fish cane in whole fished processing manufacturing stages. Every worker needs to spend about 15 to twenty minutes to finish every fish cane , till to delivery to any supermarket to sell. If this fish canes manufacturing factory can apply manufacturing robots to help them to finish any one working tasks , every robot can only spend five minutes to finish whole fresh fish cane manufacturing process. Thus, every robot can

help this factory save 10 to 15 minutes time to finsh every fish cane manufacturing process. IN fact, time is money, because when every robot can help this factory to reduce 10 to 15 minutes time to compare human worker. Then, this factory can finish about 20 fish canes in one hour if it can use robot to help it to manufacture fish canes. Otherwise, if this factory still use human workers to help it to manufacture fish canes, then it can finsh about 3 to 4 fish canes in one hour. SO, the manufacturing efficiency ensures that robots must help this fish manufacturing factory to raise fish canes number more than human workers. So, in robotic behavioral economy view, manufacturing robots must help this fish canes manufacturing factory to raise fish canes manufacturing number and deliver increasing number to supermarkets to prepare to sell every day. Robots can help this fish canes manufacturing factory bring manufacturing time saving, rising manufacturing efficiency, improving performance and reducing wages expenditure long time advantages in micro economy view. However, manufacturing robots can also bring disadvanages to society, e.g. increasing unemployment ratio, increasing crime rate,

this factory workers will lose jobs and income, they need earn social welfare from government and increasing government finance pressure in short time, even long time in macro economic view.

Stanford University graduate program in economics, Scott lecturer explained that "in demand and supply economic theory for robots supply and demand case, robots supply number increasing may influence human workers demand number decrease. It sometimes calls " the efficient frontier".

No specific human beings were mentioned in any of economics classes. As robots supply and demand in market case, They ( robots) may be purely theoretical " agents" who reached to the most reasonable sale prices in order to persuade any one businessman buyer to make manufacturing robot buying decision whether robots can help him / her to bring how much saving time , saving money, saving cost, improving performance, efficiency economic benefit before he/she plans to reduce workers number when he/she decides to apply robots to replace human workers in his/her factory or office or any service department, e.g. cinema ticket sale service, shopping center customer service, shopping center cleaning , supermarket customer service etc. service or sale tasks. When robots can replace human to do any one of these tasks in any organizations. So, robots may be human worker agents who reached to prices the way robots would react to a software

command. There was nothing that explained why some people thrived and others did n't or why truly brilliant, hardworking people could fail when much lazier folks succeeded." Having been admitted to the Stanford University graduate program in economics, Scott lecturer hoped to get his answers there.

How robots influence our future social changing? Using the right technology can be a boon to your business in this economy. For internet example, it is easier than ever to find well-matched customers all around the world, to stay in contact with them, and to more quickly design the products they want. If you focus solely on being cutting -edge, though you risk letting the technology

take over what should be very robust relationships with your customers , employees, and colleagues. IN nowaddays society, technoligical advances and cutomation, personal

relationships in business are more crucial than ever. I mean that robots can not replace human to serve clients to let them to feel more comfortable and passion more easily. For shoe shop case example, if the shoe shop apply one robot to serve its clients to replace human shoe salesperson to serve its shoe customers. Robots ensure that they can not persuade every shoe potential buyer to make shoe buying decision more easily when robots need to contact every shoe potential buyer. The reason is simple, because robots can not touch any one shoe buyer individual emotion very easier.

If the shoe buyer needs the robots to help him/her to choose any right shoe styles when he/she can not feel himself / herself can make the most right shoe style choice decision. The robots can not replace human shoe salesperson to make shoe style choice judgement more easily. They must need longer time to analyze whether which shoe style may be the most suitable to the shoe buyer. Otherwise, human shoe salesperson may attempt to make the most right shoe style choice decision to help any one shoe buyer to chooce the most right style shoe because he/she owns shoe style sale experience, shoe style knowledge, the most important reason is that they can feel every shoe customer individual emotion to touch whether he/she will feel comfortable or happy when they attempt to help every shoe customer to seek the most right shoe style in every shoe customer whole shoe searching processing. Othwerwise, serving robots are only one machine, they can not touch or feel every shoe customer individual emotion whether he/she feel comfortable or unhappy or happy when they need to contact them in whole shoe searching processing.

Hence, I believe that some tasks robots can

not repalce human staff to do very easily. Otherwise, robots may bring disadvanatges to let any one businessman to loss his/her customers, due to robots can not touch every customer

emotion to compare human staff in service tasks more easily. Robots serving customer behaviors may cause money lose and customers number lose to the shop in micro economic view.

Intellectual human economic behaviors

What does intellectual human economic behaviors mean ? I believe that when we choose or decide to do intellectual behaviors, then our societies will be influenced to bring economic growth in consequence.I shall attempt to indicate pollution case to explain how and why eithet our intellectual or foolish behaviors may bring economic growth or recession in consequence as below:

On one hand, for air pollution social case aspect example, if we only consider to buy cars to drive for working aimr or holiday leisure aim. Then, our societies air will be polluted. Our health will be influenced to bad. Our car driving behaviors may cause global environment air pollution serously. In long tiem, global air pollution will bring our bodies health to be bad. Although, ourselves car driving behaviors may bring our driving travelling leisure enjoyment and comfortable feeling in short time, also we so not need to pay public transport fare often, but we need to compensate ourselves health economic intangible loss due to air pollution , when cars number increases, dirty air will cause ouselves health to become bad.

In the result, we will need to pay more medical expenditure when we are old age, due to ourselves bodies will become bad, due to we breathe global dirty air every day, due to ourselves cars pollute air in long time, e.g. 10 to 20 years, even 30 more without limited air pollution environment. So, driving cars behavior may be one kind of human foolish behavior and our foolish behavior may bring ourselves future long time medical expenditure absolutely.

One the other hand, water pollution social aspect, if we often keep much rubblish to pollute sea, oil exploration porcessing pollute ocean , ships gas pollute ocaen, then fishes will eat polluted food and drive dirty water, due to global ocean is polluted.

In fact, because human only to conside how to buy boats to carry on leisure enjoyment activities, or catch cruises to travel on the sea. Also, oil manufacturers only consider researching anywhere to find new oil exploration places to manufacture oil product, when their oil exploration processes pollute ocarn . Consequently, global fishes drink polluted warer or eat polluted food. They will have poison. SO, human will have high chance to eat poison polluted fishes, due to fishes are poison or are polluted.

So, human is doing foolish activities, we only hope to find oil exploration places to pollute ocean or we only spend money to buy ticket to catch ships to travel anywhere in global ocean. All of these human foolish behaviors will bring pollution to global ocean. On consequently, we will need to compensate to eat polluted or dirty or poision fishes, ourselves bodies health will be bad. In long time, we need have high chance to pay medical expenditure when we are old. So, pollution case may be one good example to explain how and why human foolish behavior may influence ourselves future need to compensate serious medical loss.

All of these human foolish behavior will bring pollution to global ocean. On consequently, we will need to compensate to eat polluted or dirty or poison fished , ourselves bodies health will be bad. In long time, we will have high chance to pay medical expenditure, when we are old. So, pollution case may be one good example to explain how and why human ourselves intellectual or foolish behaviors may influence future long time economic loss or economic growth or recession in micro and micro economic view.

On another water pollution aspect hand, if we often keep rubbish to sea, oil exploration processing pollutes ocean and ships' gas pollute ocean, then fishes will eat polluted food and drink dirty water, due to fishes will eat polluted food and drink dirty sea water because the global ocean is polluted seriously.

In fact, because human only consider how to buy boats to carry on any leisure water activities, or catches cruises to travel on the sea. Also, oil manufacturers only consider any where to find oil exploratin places to manufacture oil products from ocean, when their pol exploration processes can plooute ocean. Consequently, global fishes drink polluted water or eat direty food. They will have poison. So, human will have high chance to eat poison fishes.

Otherwise, such as pollutin case, it can infuence inflation or deflation. Consequently, the reason indicates supply and demand theory. If air pollution is serious, then we will consider health issue, global cars demand number may be influenced to reduce, when global cars number demand will reduce, global car prices and supply number will need to change to fall down in order to attract or persuade global car consumers choose to make car purchase decision.

Hence, global car manufacture number and car price will be influenced to reduce, due to global air pollution issue. Consequently, deflation will occur because when the country citizen usually does not spend much extra saving money to buy car expensive goods. Money value will be low. Otherwise, if global cair pollution is not serious, human considers to buy cars to enjoy driving leisure lives. So, global car demand is influenced to increase , also global car price will also influenced to increase.

Consequently, gobal human will choose to buy cars to drive. Due to we accept to spend extra saving to buy expensive car goods. Car sale price and supply may be influenced to rise up. Money value is influenced to reduce. Inflation may be influenced, due to global car consumers number increases, we would not have extra money to spend easily. Car expensive goods expenditure influences our spending habit to avoid to make car purchase decision more easily. So, human intellectual or foolish activities may bring inflation or deflation consequency in possible indirectly in macro economic view.

On conclusion, above pollution case explain that how and why human intellectual or foolish economic behaviors may bring inflation or deflation consequency as wll as economic growth or recession consequency as well as any goods demand and supply increasing or decreasing consequency. It implies that human behavior may have indirect relationship to influence any goods demand and supply number to either increase or decrease result as well as any goods price will be influenced to increase or decrease in micro and macro economic view.

The relationship between social change and human behavior

Why does economic changes may influence human individual behavioral change? I shall attempt to indicate shopping behavior and staying at home behavior to explain their case and effect relationsip as below:

Human behavior can be influenced by economic change or economic change can be influenced by human behavior? Why does recession may influence consumers reduce shopping desire? In social recession suitation, it is possible that many people lose jobs suddenly, due to businessmen lose many customers. They need to make decision to reduce employees number in order to continue to keep businesses. Consequently, many firms ( organizations) their employees may lose jobs. When they have much time, due to lose jobs, they will feel to avoid to spend too much time and money to go to shopping often. Many losing jobs people, they will often stay at homes.

So, they will reduce time to go to shopping, then non essential products won't their preferable choice purchase products. Hence, recession will change many losing jobs people their shopping or consumption desires to avoid to buy non essential products often . Usually when economic boom, many people have jobs to do because consumers number must increase when many people have jobs to do. Then, many people can accept to spend money to buy non essential products often. Many people feel spend time to go to shopping can satisfy their purchase of any kinds of new products useful psychology or desire. So, recession is one good example to explain it can influence many people do not like often to leave homes to go to shopping easily. Many people like to stay at homes, becaue they feel worry about spending too much shopping time when they leave homes. Their staying home time is one good negative shopping behavior example. So, economic change may influence human individual behavior changes , they have direct cause and efect relationship in behavioral economic view.

May human behavior influence economic change? Is it possible that human behavior may bring the country social economic change in macro economic or micro behavioral economic view ? I shall indicate publishing industry example. Do you feel that if there are many students feel learning is very important when they read many books or many of students feel interesting to read or they have reading new books in habit, then it is possible that the country will have many students like to spend time to go to any book shops to choose the books, they feel that they can help they learn new knowledge. Then the country will increase students number, they often spend time to visit any one book shop every week. Their visiting book shops behavior which may become their habits. So, the country will increase students number, they often spend time to visit book shops. Also, it implies that visiting book shops behaviors may be their behavioral habits.

So, when the country has many students often spend time to visit book shops , their visiting book shops behaviors may help any one book shop to raise books sale chance. So, the country's student individual often visiting book shop behaviors, their habitual visiting book shops behaviors must may assist help any one book shop to increase books sale number absolutely.

Consequently, any one book shop , its books sale bumber must be influenced to increase to increase because the country will have many students like or feel need visit book shops habit in order to choose any suitable books to buy to read at home in order to raise themselves learning effort. When the country has many bok shops often have many students visit their book shops, then their books sale number may be influenced to increase. It explain why student individual visiting book shop behavior may help any one book shop sale number increases also.

How human productive behavior may influence economic development

May any country which citizen behavior assist themselves country development? It is one cause and effect economic question. I mean that if the country itself citicen can not concentrate mind or energy to choose to do one kind of industry in order to let themselves country can bring the most benefit, then whether the counry itself economy can bring the most serious economic benefit. I shall attempt to indicate these countries themselves indistry choice to explain whether these countries themselves citizen productive behavior may help themselves countries to achieve the largest economic benefits. I shall indicate as below:

New Zealand farmer individual wine productive behavior

For New Zealand country example, this country concerns itself effort is foucs on farming agricultural aspect. So, this country has many farmers concentrate on farming agricultural aspect. May New Zealanders choose to spend time to produce different kinds of wines, e.g. wine or red grape wine is for the people are eating meat, or they are eating dinner.

When these New Zealanders their behaviors choose to do farming or agriculture to grow and produce different kinds of taste of white or red grape wine drinking products job. Themselves grape agriculture behavior will influence these New Zealanders themselves, they can learn how to improve different kinds of grape wine drinking products in order to achieve every kinds of white or read grape wines taste improving aim during their white or red grape producing process.

Why can New Zealander every individual white or read grape wine producers improve their white or read grape wine taste more easily? In behavioral economic view, it can explain that why any one New Zealander white or read grape wine producer can be encouraged or excited or persuaded to concentrate nervous and energy and effort to learn how to improve their white or red grape wine products easily.

In fact, New Zealand is one agricultural food export country. It has good natural environment resource , e.g. land, seed to provide any one farmer to produce themselves any kinds of agricultrual food products, e.g. fruit, or wine food products. Because New Zealanders know themselves country has enough natural resource . So, in common, many New Zealanders choose to attempt to do farming agricultural jobs in order to export themselves any kinds of fruit or meat or wine products to overseas or sell to domestic in order to earn profit.

So, when these New Zealand farmers number has been increasing every year. This country farmers will feel themsleves competition between this New Zealand farmers themselves are serious due to they may feel New Zealanders choose to do agriculture businesses in order to export themselves different kinds of farming food to overseas or sell to local to earn profit.

Hence, when many New Zealand farmers feel that farmers number has been increasing every year. They will feel themselves competition is serious. They must need to spend much time and nervous and effort to research what method is the best how to produce the best taste of white or red grape wine products in order to let local or overseas wine buyers to choose to buy his/her producing white or read grpae products to drink.

Hence, in competition psychological view, may influence many New Zealand white or reaad wine producers had been beginning to change their learning behavior on researching what method is the best in order to produce the best quality of taste red or white wine products to sell in order to attract overseas or local white or read grape wine drinkers to choose to buy his/her wine products. Their behavior will focus on learning how to raising or improving white or read grape wine taste method more than only focus on producing a large number white or red grape wine products. They believe wine quality is more important to compare wine producing number. So, New Zealand wine producers themselves wine producers behaviors have been changing on concentrating on researching wine quality method aspect more then wine producing number aspect in behavioral economic view.

America high technological productive behavior

For America example, US is one high technological country, it owns many high technological knowledge talent inventors, e.g. computer science inventors. Hence, US must attract many diferent countries owning high technological computer inventors choose to go to US to develop their computer science profession career. Also, it seems that when many computer science inventors or professions choose to go to US to develop themselves computer science new career. In behavioral economic view, due to their leaving themselves countries choice, which may bring influence themselve country job behaviors need to be changed. They must need to adapt US new live. Because they will forgive their past computer science job. These computer science professionals need to spend time to adapt US new lives. They " past computer science job behaviors" will need to be changed to their new US any computer employer's new computer science job model.

Because their traditional computer science jobs needed to be forgot in their themselves countries. They will feel their old computer science job knowledge and behavior needed to change in order to let their US any one new of computer company employer feels satisfactory to accept their new working behavior in any one US computer organization.

So, on the other hand, many US computer company employer will feel that they must need time to accept any one new overseas computer science professions their working behaviors, their working attitude daily, because these foreign comouter science professional, their past computer working behaviors and working attitude must be different to US domestic computer science professions.

In behavioral economic view, these overseas computer science professions, their working behaviors and attitude must be needed to change in order to adapt any one US new computer company itself domestic or local computer science professional stafs themselves daily working behaviors and attitude because these overseas and local computer science professionals must need to team work together.

In behavioral economic view, it is only one way that foreign computer science professionals must need to change themselves past country traditiona daily working behaviors and attitude in order to cooperate with these US local computer science professionals in teams more easily.

Consequently, if these foreign compute science professionals can change their past working behaviors and attitude to let any one US local computer science professional feels to cooperate with them easily in short time. Then, the US computer company itself whole computer professional teams themselves efficiencies will be influenced to raised or improved by the changing past working attitude and working behaviors of these foreign computer science professionals. So, in behavioral economic view, only if US any one computer company hopes itself computer teams themselves efficiency can be raised or improved when it decides to employ foreign computer science professionals and US domestic computer science professionals. They need to work in teams together. They must need to let these foreign computer science professionals to know how to change their working behaviors and attitude to let their domestic computer science professionals feel easy to work together. Then, the US computer company itself whole team efficiency must be rasied or improved easily in short time.

● China share market investing behavior

For China share market example, economic development depends on financial market. Because if many Chinese have interest to invest to carry on shares buying and selling activities in orde to learn how to earn shares interest and share profit when the China shareholder can make decision to sell himself/herself shares in the the high price, then he/she can earn money when he/she can sell the China company's shares in the high sale share price position.

If China has many Chinese like to spend time to carry on investing shares activities. Themselves shares buying and selling behaviors will influence China has many companies can increase fund from many Chinese shareholders in order to have enough money to expand or develop themselves businesses in China in long term.

Consequently, when China can have many Chinese like to attempt to carry on buying and selling shares investing behaviors in China share market. Themselves buying and selling shares behaviors can help many Chinese companies have effort to increase enough money or capital in order to continue to do their businesses in long term absolutely. So, it explains why when many Chinese become shareholders , they can assist China will have many companies continue to develop their businesses if many Chinese like to carry on shares buying and selling investing behaviors in long time in China financial investment market nowadays in behavioral economic view.

Why has any individual country have many people invest share behavior which can influence the country's macro consumption desire?

I shall apply shares market buying and selling investment behavior to explaiin why shares investment behavior which may impact the country's overal consumption desire as below:

In behavioral economic view, I assume that when the coutry has many people have interest to attempt to carry on shares buying and selling investment behavior, then their frequent shares buying and selling behaviors which may bring negative consumption desire or shopping desire of these shares investors their consumer behavior.

The reason is simple, when the country has many share buyers number suddenly been increasing rapidly. Consequently, these large group share investors must need to spend much time to research any kinds of company shares variations, whether when their share prices will rise up of fall down in order to achieve buying the company's shares in the lowest price and selling the company's shares in the highest price level in order to earn profit.

Basic on this reason, they must need to spend much extra time to research share prices changing behavior every day, e.g. one working person will wait to leave his/her job, after he/she can spend time to gather data to research the day's share price changing behavior after dinner. So, the working person's right time may be his/her share price market research behavior. Before he/she may spend his/her night time to go to shopping after dinner, but nowadays, he/she will fogive to do his/her shopping behavior before dinner or after dinner at hight sometime. He/she will make decision to spend much night time to turn on computer to click on share market website to research his/her share purchase choice to investigate whether his/her share price whether it rises up or falls down at the moment in order to make his/her share buying or selling decision at ever night time.

I mean the when the country has many people are share investors, their shares investment behavioral spenging time which will influence many shops lose customers at might often because the country will have many people feel need to spend night time to turn on computer or watch television to investigate share price variation. So, the country will have many people / share investors choose to stay at home in order to carry on share price variation investigation behavior, they need to listen share market update news from radios or watch the share market update news from computer or TV at home every night. Consequenly, they must reduce times to leave themselves homes at night. So, their shopping behavior also will be reduced. Because these share investors feel need to spend time to investigate share price variation news at homes which can bring economic benefits ( high opportunity benefits) when they choose to forgive to leave homes to go to shopping times ( opportunity cost) every night.

On conclusion, it seems that when the country has many people are share investors, then their share price investigating behavior may bring negative shopping emotion at night. Consequently, the country's any one shop may lose many customers from this share investor consumer group in behavioral economic view. Hence, when the country's share investors number had been increasing rapidly, it will influence any shops lose many customers from this share investing customer group at night frequently in short time, even long time in behavioral economic view, because their shopping desires or shopping emotion will be brought negative feeling when they make decisions to spend much time to listen radios or watch TV or computers share price update nes at night. Hence, share market will bring negative impact to influence consumer shopping desire or negative shopping emotion in behavioral economic view.

Can technology influence human shopping behavioral change?

Nowadays, technological development has reached mature stage, whether technological mature stage may bring positive or negative shopping emotion influence to global consumers. I shall aplly internet inventin or ecommerce shopping channel tool to explain whether internet technology can bring postive or negative influence to global consumer behavior in behavioral economic view.

Internet is a good technological tool, it brings e-commerce business chance. In fact, commonly, global has have many businessmen choose to use internet channel to carry on their products transactions between global online-buyers and their electronic websites. So, global many shoppers had begun to feel online shopping is more convenient to compare visiting shops shopping. Their shopping behaviors have been changed from internet technological tool. Global has many shoppers choose to buy any products from any overseas or local businessmen their web stores. They only need to spend time to find any businessmen their webstores to choose the most suitable products to pay visa to buy from their webstores. at homes. So, in general, global had have may shoppers had changed their shopping behaviors from visiting shops to visiting webstores at homes often.

So, it seems that internet technological tool had influenced global many shops disappear, but internet webstores will be replaced their actual shops on streets. Some of businessmen either they choose webstores to replace shops or choose websotes and shops both or still keep shops only. Hence, internet tool influences global businessmen have three kinds of products sale channels to let globa local and overseas consumers to choose how to buy their products. However, in fact, many of global shoppers, youngers and olders had begun to accept to buy any products from webstores. They feel to spend time to leave homes to visit shops , their shopping behaviors will be wasted time to not essential part to their daily lives. Hence, since internet technological invention, it had changed many consumers their traditional visiting shops shopping habit to change to buying products from webstores channel.

However, on the one hand, internet creates webstores ecommerce shopping channel to let global many consumers do not need to leave homes to go to shopping. It brings negative visiting shops shopping emotion to global general consumers nowadays. But on the other hand, it also brings positive visiting internet webstores shopping emotion to global general consumer nowadays. So, it seems that global many consumers feel that they often do not need to spend much time to go out shopping. Many global consumers feel convenient and enjoy to choose any products to buy from different internet webstores, when the online buyer chooses the most suitable product, he she only needs to pay visa card to buy the product from the online seller's webstore conveniently at home.

Hence, online shopping can bring economic benefit to online buyers, e.g. avoiding walking time or spending transport fare to visit the shop to go to shopping, shortening or reducing shopping time to do another important matter.

On conclusion, global many consumers began feel online shopping can bring more economic benefits on shortening shopping time, avoiding transport fare spending aspect. So, online shopping will be popular shopping behavior for future long time. It may encourage global many shoppers can make rapid shopping decision in short time in order to carry on any products buying transaction to global any one online shopper in short time easily in behavioral economic view. So, global many businessmen had begun to build themselves one attraction webstore in order to persuade different countries consumers to choose to click themselves webstores from internet channel to buy any kinds of products in short time easily.

So, internet technology had changed consumers traditional shopping behaviors to build positive online shopping emotion as well as raise online sellers' any products sale chance easily in behavioral economic view.

Why and how human behavior may influence the country's economic growth or recession?

When one country has many people choose to do the same matter for one period, whether their behavior may influence the country's pvera; economic growth or recession . I shall attempt to indicate cases toexplain their relationship as below:

For flowing rubblish behavioral case example, do you feel that when the country has many people often flow rubblish on the streets, instead of their flowing rubblish behavior may bring streets dirty? But, their flowing rubblish behavior may explain that this country has people may have enough money to buy food to ear, or enough cloths to wear, enough bottles of water to drink, even they may have enough money to buy new television, radio, refrigeraters , washing machines, desktops or laptops electronic home products from old to new to use in order to satisfy their living needs. So, when they flow old electronic home products, their flowing old home electronic products behaviors may seem that they have enough money to buy other new home electronic products to replace old home electronic products to use at homes.

However, it seems thaat this country ought have many people have jobs to do. So, many of them, they can easy to make purchase decison to flow any old home electronic products and buy any new home electronic products to use . Because this country has many people have jobs to do. So, they can often not use old home electonic products to become rubblishs to flow on streets after they had bought any kinds of new home electronic homes.

In fact, it also implies that this country's economy grows rapidly. So, many businesses can glow up rapdly. When they expanded their businesses, they must need to increase employees number in order to let they help themselves to raise productivity or serve their clients absolutely. So, when the country has many businesses can grow up, it seems that its economy must be better or it is improved to compare past. Due to many different kinds of home electronic products had been often bought to use by this country people in this period. So, this country's any streets can be observed that expensive electronic home products were flowed on streets anywhere. then, this country will have many electronic home products sellers can sell their home electronic products very easily. When this country has many people can find any kinds of jobs to do easily. So, due to unemploymen rate had been decreasing.

In behavioral economic view, as this many electronic home products rubblish country case, we can observe this country may have many people have jobs to do. So, consumption number has been increased long time. So, cheap food, or expensive home electronic products may be rubblish on any streets. This country's people , their flowing rubblish behaviors may be explained that many of people have enough jobs to do, so they have ability to buy any good taste food to eat or buy any kinds of expensive electronic home products to use. So, this country's economy may be improved for this long period. So, in behavioral economic view, when this country can have many electronic home products rubblishs are flowed on anywherer in streets frequently. It seems that this country will have many people have jobs to do, so it causes they often change old home electronic products or replaced them easily, when they have enough income to spend to buy any kinds of new home electronic products to use at homes easily. Moreover, their flowing old electronic home products behaviors also indicate that this country has many people their salaries may be increased in possible from their emplyers. When this country can have many different kinds of home electornic products are sold. It means that this country's electronic home products needs or demand had been increasing, due to many people have jobs to do and income increases to excite their living of needs also improve. Consequently, this country may seem have better economic improvement. We can observe from this country's electronic home products rubblish increasing income in theis period.

On conclusion, this country ought experience economic growth at this period. So, " flowing expensive electronic home rubblish increasing number " may seem that this country's economic growth is rapidly in this period, due to many people have jobs to do as well as salaries increase in this period.


Technology how impacts human behavior changing?

Technology how influences human behavior to bring changing? For example, online share purchase and sale transaction from smart phone brings share investor can do share buying or selling transation in any where and any time conveniently, non manual driving auto vehicle, bring car owner feels comfortable and spends free time to do other matter, e.g. reading, listening mucis in himself or herself car freely. electrical energy vehicle can help car owner to reduce air polluton and it can brings the drivers do not feel drive long time in any journeys in order to avoid air pollution for environmental protection responsible car drivers in our societies. Thus, they will drive long time in any journeys when they can drive electronic energy cars to replace oil energy cars.

However, online technology can also bring consumers can choose to stay at homes to buy any things from seller individual online webstore conveniently. Such as online technology can bring shoppers do not need to spend much time to visit shops to buy any things. They can choose any kinds of products from any online sellers individual online webstores conveniently at homes. Online technology excite busy consumers can make purchase decision easily as well as it can help online sellers sell any kinds of products from internet easily.

In behavioral economic view, technology can change human behavior to be improved, it can let human feels comfortable, more free time ro use, rapid making any decisions, such as apply smart phones to make share purchase or sale transaction decision, online shopping decision, even travelling any where decision in short time, when the traveller finds the most cheap hotel accommodation room price and air ticket price frm any travel agent online tourism webstore, then the potential travel customer can follow the online hotel accommodation price and air ticket price data to make decision when to buy the air ticket from the airline travel agent or make decision when to prebook which hotel accommodation room to go to the country to travel from online travel agent tourism webstores. So, technology can encourage global any country travelers to make anywhere to trvel rapidly. If the traveler can find the country's general hotel rooms and airline tickets prices had been decreasing more sightly. The traveler may make travel decision to choose the country to travel in short time, then he/she can prebook the country;s any hotel room and airline ticket to pay by visa fraom the country's any hotel and airline travel agent webstores., before one week, even one month or more easily. Hence, online technology can also encourage traveler individual frequent travel times to be increased, due to global travelers can find any hotel rooms and airline tickets prices from internet conveniently at homes. They do not need to spend time to visit any airline travel agent to enquire travel choice country's hotel rooms prices and airline ticket prices. They can compare global travel of countries choices ' all hotels rooms and airline agents air tickets prices to make prebook airline seat and hotel room decision before one week, one month even six months early.

On conclusion, online technology can encourage global travelers can make travelling any where and when traveling time desicions easily. It can excite tourism industry develops in long time. Also, such as electricity cars invention can encourage environment protection car owners do car purchase decision easily, because they can choose to drive electronic energy cars to replace oil energy cars in order to avoid air pollution occurs easily. So, electronic cars can increase electronic car purchasrs number, due to many of environmental protection attitude of car owners can choose to drive electricity cars to bring air cleans, even non -manual driving cars can encourage lazy driving and free time driving car owners to choose to buy non-manual ( artificial intelligent) cars to drive , because they can spend much free time to read, listen music or do any matters in themselves cars, they do not need to drive cars, robotic (AI) auto driving machine is such one non-manual driver to help them to drive themselves cars confidently. So, non-manual driving cars can attract lazy and enjoying free time driving car owners to choose to buy to replace traditional manual cars to drive easily. Moreover, online share transaction can help any share investors to make share buying and selling decision in short time easily. When they can apply smart phones technological tool to carry on share buying and selling activities easily. They can observe any share rising or falling price suitation from smart phones in any where any any time easily. So, smart phone technology can help global any shareholders to make share purchase and sale transaction easily. So, technology can encourage human makes decision in short time rapidly.


How and why employees behaviors may influence economy development?

In behavioral economy view,I believe the country's any organizational employees behavior may bring indirect relationship to influence the country's long term economic development. I shall indicate past manufacture industry social development period to explain their relationship. For many countries' past business activities had belonged to manufacturing industry, such as US, UK past before 1980 year, it focused on steel manufacturing and steel manufacturing related machine products. So, US, Uk developed countries manufacturing industries may be past main country's economic income sources. I assume US , UK past had one million number different kinds of industries. They ought had about seven houndred thousand number organizational businesses were belonged to manufactured industry. They may include:

Steel manufacturing and steel related machine manufacturing, e.g. vehicle manufacturing, home appliances, e.g. washing machine, television, radio, refrigerate cooler, heater, air condition etc. different kinds of different kinds of steel -related manufacturing machine, they were manufactured from US, UK steel machine manufacturers. So, US, Uk the other three hundred thousand number industry may be general service industry, e.g. hotel service, restaurent, cinema, public transport service, tourism lesiure , wine bar, supermarket etc. different kinds of non-manufacturing industries business organizations were operated in UK, US past before 1980 year.

So, in UK, US developed countries industry development history, they ought have high percentage of businesses belonged to steel related manufacturing machine and steel products. Also, in the past before 1980 year, US, Uk business employers , they employed many workers are manufacturing workers. They needed to spend long time to work in factories. They were skillful workers, and they are trained to manufacturing cars, washing machine, television, heater, etc. even steel itself different kinds of steel related products to prepare to deliver to their shops to sell to US, Uk local or overseas clients.

So, I believe that past UK, US ought employ many employees, they belonged to skillful manufacturing workers, manufacture increasing steel machine or steel related machine number of products rapidly daily. So, if UK, US had had many of these manufacturing factories owned high skillful workers, then their manufacturing steel-related machine or steel both kinds of products number must be influenced to raise rapidly. Consequently, their steel machine manufacturing products would been exported to overseas or would been sold to local both markets , they may be influenced to raise sale number. They ( these manufacturing workers) needed to be trained to know how to manufactur these different kinds of machine products in the efficient teams and they ought to be trained to raise their efficiencies in order to shorten time to manufacturing many kinds of steel related manufacturing machine or steel itself products rapidly. So , if their efficiencies and manufacturing performance was improved, these US, UK any one manufacturing worker and their teams ought achieve raising productivities significantly.

Hence, when past UK, US manufacturing industry development period, if these two countries' any manufacturing factories could have many manufacturing workers could be trained to be skillful and proficient manufacturing workers. Then, in past every day to these factories workers, they ought help their steel or steel related manufacturing employers to raise any kinds of machine or steel products number in every team. So, when past in the manufacturing industry development, US, UK could have many factories' manufacturing workers themselves steel or steel related machine products manufacturing skill could be trained to to improve to any kinds of these machine or steel manufacuring products quality as well as their products number could be influenced to raise by themselves skillful improvement significantly every day.

Then, what would be influenced to occur to past UK, US manufacturing industry period? In behavioral economic view, when these two manufacturing industry developed countries, such as UK, US , if they had many factories workers can be trained to improve their skill in order to achieve any kinds of steel or steel-related machine products quality could be improved as well as products manufacturing number could be also increased absolutely.

In consequence, past UK and US both countries ought increase themselves any kinds of steel and steel related machine products number to be supplied to themselves local shops to let local clients to choose any one kind of machine manufacturing products to buy easily as well as they could also export to supply overseas any countries to buy their different kinds of steel or steel related machine products to let overseas steel or steel related manufacturing machine product buyers, they can have many of these different kinds of these steel or steel-related different kinds of manufacturing machine from UK and UK these both countries easily to compare other countries.

On conclusion, I believe that past US, and UK macro manufacturing industry income GDP would increase significantly. So, they would have good economic growth performance because when many of these manufacturing workers themselves manufacturing effort could be improved. So, it explained when employees manufacturing abilities can influence economic growth indirectly.

Robots invention whether they can help organizations to raise efficiencies or inefficiencies?

In behavioral economic view, in any organizations, when the organization hopes its worker teams can raise efficiencies , the organization may choose to increase more workers number and/or it can provide training to improve these workets themselves skills in order to raise their efficiencies. For one warehouse example, when the warehouse increases many goods , they are needed to delivered these goods from the shelves to the delivering destination locations. If this warehouse supervisors feel these workers themselves goods delivery speeds are slow, which is possible due to this warehouse's workers number is not enough. So, this warehouse supervisor ought increase workers number in order to increase their goods delivery speed in order to deliver goods from the shelves to every indicated goods delivery destination in order to let any one lorry driver can transport the right kinds of goods and ensure the accurate goods number to transport to any one client home rapidly.

However, if this warehouse supervisor planed to buy several warehouse goods delivery robots to assist these warehouse workers to find the right kinds of goods from shelves and then deliver to the right destination location in the warehouse. So, these warehouse orkers can concentrate on counting the accurate goods number and ensuring the right kinds of goods in order to prepare to let lorry drivers to transport these goods to these goods of buyers themselvers homes rapidly. Consequently, in the first step, robots can concentrate on finding th right goods from shelves and delivers them to the right goods transportation of location destination. Then, in the second step, these warehouse workers can concentrate on counting the accurate goods number and ensuring the right kinds of goods in order to prepare to put them to the lorry. Consequently, when warehouse robots and warehouse workers can cooperate to work together, the most important, robots, can deal on finding the right kinds of goods and deal on delivering the accurate number of goods of job duty as well as these warehouse workers can only concentrte on counting the right kinds of goods number in order to avoid it has none any mistake of wrong kinds of goods and inaccurate goods of delivery number to be transported to the lorry and to deliver to any one buyer's home.

So, it seems that warehouse robots ought help any one warehouse worker to raise himself efficiency and avoid goods delivery of mistake occurrence easily as well as their help to warehouse workers that can let any one goods buyer feels their goods can be delivered to their homes rapidly. Moreover, warehouse robots can also help these warehouse workers to raise efficiencies because warehouse robots can help them to shorten goods delivery time between any one shelf and any one goods delivery destination of location in the warehuse because robots may help them to find the right kinds of goods from the right shelf in the short time. So, any one worker does not need to spend long time to seek anywhere is the right shelf location for the kind of goods when the kind of goods are needed to deliver to the buyer's home from lorry. Warehouse robots can help them to do this aspect of " finding the goods from the right shelf in short time job duty". So, any one warehouse worker only needed tospend less time to do the counting of any right kind of goods number and ensuring the right kind of goods job duty. Consequently, this warehouse 's any one worker, his any one kind of goods delivery time may be reduced, because robots' assistance and they may have more confidence to avoid mistake to deliver the wrong number of goods and/or the wrong kind of goods to any one goods buyer's home.

On conclusion, it seems that warehouse robots ought may help any one warehouse worker to raise efficiency for any one team in the warehouse as well as the warehouse any one supervisor does not need to spend much time to observe any one worker individual performance for " goods delivery job duty aspect" because their goods delivery job duty that had been replaced to do by these several warehouse robots. Robots can achieve the more accurate of right kinds of goods and the right number of goods delviery job performance to compare any one of human warehouse worker themselves right kinds of goods of delivery and right number of goods of delivery job performance. So, when robots can participate to cooperate with this warehouse's any one worker to do their goods of delivery job duty in this warehouse every day. Then, robots can raies any one of supervisor individual confidence in order to let they do not need to spend time to observe any one of worker individual whose goods of delivery job performane. They can concentrate on supervising any one worker whose goods transport to lorry in the final step in order to avoid to deliver wrong goods number and / or wrong kind of goods to any one goods buyer's home every day. Consequently, this warehouse's overall teams of their delviery of goods performance many be improved by robotss' participatin to goods of delivery task as well as this warehouse's oveall teams themselves efficiencies may be influenced to raise by robots' goods of delivery task participation.